ALLE · ZEIT · WACH · 1842

Springer Series in Wood Science

Editor: T. E. Timell

M. H. Zimmermann
Xylem Structure and the Ascent of Sap (1983)

J. F. Siau
Transport Processes in Wood (1984)

R. R. Archer
Growth Stresses and Strains in Trees (1986)

W. E. Hillis
Heartwood and Tree Exudates (1987)

S. Carlquist
Comparative Wood Anatomy (1988)

L. W. Roberts/P. B. Gahan/R. Aloni
Vascular Differentiation and Plant Growth Regulators (1988)

C. Skaar
Wood-Water Relations (1988)

J. M. Harris
Spiral Grain and Wave Phenomena in Wood Formation (1989)

B. J. Zobel/J. P. van Buijtenen
Wood Variation (1989)

P. Hakkila
Utilization of Residual Forest Biomass (1989)

J. W. Rowe (Ed.)
Natural Products of Woody Plants (1989)

K.-E. L. Eriksson/R. A. Blanchette/P. Ander
Microbial and Enzymatic Degradation of Wood and Wood Components (1990)

R. A. Blanchette/A. R. Biggs (Eds.)
Defense Mechanisms of Woody Plants Against Fungi (1992)

S. Y. Lin/C. W. Dence (Eds.)
Methods in Lignin Chemistry (1992)

G. Torgovnikov
Dielectric Properties of Wood and Wood-Based Materials (1993)

F. H. Schweingruber
Trees and Wood in Dendrochronology (1993)

P.R. Larson
The Vascular Cambium: Development and Structure (1994)

M.-S. Ilvessalo-Pfäffli
Fiber Atlas: Identification of Papermaking Fibers (1994)

F.H. Schweingruber

Trees and Wood in Dendrochronology

Morphological, Anatomical, and Tree-Ring Analytical Characteristics of Trees Frequently Used in Dendrochronology

With 563 Figures

Springer-Verlag
Berlin Heidelberg New York
London Paris Tokyo
HongKong Barcelona Budapest

PROFESSOR DR. FRITZ HANS SCHWEINGRUBER

Research Group: Tree Ring and Site
Department of Forest and Climate Ecology,
The Swiss Federal Institute of Forest
Snow and Landscape (FSL)
Birmensdorf, Switzerland

Series Editor:

T.E. TIMELL
State University of New York, College of Environmental, Science and Forestry,
Syracuse, NY 13210, USA

Translation: SUSAN JOHNSON
Text-Editing: T.E. TIMELL

"Trees and Wood in Dendrochronology" is an expanded version of a lecture presented by the author to the Botanical Institute, the University of Basel.

Cover: Transverse section of *Pinus lambertiana* wood. Courtesy of Dr. Carl de Zeeuw, SUNY College of Environmental Science and Forestry, Syracuse, New York

---

ISBN-13: 978-3-642-77159-0   e-ISBN-13: 978-3-642-77157-6
DOI: 10.1007/978-3-642-77157-6

Library of Congress Cataloging-in-Publication Data

Schweingruber, Fritz Hans. Trees and wood in dendrochronology: morphological, anatomical, and tree-ring analytical characteristics of trees frequently used in dendrochronology/F.H. Schweingruber; [translation, Susan Johnson].
(Springer series in wood science)
Translated from German.
"An expanded version of a lecture presented by the author to the Botanical Institute, the University of Basel" – T. p. verso. Includes Bibliographical references (p. 386) and indexes.
ISBN-13: 978-3-642-77159-0
1. Tree-rings. 2. Dendrochronology. 3. Trees – Anatomy. 4. Wood – Anatomy. I. Title. II. Series.

Softcover reprint of the hardcover 1st edition 1993

31/3020-5 4 3 2 1 0 – Printed on acid-free paper

Preface

The science of dendrochronology has grown significantly in the past 20 years. In the 1950s and 1960s, interest in the subject was limited to only a handful of scientists who perceived in dendrochronology a "l'art pour l'art". Today, however, specialists from many different fields recognize and are pursuing the problems of dendrochronology. Tree-ring research has acquired a permanent role in the various sciences of archeology, history, geology, ecology, and climatology.

The founders of dendrochronology themselves were of varied scientific backgrounds and interests. For example, A.E. Douglass in the United States was an astronomer, B. Huber in Germany a forest-biologist, and F.N. Shvedov in Russia a climatologist. Today the spectrum is even broader. Many dendrochronologists are authorities in mathematics, archeology, history, forestry, botany, wood technology, ecology geography, geology, etc. It is, therefore, understandable that it has become almost impossible for one individual to encompass the entire field. Bitvinskas (1974), Fritts (1976), Schweingruber (1983), and Mitsutani (1990) have attempted, each guided by his own interests, to provide at least an overview of the field. Recently, individual aspects have been presented by groups of authors in books edited by Fletscher (1978), Hughes et al. (1982), Jacoby and Hornbeck (1987) and Bradley and Jones (1992).

It is very likely that in the future summaries covering each branch of dendrochronology will be published. Indeed, the initial steps have already been taken by the present work on dendrochronology and by the book on methods of dendrochronology edited by Cook and Kairiukstis (1990). Anticipated as well are studies relating to archeology, building history, art-science, environmental science, etc. Considering the rapid expansion of dendrochronological knowledge, it must be emphasized that the tree species presented in this work are limited to studies that are currently available. In principle, all tree species, and practically all shrubs found in seasonal climates may, in the broadest sense, be dendrochronologically analyzed. Yet, of the over 1000 European tree and shrub species, only an estimated 15 have been studied, and only a few of these studies are of any major importance.

The purpose of this work was not to issue a comprehensive list of species already tested in dendrochronology. Instead, its purpose is to display the dendrochronological potentials exhibited by individual species. The work is also by no means intended as a textbook of dendrochronology; rather it is a summary of the important well-known morphological, ecological, and anatomical characteristics of trees and their relationship to the relatively unfamiliar studies found in tree-ring research.

With this publication I would like:

- To contribute to mutual understanding; it is imperative that professional and language barriers be torn down and new paths be laid.
- To show that dendrochronological problems in the sciences of ecology, climatology, and history are the same worldwide and that in different regions and continents only the tree species vary.
- To draw attention to gaps in research. Is it not, for example, astounding that deciduous trees have worldwide received only limited attention?
- To point out that because the most important decisions take place while choosing samples, a basic ecological understanding is vital to dendrochronological research.

The book is a byproduct of wood anatomical and dendroclimatological projects of the last 20 years. On collection trips through many countries of Europe, in North America, Russia, and Japan, I made photographs and contacted scientific colleagues to evaluate the dendrochronological significance of the described species. The book grew slowly. First I made some species descriptions for classes in dendrochronology at the University of Basel, later followed by a small German textbook, and finally the present book.

I am fully aware that each species has not been presented in equal detail. Language barriers and a deficient literary awareness sometimes prevented my access to desirable references. I also realize that the species from my area of experience have been much more thoroughly presented than those species with which I have never worked or ever seen.

Many colleagues provided me with materials or helped me scientifically. I am grateful to all of them. I would particularly like to thank Rosmarie Hirzel-Müller and Miroslav Sebek, who contributed the drawings of branches and seeds; Paul Scherrer and Pierre Sulser, who together took care of the photography; and Padruot Nogler and Ernst Schär, who provided the X-ray films for the presentation of the tree-ring sequences, as well as Konrad Häne, who made the stamps available, and Charles von Büren and Edith Stamm-Leuenberger, who proof-read the German work. My special thanks go to Margrit Wiederkehr, who turned my rather jumbled manuscript into a legible text, and Doris Pichler who mounted the photographs and texts along with making hundreds of small additions. Oscar Hugentobler, Walter Keller, Enrico Marcet, Georg Miehe, Fidel Roig and Tore Timell were helpful in making corrections and providing specialized information.

I very much appreciate Susan Johnson's excellent translation and Tore Timell's very careful editing.

Many thanks to the Swiss National Science Foundation. The present work is a byproduct of his support.

Birmensdorf, Summer 1993 Fritz Hans Schweingruber

Contents

Introduction

The description of each tree species follows the same pattern.

The exterior form of the tree is illustrated concisely, without text. A paragraph characterizing the sites explains, above all, the soil, climatic, and ecological conditions. Specific ecological aspects supplement the sections covering the distribution of the particular tree species. A map illustrating the geographical areas of distribution is of significant importance. It is also possible, in connection with a climate atlas, to interpret these maps ecologically. Microscopic representations, mostly cross-sections, define the tree-ring sequences given at the end of the description of each species. These representations are copies from X-ray photographs. Based on literary studies and personal observations, I have attempted to present each species in such a manner as to make it

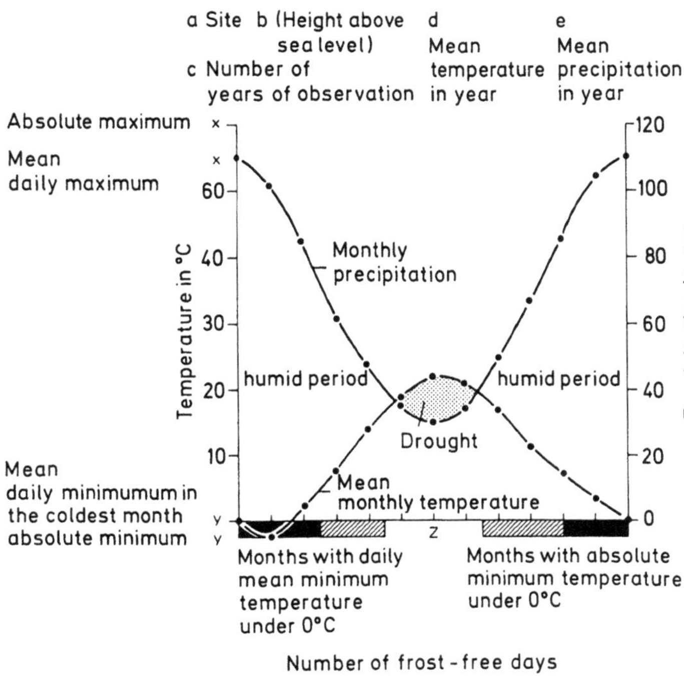

Fig. 1. Explanation of a climatic diagram and meaning of the designations

adaptable to chronological, ecological, and climatic studies. I have included brief summaries of previous dendrochronological studies to show how the species have been used in the past. An introductory chapter recapitulates particular wood anatomical and dendroecological basics. Of special importance are the climate diagrams, which, with only a few exceptions, have been taken from the standard work of Walter and Lieth (1960–1967).

Many illustrations are derived from standard works. The distribution maps were taken from the works by Meusel et al. (1965–1978), Fowells (1965), and Dallimore and Jackson (1974), the tree silhouettes are from Phillips (1979) and Hosie (1973). The illustrations from these works are not bibliographically cited in the text. Dendrological evidence and the descriptions of site conditions are mainly from Brockman (1968), Lamprecht (1980), Zoller (1981), Krüssmann (1972), Fowells (1965), Dallimore and Jackson (1974), and Mayer (1984). All the photographs without given source are from F.H. Schweingruber.

Factors Influencing the Development and Form of Stem Cross-Sections

The Reaction of Trees to Environmental Influences

Tree growth and specifically cambial activity are significantly influenced by climate and weather. Yet the peculiarities of each species and restricted biotic and abiotic factors often surpass the climatic effects.

Climate

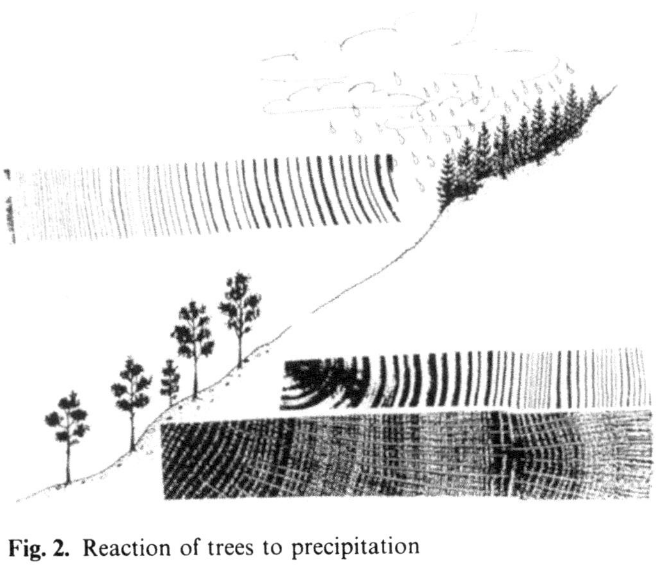

Fig. 2. Reaction of trees to precipitation

Region of High Precipitation

The sequences of tree ring widths vary only slightly, but those with a maximum amount of latewood vary considerably.

Region of Minimal Precipitation

The tree-ring sequence is extremely variable and indicator years are common.

Soil

The drainage, depth, and nutrient supply of the soil determine the rate of growth considerably.

Site

Fig. 3. Deep-rooted, fine subsoil, moist. Large trees with a compact crown and wide tree rings

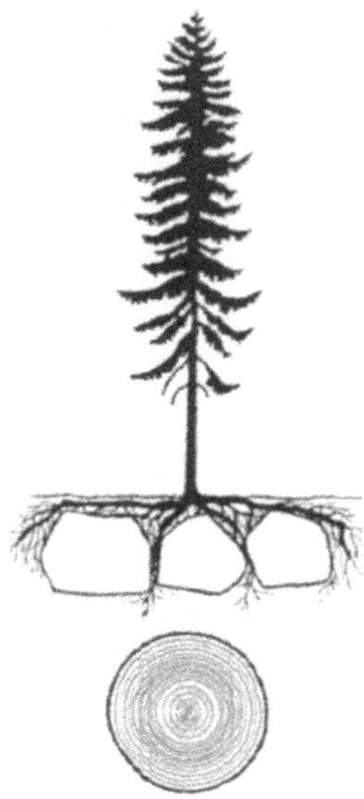

Fig. 4. Shallow-rooted, rocky subsoil, dry. Small trees with a loose crown and narrow tree rings

Fig. 5. Shallow-rooted, bog, and high ground water surface. Small trees with a thin crown, frequently yellowed needles, and narrow tree rings

Light

Tree growth clearly responds to variations in availability of light. Trees cleared by fire, storm, or forest management will in the following year produce notably wider tree rings. A tree growing beneath a dense canopy receives only poor illumination, resulting in a drastic reduction in ring width.

Fig. 6. Two dominant and one suppressed fir. The tree rings of the dominant firs are wide, those of the suppressed fir are narrow

Fig. 7. A dominant fir, a fir cleared 5 years ago, and a fallen fir. The sudden supply of light to the formerly suppressed fir results in an instantly improved growth rate

Site Changes Due to Wind, Snow, and Unstable Soil

In coniferous species, site changes produce compression wood on the lower side. In deciduous species site changes produce tension wood on the upper side.

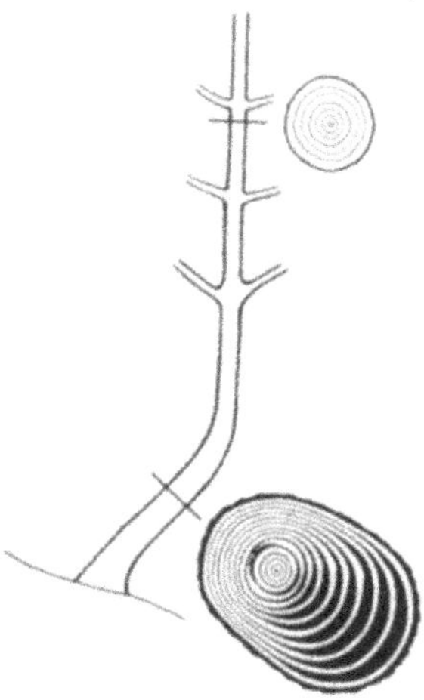

Fig. 8. Saber butt, continuous compression wood formation

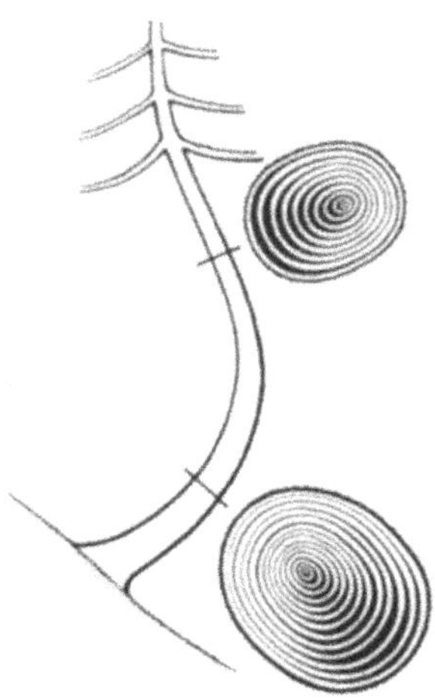

Fig. 9. Saber butt with overcompensation, continuous compression wood formation on the upper and lower sides

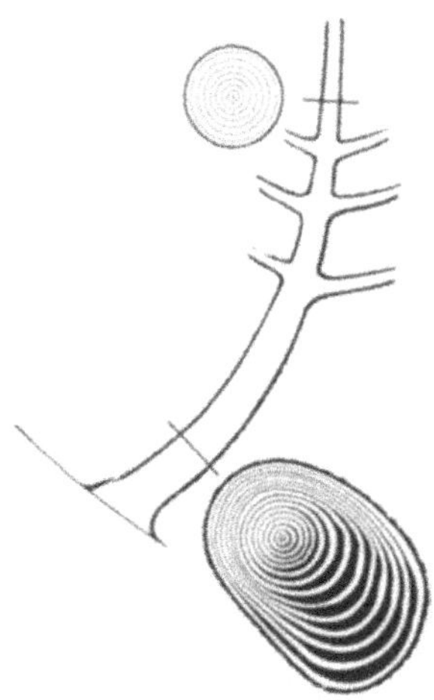

Fig. 10. Bayonet growth, sudden site change with abrupt initiation of compression wood

Wind

A wind constantly blowing from one direction results in the deformation of the tree-crown and formation of an eccentric stem cross-section.

Moderate Wind

Fig. 11. Very slender, slightly one-sided crown. The wind blows only against a small surface area. The trunks are slightly eccentric in form, but exhibit no trace of compression wood

Fig. 12. One-sided crown. Branches on the windward side die away. The wind blows strongly against the tree, resulting in eccentric trunk, cross-sections, and formation of compression wood

Fig. 13. Tree crown and stem cross-section of a tree growing in a wind of varying intensity

Heavy Wind

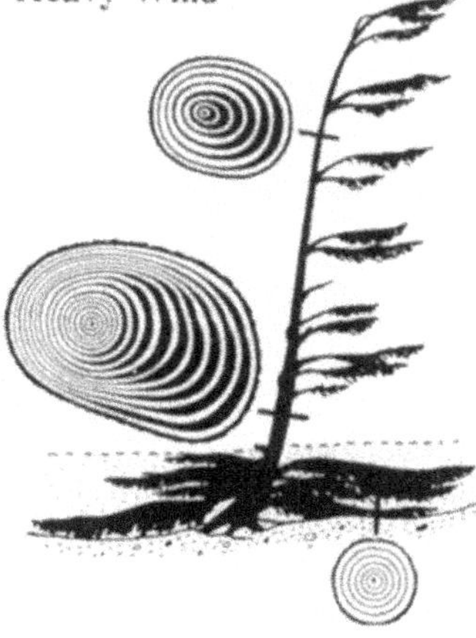

Fig. 14. Heavy winds result in small, flag-shaped trees with extremely eccentric stem cross-sections. Conifers form compression wood on the leeward side

Fig. 15. Extremely heavy winds result in dwarf or stunted trees growing along the ground

Snow: Creeping Snow, Avalanches

The crowns and growth rings of trees growing in forests located on ecological boundaries, i.e., in forests bordering the timber line, clearly show the results of climatic, edaphic, and microclimatic factors.

Fig. 16. a Young trees on snow-creep inclines. **b** Young trees on avalanche and rock-fall inclines

Fig. 17. Old trees in the path of avalanches. The tree tops may break off, resulting in new top shoots. The moment at which this event occurs is detectable in the growth rings as an abrupt reduction in the growth rate of the tree

Young trees are bent back and forth and form compression wood on different sides of the trunk, relative to the particular location of the tree.

Bent trees display the extreme dynamics of the regeneration process in crown and tree ring formation: growth reduction at the base of the trunk, compression wood develops on the lower stem side. The top bends upward, and the original dormant buds begin to sprout on the former stem and branches. Newly formed, vertical shoots contain no compression wood.

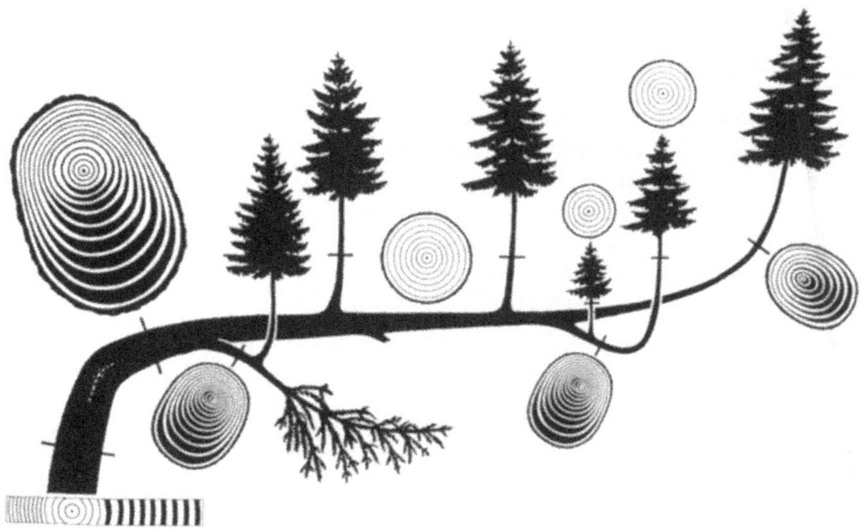

Fig. 18. Harp spruce in a wind-swept location or avalanche path

Defense Mechanisms Against Biological and Mechanical Damage

Cell contents and cell walls provide the nutrient base for all decomposing organisms, such as fungi and bacteria. Trees protect their energy-rich tissue with a resistant layer of cork in the outer bark. Damage to this layer is, nevertheless, quite frequent. Thus, over time, trees have developed vital defence mechanisms that protect the undamaged tissue.

Resin Formation

The wood and bark tissue of many softwood trees is intersected by a system of horizontal and vertical resin canals. On injury, a slightly fluid resin is exudated into the wounded area, which protects the living tissue from further damage. This layer of resin establishes an almost impenetrable defence against fungal infection.

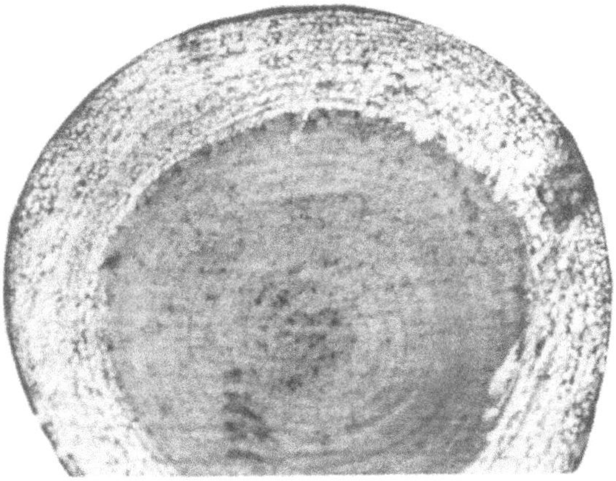

Fig. 19. Cross-section of a freshly cut pine stem. The living resin duct cells in the sapwood exude resin

Fig. 20. Cross-section of a freshly cut fir stem. Resin canals are absent in the wood. The living resin canals in the bark exude resin

Heartwood Formation

Physiological changes in the tissue cause cytological, chemical, and anatomical, alterations, specific to the individual species, to occur within the tree, such as reduced physiological activity, dying cells, diminished sugar and starch content, a lowering of water and increase in gas content, and a reduced permeability due to formation of thyloses, aspiration of pits, encrustation of the pit border, and the formation of phenolic compounds.

A reduced supply of easily decomposed material and a reduced permeability, an impaired physical condition (dryness), and fungicidal materials reduce the ability of decomposing organisms to act. It is possible to differentiate between three basic types of heartwood formation occurring in both coniferous and angiosperm species.

Dark-Colored Heartwood

The formation of dark-colored heartwood is in part facultative (ash).

The protective effects are great for example in ash, pine, larch, *Fitzroya*, *Sequoiadendron*.

Light, Dry Heartwood

Sapwood and heartwood are indistinguishable by color.
The protective effects are moderate, for example spruce, fir.

No Detectable Heartwood

The entire tree is moist.
The protective effects are low, for example beech, birch, maple.

Wet Heartwood

Wood with dark-colored heartwood and wood with normally dry heartwood occasionally exhibit a pathological wet heartwood, for example in fir.

Heartwood and Sapwood

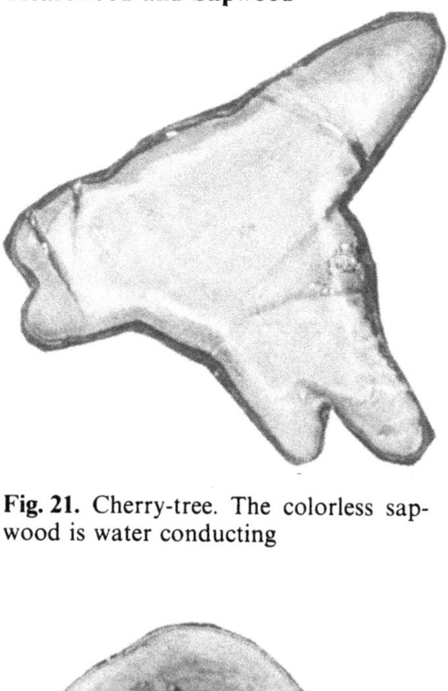

Fig. 21. Cherry-tree. The colorless sapwood is water conducting

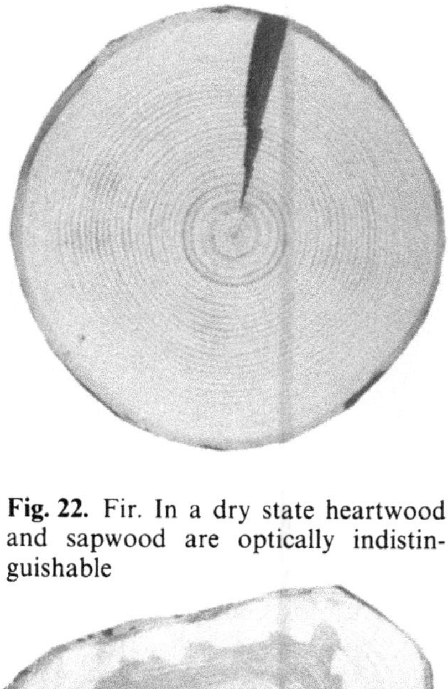

Fig. 22. Fir. In a dry state heartwood and sapwood are optically indistinguishable

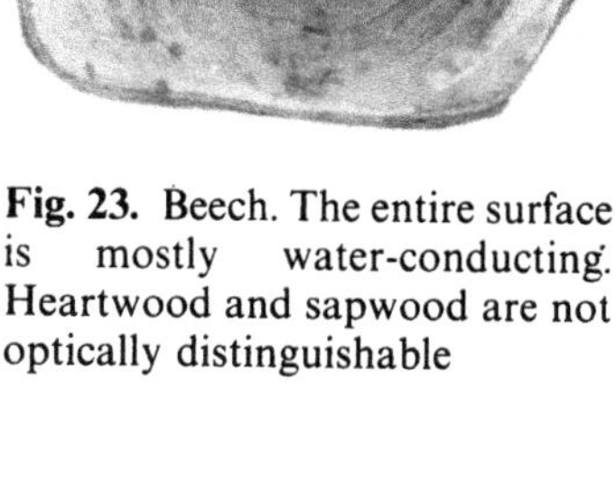

Fig. 23. Beech. The entire surface is mostly water-conducting. Heartwood and sapwood are not optically distinguishable

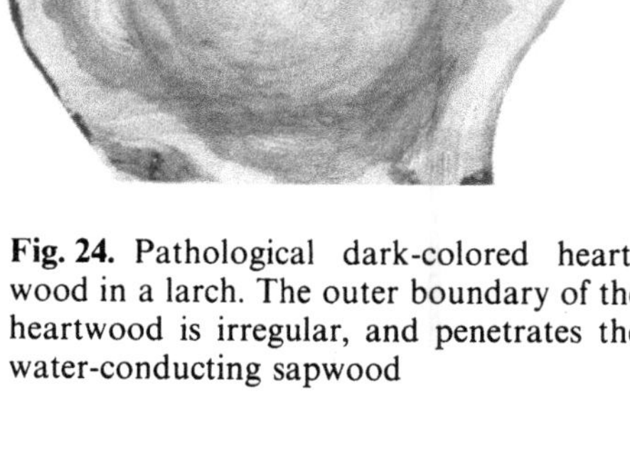

Fig. 24. Pathological dark-colored heartwood in a larch. The outer boundary of the heartwood is irregular, and penetrates the water-conducting sapwood

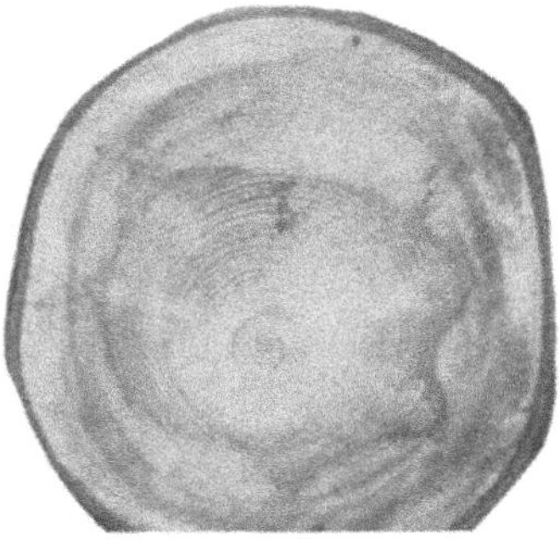

Fig. 25. Pathological wet heartwood in a fir subjected to fir decline. The wet heartwood with stagnant water penetrates the water conducting sapwood

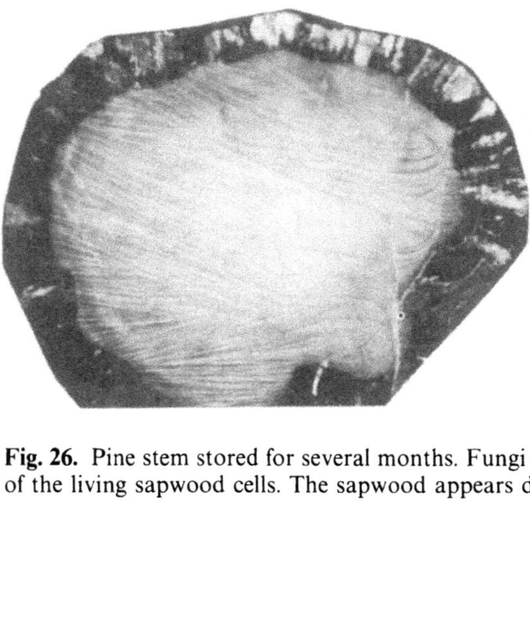

Fig. 26. Pine stem stored for several months. Fungi and bacteria feed from the contents of the living sapwood cells. The sapwood appears dark

Fig. 27. Pine stem stored for many years. The cell walls of the sapwood have been decomposed by fungi. The heartwood contains fungicidal material which deters decomposition

Compartmentalization

The spreading of fungal infection is restrained, above all on rays, on the tree-ring boundary, and on the base of the dead branches.

Radial dispersion of rays highly restrains the tangentially directed growth of fungal infection.

The radial direction of fungal growth is restrained, especially at the growth ring boundary.

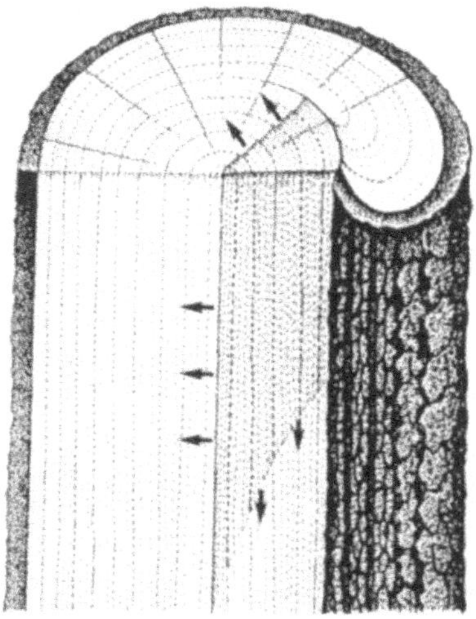

Fig. 28. Schematic presentation of compartmentalization

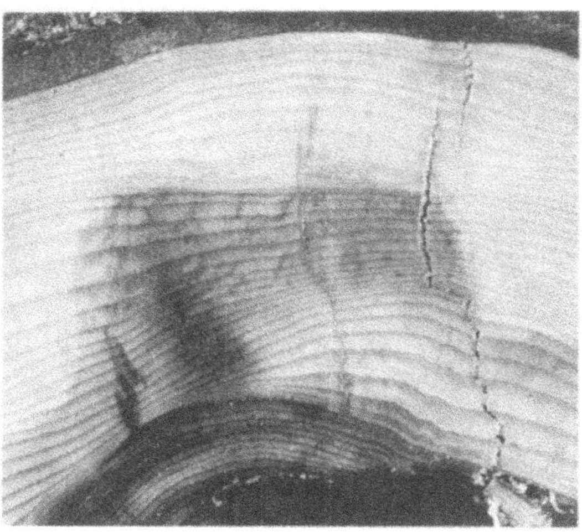

Fig. 29. Red rot in a spruce stem. The sapwood is attacked by fungi from the inside. Further spreading of the fungal infection is restrained, above all, at the tree-ring boundary and the wood rays

Injuries are healed. The new tissue possesses a protective cell wall layer, immune to the decomposing organisms found in the old wounds.

Longitudinal direction: The organisms spread quickly, especially along the pith. Longitudinally, only a few barriers exist against the spreading of fungi.

Fig. 30. Dead, decomposing pine stem. Further spreading of the various fungi is impeded at the growth-ring boundary and the rays of the dead tree

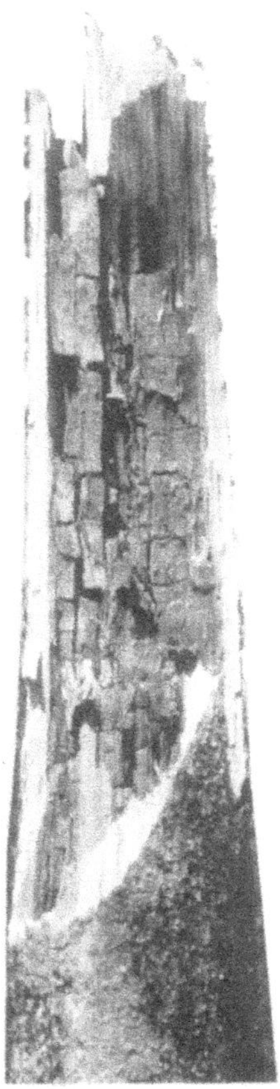

Fig. 31. Red rot in a spruce stem

Healing

At the spot where potential wood-decomposing organisms could enter the tree, for example through an injury or at the base of a dead branch, the tissue grows exceptionally fast, attempting through a process of healing to protect the open spot, which until then had been only chemically protected, from further infection.

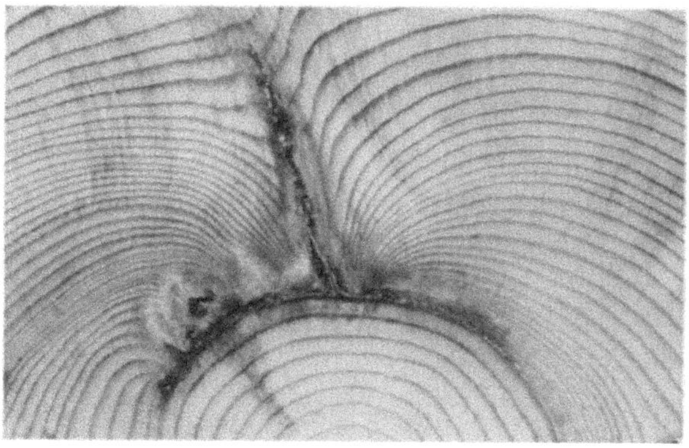

Fig. 32. Cross-section of a healed pine. Cambial activity on the injured area has been increased. In the next few years the small wounds will be completely overgrown

Fig. 33. A healed Douglas-fir. The annual growth is easy to identify in this species

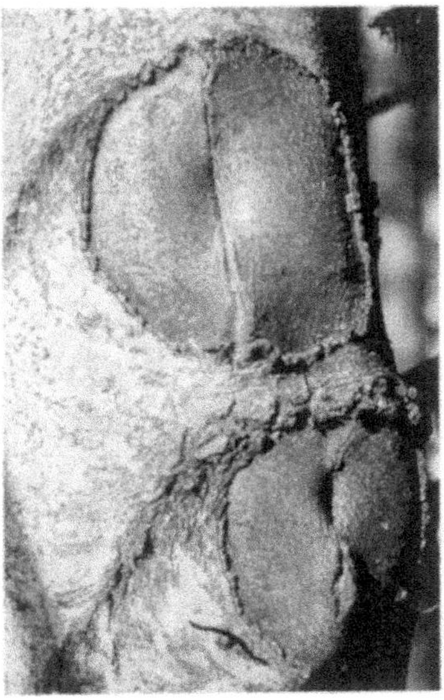

Fig. 34. Completely healed branch of a beech

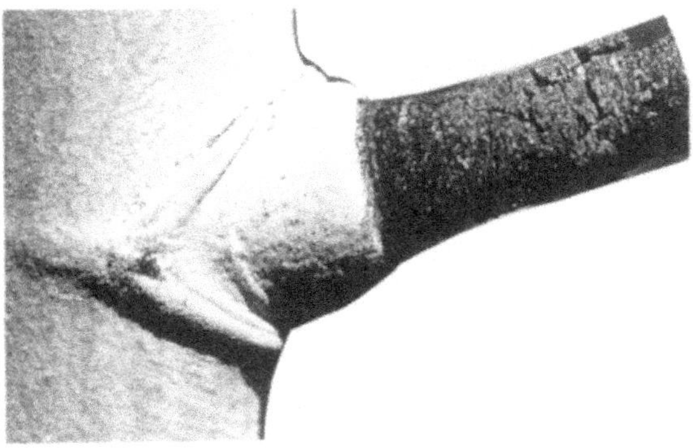

Fig. 35. A healed dead poplar branch. The healed area is immune to fungal growth in the dead branch

Macroscopic Wood and Tree Ring Features

In wood identification, macroscopic characteristics are visible to the naked eye or with the aid of a magnifying glass enlarging up to 20-fold. Providing the wood is fresh and dry, it is possible to identify macroscopically the majority of woods and trees in the temperate zones. This procedure, however, requires experience. People who work with wood draw extensively from a knowledge of species based on macroscopic characteristics.

Guides to identifying wood through the use of macroscopic characteristics have been published for the Central European region by Vorreiter (1949), Trendelenburg, and Mayer-Wegelin (1955), and Bosshard (1982), and for North America by Panshin and de Zeeuw (1980). Descriptions of different kinds of wood can be found in Begemann (1981–1987), Sell (1989), and Dechamps (1983).

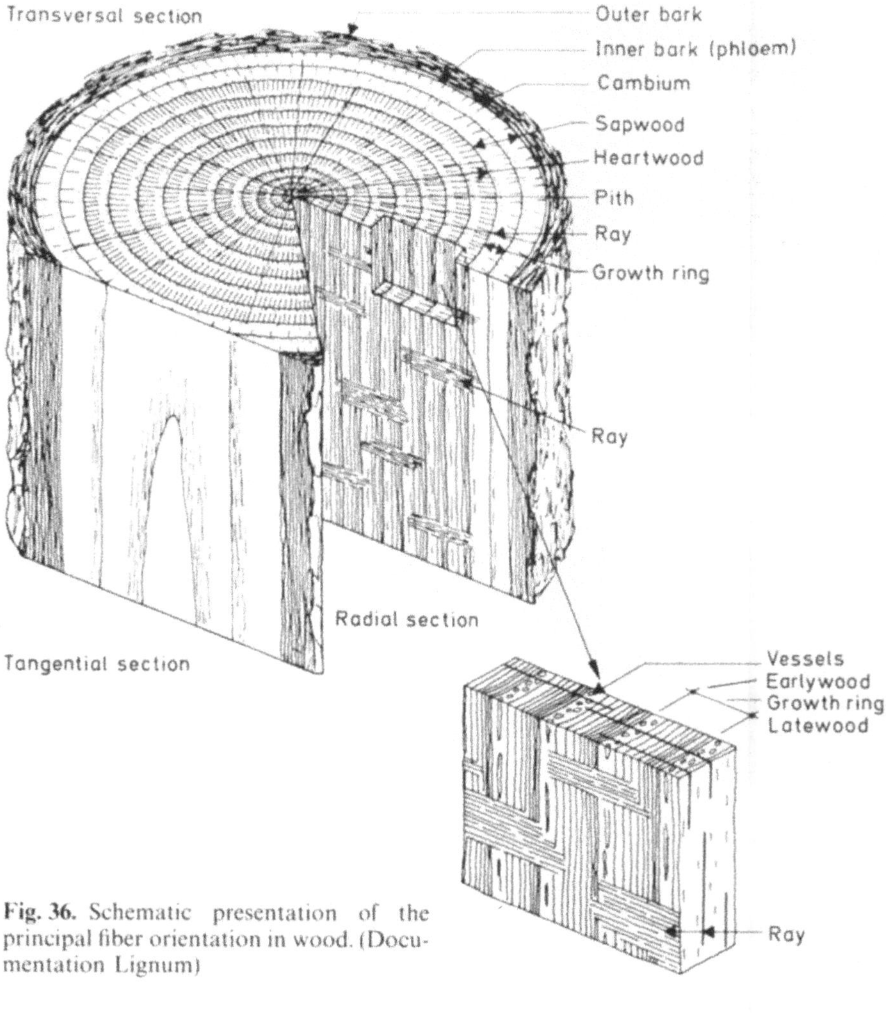

Fig. 36. Schematic presentation of the principal fiber orientation in wood. (Documentation Lignum)

Macroscopic Tree-Ring Characteristics

The macroscopic characteristics of a tree-ring sequence are of fundamental importance. Through them it is possible to obtain information on the age and overall condition of a tree. The most important characteristics are the width of the growth rings, latewood width, density variations, callus tissues, and healing tissues.

Tree ring and latewood widths are specific for individual species and sites. Event years and abrupt changes in the tree's growth rate are characteristic and ecologically significant. Abrupt changes in growth rate reveal changes occurring over more than 3 years.

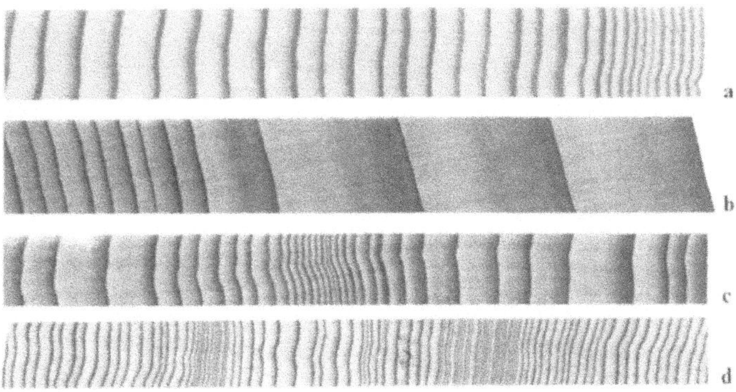

Fig. 37. a Decrease in later years, *Abies alba*. **b** Recovery in later years, *Abies spectabiliis*. **c, d** Decrease and recovery, *Abies spectabiliis*, *Fitzroya cupressoides*

Conspicuous or odd tree rings (event years) are a result of short-term, 1- to 3-year, changes. If event growth rings appear frequently in trees from one particular site or region, they are then termed "pointer years".

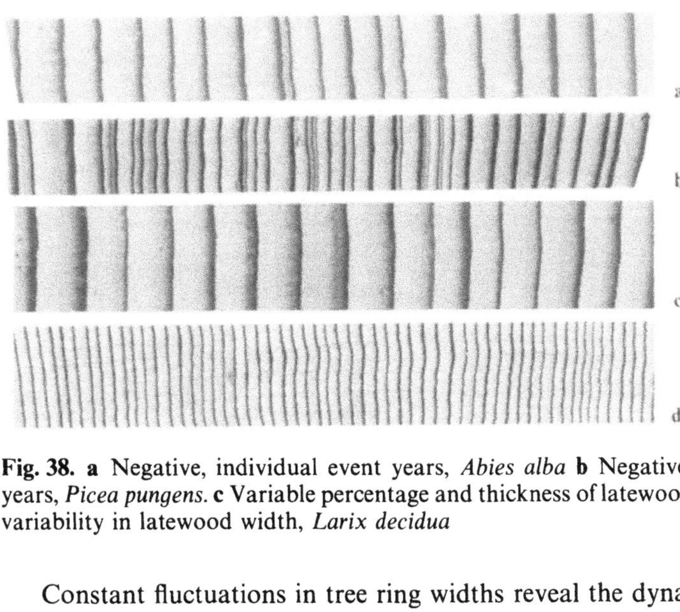

Fig. 38. a Negative, individual event years, *Abies alba* **b** Negative and positive event years, *Picea pungens*. **c** Variable percentage and thickness of latewood, *Picea abies*. **d** Low variability in latewood width, *Larix decidua*

Constant fluctuations in tree ring widths reveal the dynamics of aging and condition.

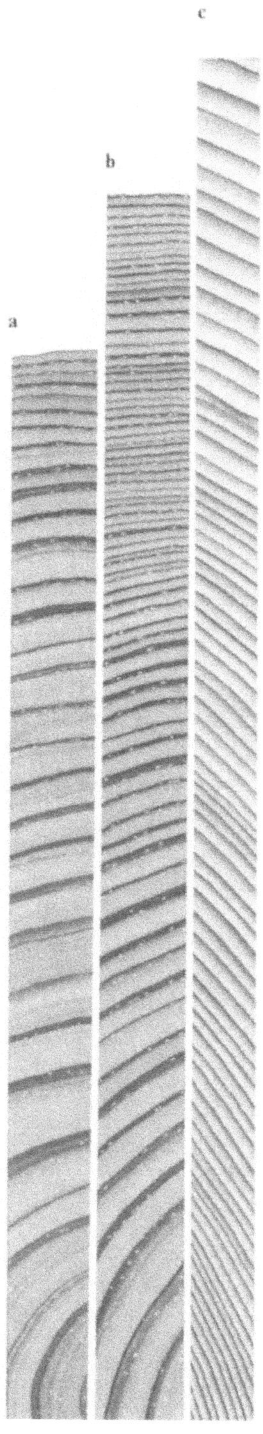

Fig. 39. **a** An abruptly reduced tree ring width. **b** A slowly reduced tree ring width. **c** An abruptly increased tree ring width

Proportions of latewood are mostly species-specific

a

b

c

Fig. 40. **a** Very low proportion of latewood, *Podocarpus*. **b** Low proportion of latewood, *Pinus cembra*. **c** High proportion of latewood, *Pinus sylvestris*

Density variation within a tree ring, unclear tree-ring boundaries and wedging rings are mainly the result of climatic influences.

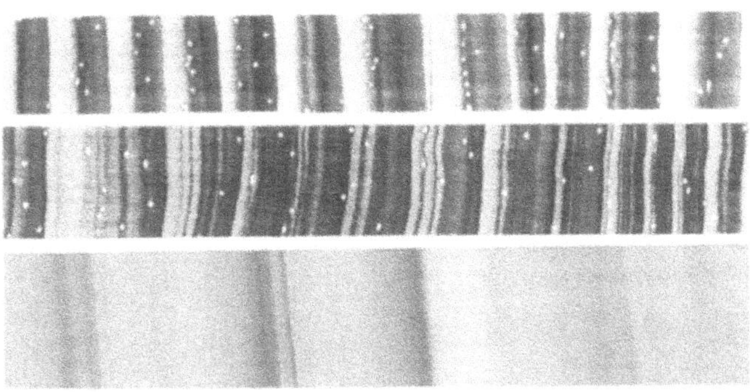

a

b

c

Fig. 41. **a** Slight to heavy variations in the width of earlywood and latewood in *Pinus sylvestris*. **b** Slight to heavy variations in width and false rings in the wood of a subtropical pine. **c** Unsharp growth ring boundaries in a subtropical Cupressaceae

Wedging Rings

Fig. 42. Yew with wedging rings

Formation of compression wood in conifers is often caused by a change in site. The growth rings in compression wood consist largely of rounded, thick-walled tracheids.

Fig. 43.a Compression wood in the stem of a young fir. **b** Extreme, short-term reaction to a change in site (*Picea glehnii*). **c** Compression wood in a fir

Fig. 44. Injury

Fig. 45. Callus formation

Microscopic Wood and Tree Ring Features

Microscopic characteristics of wood and tree rings are visible with the aid of microscopic sections or with the aid of surface preparations using highly enlarging magnifiers or a stereomicroscope. In most cases, an exact wood identification is only possible by referring to the microscopic features. This is especially true for old, heavily eroded, or decomposed wood. The following illustrations present features appearing in cross-sections. For further references on wood anatomical identification guides see for Europe: Grosser (1977) and Schweingruber (1978, 1990), for North America Panshin and de Zeeuw (1980), and for New Zealand Meylan and Butterfield (1978).

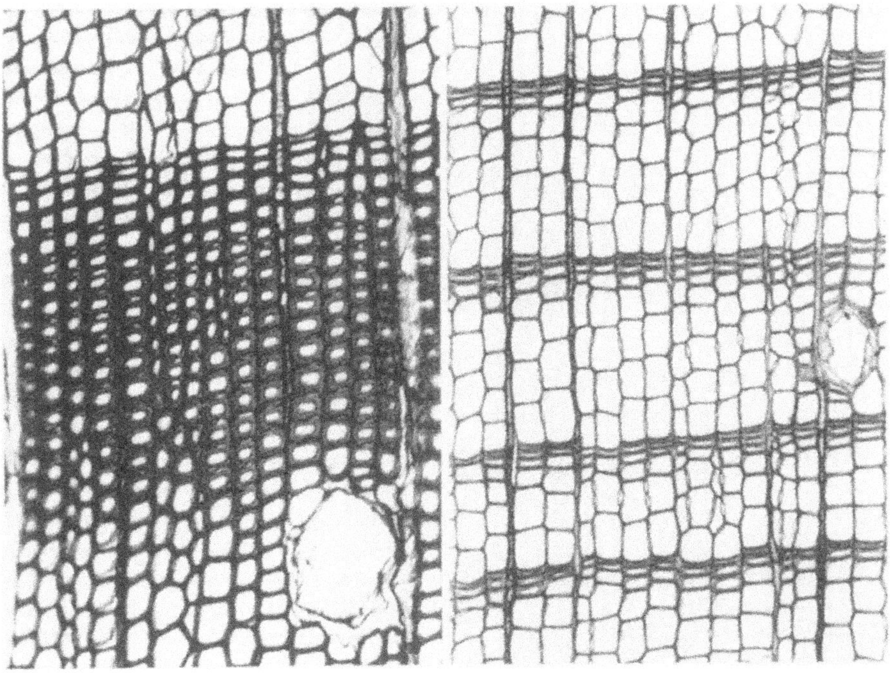

Fig. 46. Coniferous wood. **a** *Pinus sylvestris* 100 x. **b** *Larix decidua* 100 x

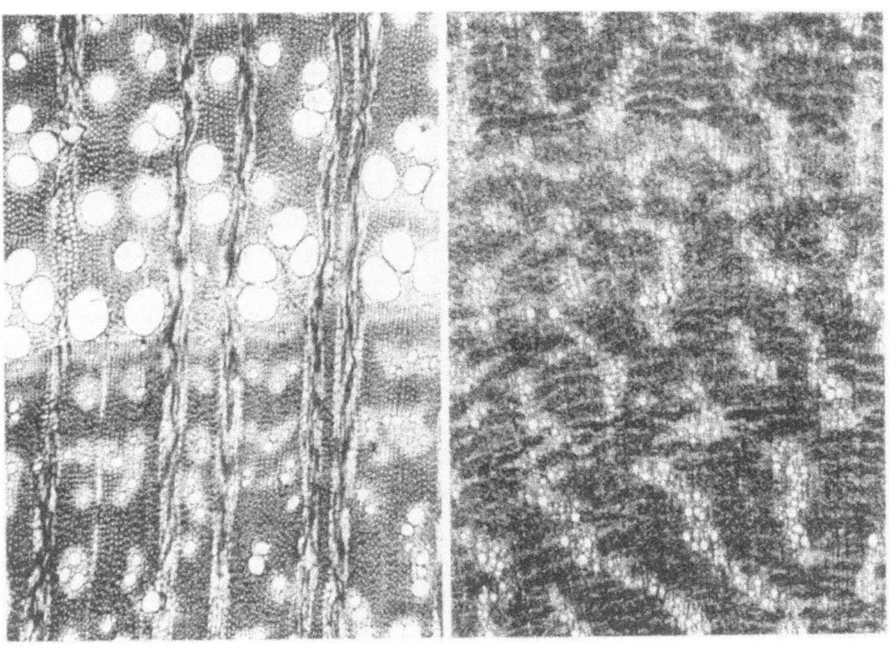

Fig. 47. Dicotyledonous wood. **a** *Tamarix gallica* 40 x. **b** *Thymelea hirsuta* 40 x

No Tree Ring Formation (all enlarged 40 x)

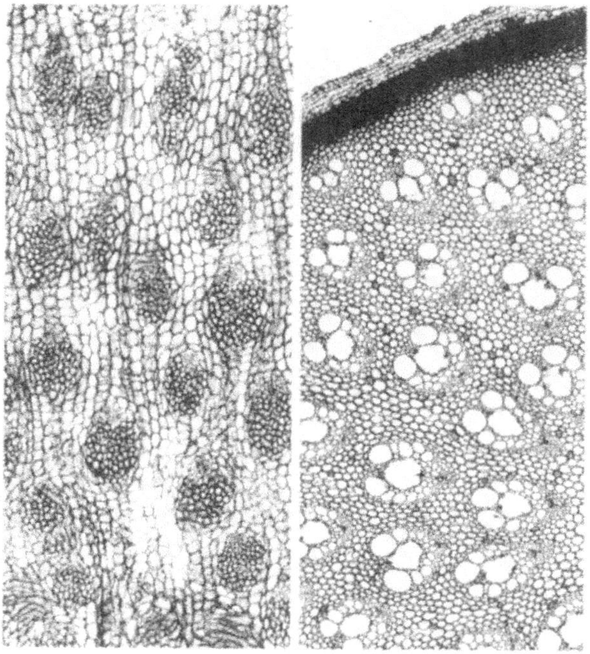

Fig. 48. a *Yucca* sp. **b** *Asparagus* sp.

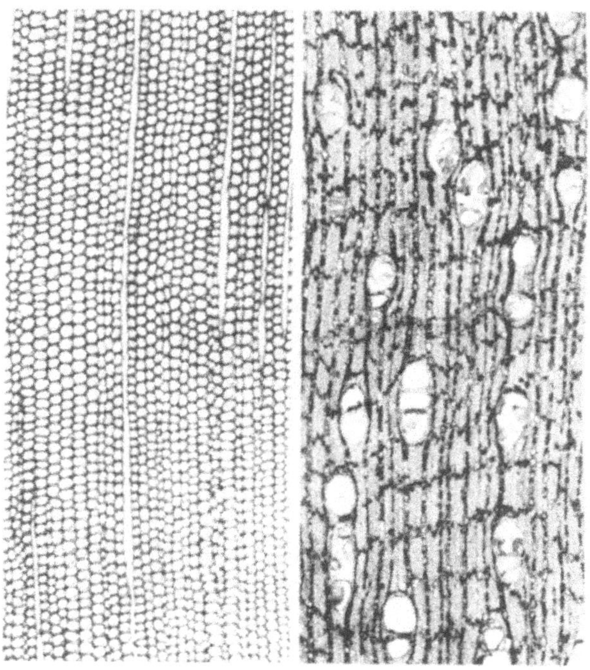

Fig. 49. a *Araucaria bidwillii.* **b** *Terminalia* sp.

Indistinct boundaries

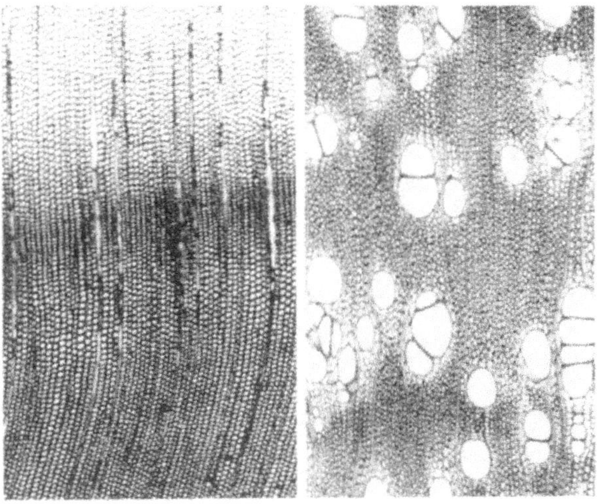

Fig. 50. **a** *Cupressus dupreziana.* The transition from latewood to earlywood is continuous. (A desert tree with contact to ground water). **b** *Acacia* sp. Tangential, thick-walled fiber bands without distinct borders suggest tree-ring boundaries.

Distinct boundaries

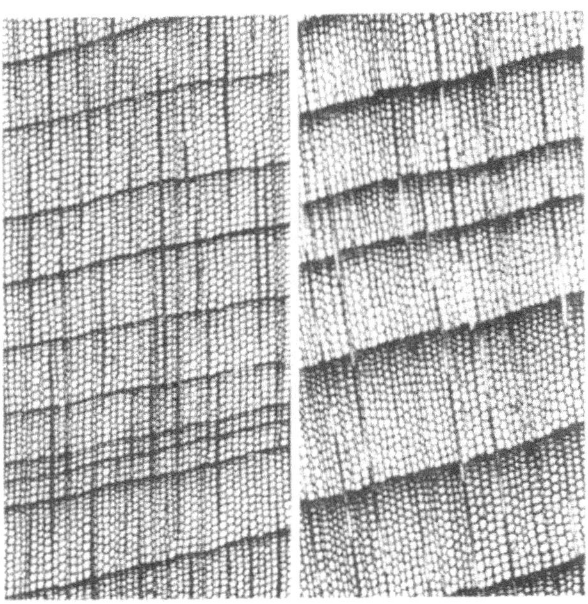

Fig. 51. **a** *Juniperus thurifera.* One to two thick-walled tracheid bands define the tree ring boundaries. **b** *Taxus baccata.* Tangentially compressed tracheids with thick-walled cells form latewood with sharp boundaries.

Vessel (pore) distribution/vessel size (all enlargements 40 x)

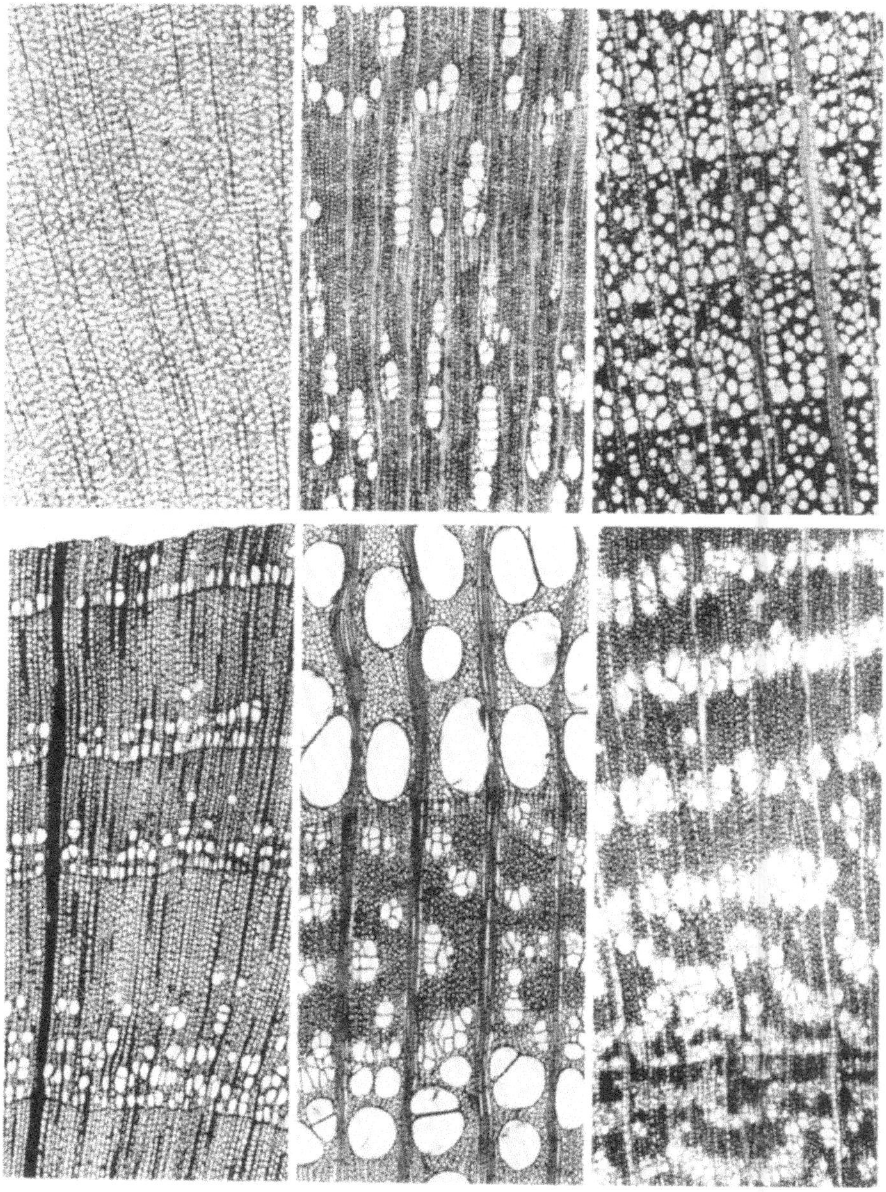

Fig. 52. a *Ledum palustre*. Diffuse-porous, very small vessels. **b** *Ostrya carpinifolia*. Diffuse-porous, medium-sized vessels. **c** *Prunus avium*. Semi-ring porous, vessel size decreases from earlywood to latewood. **d** *Myrica gale*. Semi-ring porous, the number of vessels decreases abruptly from earlywood to latewood. **e** *Morus alba*. Ring-porous, vessel size, varying between earlywood and latewood. **f** *Phlomis* sp. Diffuse-porous, tangential bands

Parenchymatic band and thick-walled fibers

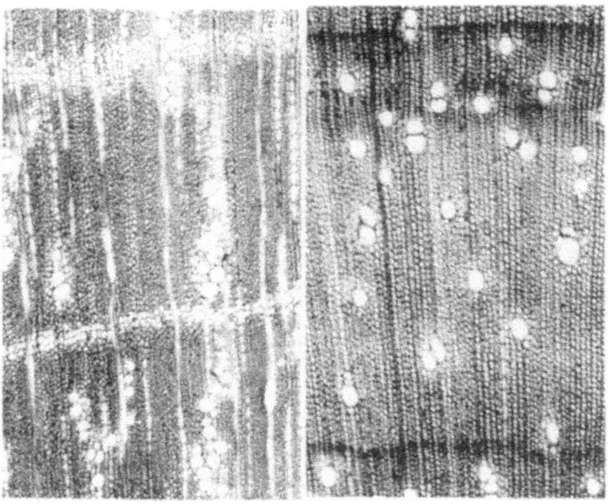

Fig. 53. a *Notholaena excelsa.* Diffuse-porous, multi-rowed tangential banded parenchyma in the earlywood. **b** *Ocotea foetans.* Diffuse-porous, thick-walled, fiber bands in the latewood

False rings (40 x)

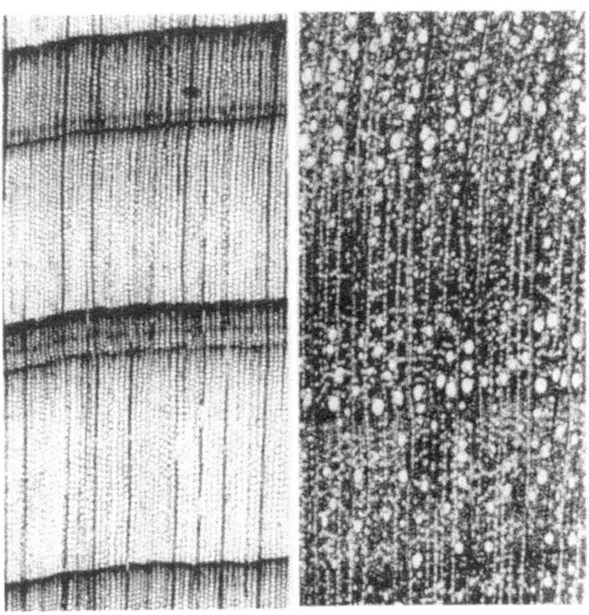

Fig. 54. a *Cupressus sempervirens.* Fiber bands varying in thickness appear in the annual growth ring. In extreme cases it is impossible to differentiate between density variations and actual tree rings. **b** *Myrtus communis*

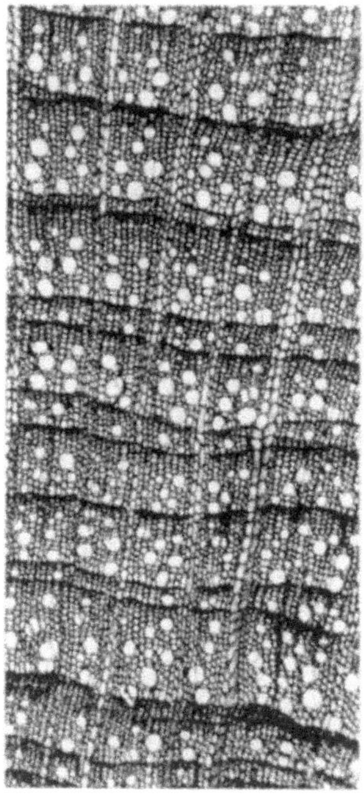

Fig. 55

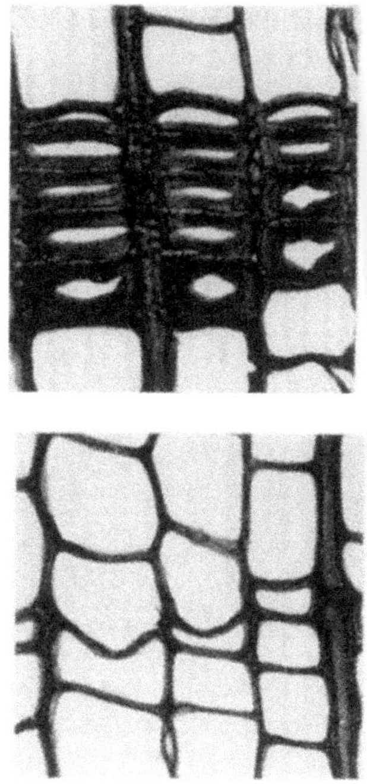

Fig. 56

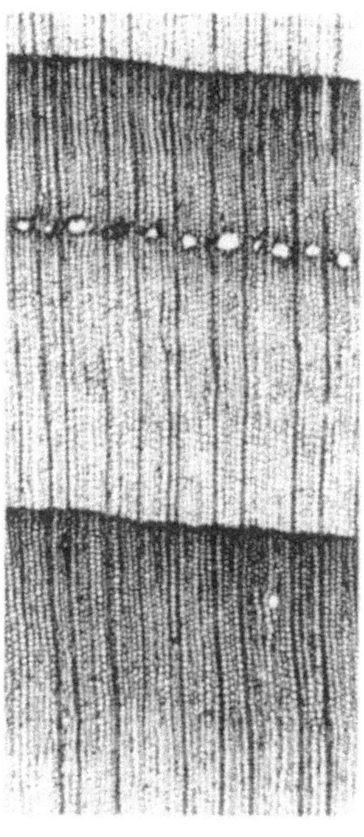

Fig. 57

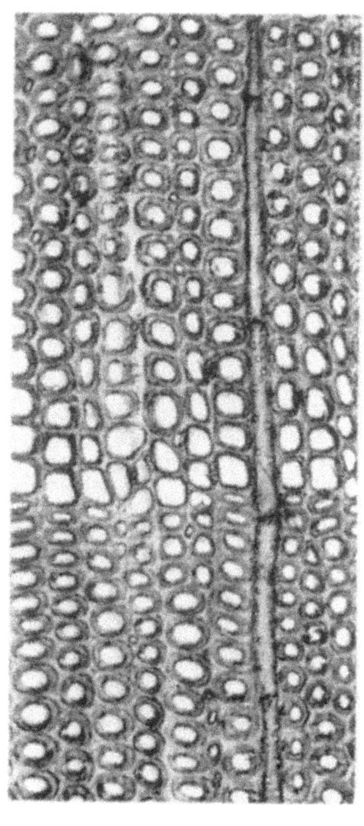

Fig. 58

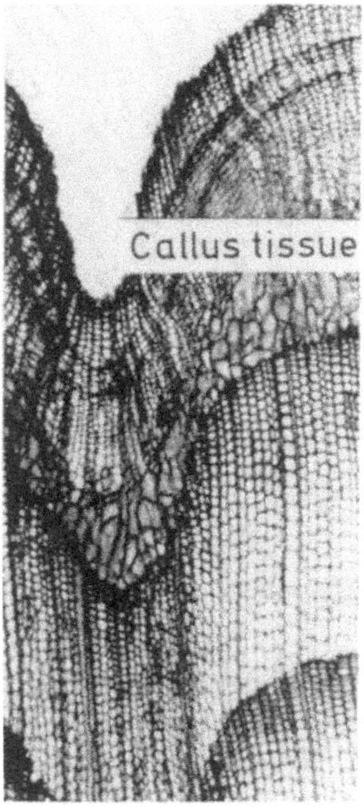

Fig. 59. *Abies alba.* Callus tissue after an injury

Fig. 60. *Suaeda pruinosa.* Anomalous secondary growth in thickness with phloem enclosures

Monocotyledoneous trees do not form tree rings. The exact age of the tree can only be determined by assessing the length of the shoot.

In the steady climate of the tropics, most species do not form tree rings. Growth zones may exist, but these zones do not correspond to annual rings.

Fig. 55. Wedging rings. *Ephedra major.* Locally the tree rings may be missing. In certain spots, wedging can occur

Fig. 56. Latewood cell walls. *Larix decidua* 250 x. *Above* latewood with very thick-walled tracheids; *below* latewood with thin-walled tracheids (larch bud moth)

Fig. 57. *Cedrus atlantica.* A tangential resin canal sequence

Fig. 58. *Abies alba.* Compression wood with very thick-walled tracheids

Tree Species in Eurasia and Japan

Abies alba Mill.

European silver fir, common silver fir, silver fir

Morphology

This evergreen conifer has a moderately broad, rather conical crown and deep tap roots. It can grow to a height of 40 m, sometimes 60 m. Trees growing in managed forests usually have trunks with a diameter of 100 cm, while trees from old growth forests can have diameters of over 200 cm.

Fig. 61. A fir forest (*Abies alba*) on an optimal site in Emmental, Switzerland.

Site, Site Conditions, Characteristics

Silver fir is a shade species that grows best in cool, humid climates with frost periods of at least 3 months.Temperatures in January average just under 0 °C, and in July rarely average lower than 13–14 °C. The stands are often located in

Fig. 62. From Hess et al. (1967)

Fig. 63. An older fir with stunted top growth (stork's nest)

cloudy regions. Silver fir flourishes in fresh, deep, humus-rich soils and is indifferent to soil acidity. In mountain zones, the species forms pure stands. On low sites, the stands are generally mixed with beech and on higher sites with spruce.

Fig. 64. A typical old fir with a flattened top. The bark of younger trees is smooth and white, while that of older trees is scaly and fissured

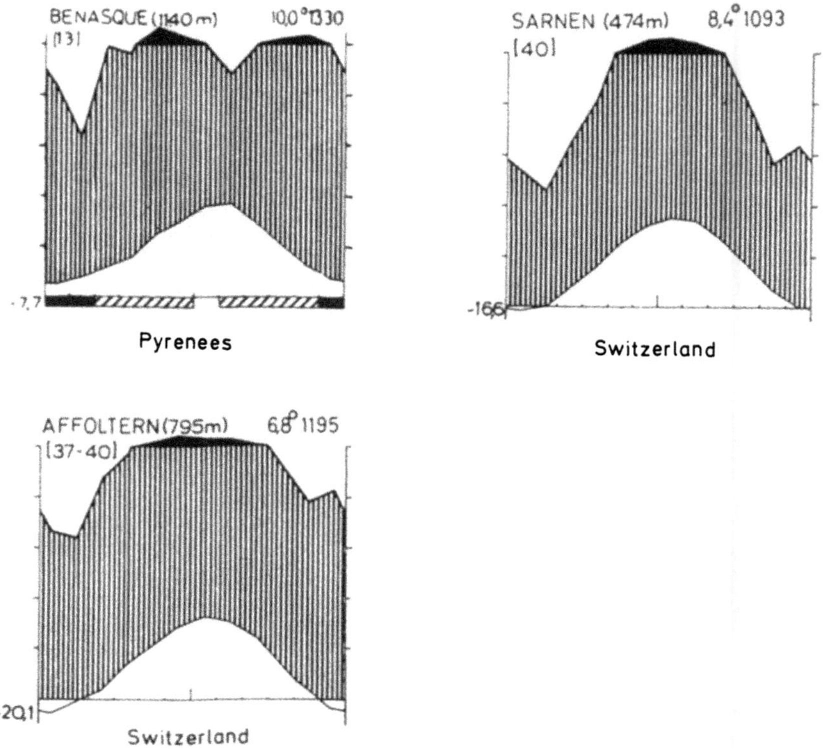

Fig. 65. Climate diagrams

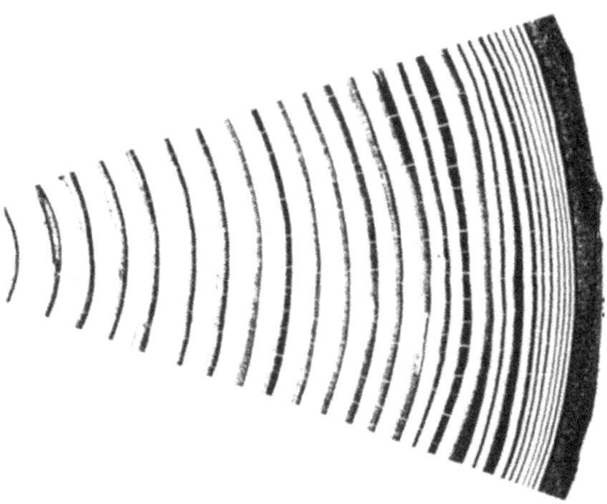

Fig. 66. A cross-section through a young fir trunk with an abrupt growth reduction

Fig. 67. Tap roots are typical of the genus *Abies*. (Köstler et al. 1968)

The range of silver fir is restricted by its high humidity demands and frost sensitivity. Thus, tree rings from frost years are generally exhibited as pointer years or are the beginning of a long-term abrupt growth reduction, for example in 1928 and 1956 in Switzerland.

In the inner-alpine distribution area, such as in Wallis, Switzerland, frost- and drought-resistant races have evolved. In dry regions, the species builds small mosaic-like stands among pine stands and among spruce and larch-Arolla pine stands at the timberline (Lingg 1986).

The species is currently being attacked by a curious disease (silver fir decline), which has resulted in the decimation of large forest areas. Stands outside the natural distribution area of the species are especially susceptible to attack. The disease is indicated in the tree rings by abrupt, long-lasting growth reductions (Schweingruber 1983).

Wood

The heartwood is white with a red shimmer. The wood has a slightly sour odor. The tree ring boundaries are distinct. The density is 320–410 kg/m^3.

Cross-section: no resin canals. Rare traumatic resin canals in tangential rows. In general a sharp transition from earlywood to latewood (Fig. 68 a–c).

Tangential section: the average ray height is high, with 15–25 cells.

Radial section: rays without ray tracheids. Taxodioid pits in the earlywood. Horizontal and tangential ray walls are thick and tooth-shaped. There is usually one, often two rows of pitting in radial tracheid walls.

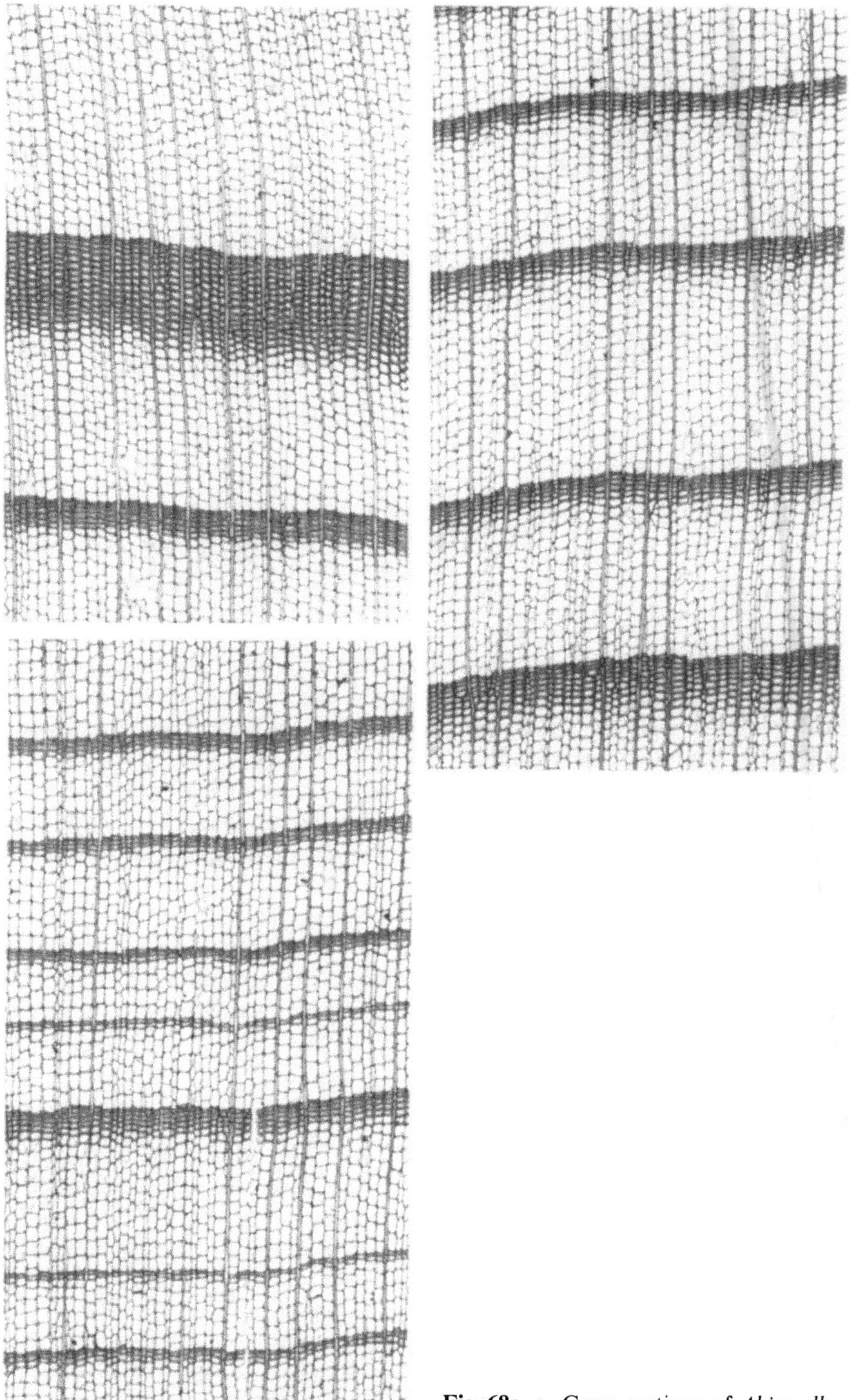

Fig. 68a–c. Cross-sections of *Abies alba*

Fig. 69. Radiographs of *Abies alba*

Distribution (see p. 42 for map)

Horizontal: in the west and central European mountains.

Vertical: in the medium altitude mountain ranges in Germany at an elevation of around 500–900 m. The upper border lies in the Bohemnian forest at around 1000 m, in the Jura mountains at 1400 m, in the Alps at 1700 m, and in the maritime Alps at 2100 m above sea level. In the Alps, the species is a relict, for example in Wallis, Samnaun, and Schans.

Use in Dendrochronology

Suitability of the Species

The wood, with its absence of resin canals and distinct latewood boundary zones, is especially suitable for chronological and climatological purposes. Samples from low sites exhibit multiple indicator years which can have many possible causes (winter cold spells, water shortage in spring and winter, and summer droughts). The fir was frequently used as building material in archeological and historic times (either for fence posts or house beams). Because the wood preserves especially well in water, fossil material also lends itself to densitometric research.

Previous Use

Prehistoric settlements in lake and swamp regions have repeatedly been dated (Becker and Giertz-Siebenlist 1970). Using recent and historical wood, Becker and Giertz-Siebenlist (1970), recorded a 1100-year chronology from southern Germany. Orcel and Orcel (1985) made regional chronologies for Switzerland available. As early as 1951, Müller-Stoll determined a similarity between recent fir tree-ring width curves reaching from Poland to France. Schweingruber (1988) established a similarity between tree-ring width curves and maximal densities from trees growing in subalpine regions from the Alps to the Pyrenees and from the Alps to former Yugoslavia. Summer temperatures can be reconstructed from the maximal densities. An ecological study from Wallis, Switzerland (Lingg 1986), showed that fir sometimes exhibits a different dendroclimatological behavior than spruce or pine, which is related to its root structure. A study of abrupt growth reductions in firs from varying regions in Switzerland resulted in new geographic and time patterns. Emmental and the canton Tessin exhibit practically none, while the central plateau and Jura mountains, the Black Forest, and the Alp foothills, on the other hand, show growth declines occurring since the 1940s. In recent years, numerous dendrochronological studies have been published, especially in Germany and France, dealing with *Abies alba*, for example by Schöpfer and Hradetzky (1986) and by Becker et al. (1990). They all indicate an increased growth rate since 1980.

Abies sibirica has, despite its wide distribution in Russia and Siberia, only seldom been used in dendrochronology (Chebakova 1987).

Abies borisii-regis Mattf. King Boris fir
A. bornmülleriana Mattf. Bornmüller's fir
A. cephalonica Loud. Grecian fir
A. cilicica Carr. Cilician fir
A. equi-trojani Aschers. et Sint.
A. marocana Trabut Moroccan fir
A. nebrodensis Mattei Sicilian fir
A. nordmanniana Spach Caucasian fir
A. numidica De Lannoy Carrière Algerian fir
A. pinsapo Boiss. Spanish fir

Morphology

In optimal climates, these evergreen trees normally form monopodial stems. On dry sites, most notably *Abies pinsapo*, the trees tend to develop cedar-like shapes. The trees can reach heights of 30 m and diameters of 1 m are common on optimal sites. The morphological variability is great, especially in hybrids such as *Abies borisii-regis* (*A. alba* × *cephalonica*) and *A. bornmülleriana*/*A. equi-trojani* (*A. cephalonica* × *A. nordmanniana*).

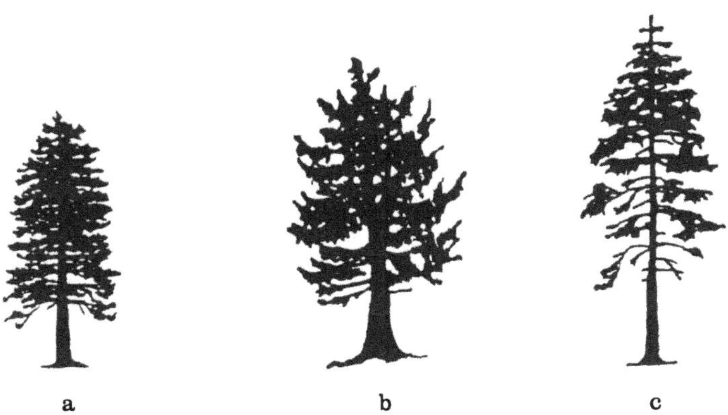

Fig. 70. **a** *A. nordmanniana.* **b** *A. cephalonica.* **c** *A. numidica*

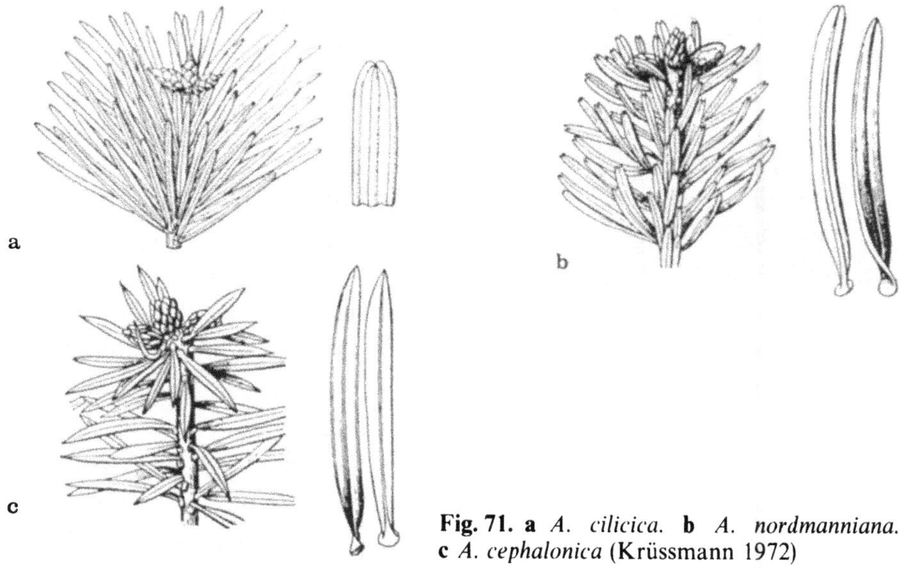

Fig. 71. a *A. cilicica*. b *A. nordmanniana*. c *A. cephalonica* (Krüssmann 1972)

Fig. 72. *Abies cephalonica* cones

Fig. 73. *Abies pinsapo* branch with cones. The horizontal arrangement of the needles is typical

Fig. 74. *Abies nebrodensis* site on Monte Scallo in Sicily. (Krüssmann 1972)

Fig. 75. A 30 to 35 m tall *Abies marocana* in a shady hollow. *Quercus ilex* and *Q. faginea* only reach the intermediate layers (Mayer 1965)

Fig. 76. *Abies marocana* on a dry slope in the Atlas mountains. (Mayer 1965)

Fig. 77. *Abies pinsapo* at timberline by Ronda (Andalusia, Spain). The limestone bedrock is overgrown with gorse

Fig. 78. Young, rapidly growing *Abies numidica.* The tapered crown and dense branching are typical

Site, Site Conditions, Characteristics

In comparison to *Abies alba*, the species from eastern Europe have lesser hydrological demands. This becomes especially apparent in plant sociological surveys (Mayer 1984). The species grow almost entirely in mountain and subalpine zones, where they profit from the high precipitation. They are, however, quite resistant to periods of summer drought, which normally last for several months. Most species grow on hard limestone but can withstand both

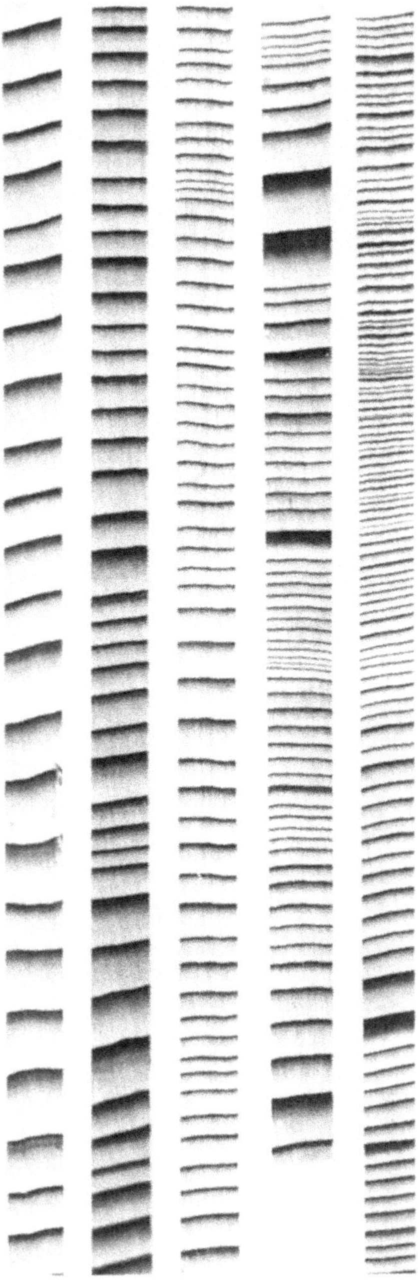

Fig. 79. Radiographs of *Abies pinsapo*

granite and acid foundations. At high altitudes they often form the forest border or are in contact with cedar or *Pinus leucodermis* (*Abies cephalonica*). At lower altitudes, the stands are accompanied by semiarid, submediterranean flora, such as *Quercus ilex*. Several of the species grow to be between 300 to 400 years old. As a result of overuse (firewood cutting, fire, and grazing) old trees are rare and the stand's structure is often ruined. *Abies nebrodensis* stands in Sicily are threatened with extinction. Only approximately 22 trees remain on Sicily's steep, rocky cliffs. However, replanting is taking place in Calabria.

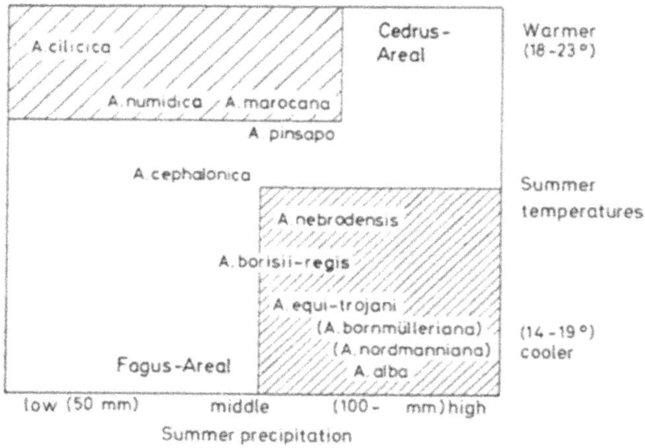

Fig. 80. Ecogram of European firs. (Mayer 1965)

Moisture and temperature conditions during the vegetation period in the mountain fir species from Mediterranean countries. *Fagus* and *Cedrus* border the sites. Species requiring humid climates (*Abies alba*, *A. nebrodensis*) clearly differentiate themselves from dry area species (*Abies marocana*, *A. cilicica*). *Abies cephalonica* and *A. pinsapo* occupy an intermediate position. (Mayer 1984, 1962).

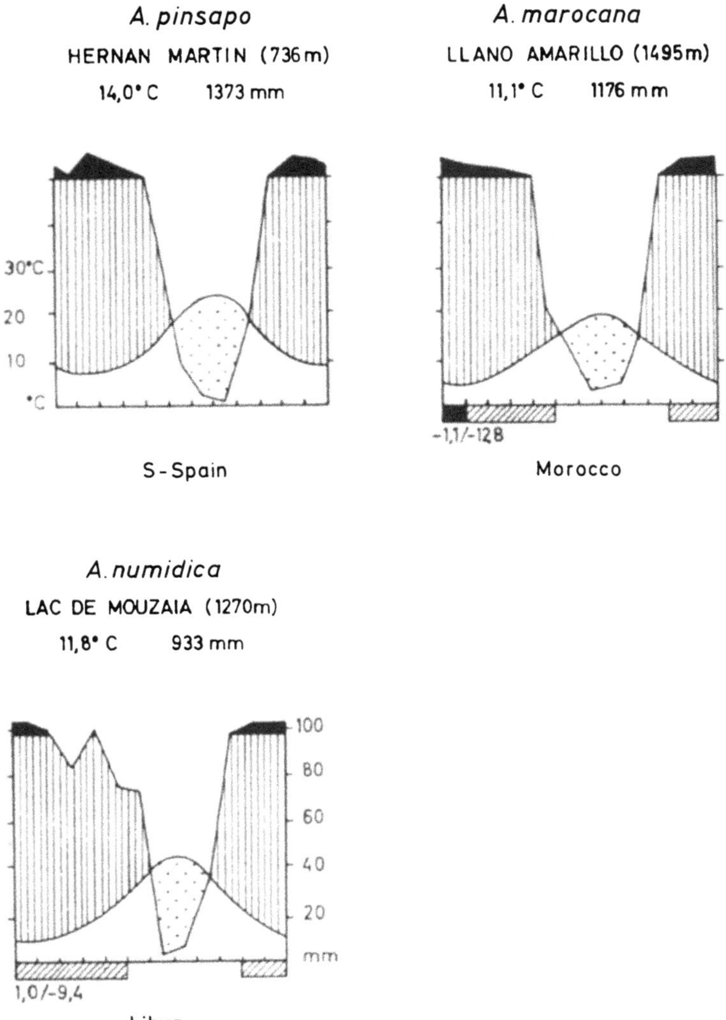

Fig. 81. Climate diagrams (Mayer 1965)

Distribution

Horizontal: the once large, united distribution area was split up during the ice age, resulting in the scattered, smaller areas found today (Fig. 82).

Vertical: *Abies cephalonica*: trees grow on the lowest sites at 400–600 m. The climax forest extends from 1200 to 1700 m. The highest examples are found at 2100–2300 m and form the timberline with *Abies marocana* and *A. numidica* at 1500–2000 m and *Abies pinsapo* at 1600–2000 m.

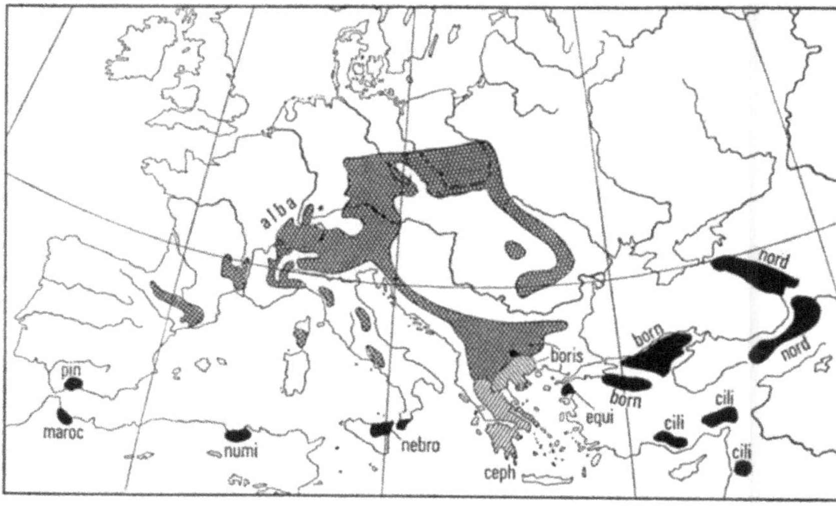

Fig. 82. Distribution of *Abies* species in Europe and Asia Minor. *Abies alba* (alba), *A. borisii-regis* (boris), *A. bornmülleriana* (born), *A. cilicica* (cili), *A. cephalonica* (ceph), *A. marocana* (maroc), *A. nebrodensis* (nebro), *A. numidica* (numi), *A. nordmanniana* (nord), *A. equi-trojani* (equi), *A. pinsapo* (pin). (Krüssmann 1972)

Wood

See *Abies alba* for description.

Use in Dendrochronology

Suitability of the Species

The other species are just as suitable for dendrochronology as *Abies alba.* Due to their broad ecological amplitudes, the species can be used in ecological studies on moisture gradients in relation to vegetation and tree growth. Considering the tree shortage in the Mediterranean region, *Abies* species are those best suited for dendrochronological studies.

Previous Use

The Mediterranean species have received minimal attention. Munaut and Serre-Bachet (1982) cited studies on *Abies pinsapo, A. marocana, A. numidica,* and *A. cilicica. Abies pinsapo* could not be fitted into the European sample grid because its maximum ring width, is restricted by periods of summer drought (Schweingruber 1987). Mayer (1965) published mean curves for *Abies marocana, A. pinsapo,* and *A. numidica. Abies nordmanniana* from the western Caucasus was dendrochronologically characterized by Brukstus and Baliciunas (1983).

Abies spectabilis Mirb.

East Himalayan fir

Morphology

This large tree can grow as tall as 50 m and in extreme cases has a trunk diameter of over 1 m. The crown is broad and conical with widespreading branches. *Abies spectabilis* has a deep root system which allows it to grow on relatively dry sites.

Site, Site Conditions, Characteristics

Abies spectabilis is the largest tree of the upper fog-forest zone on the southern slopes of the Himalayas and of the temperate moist valleys of the interior of the Himalayas. It is found growing in a wide ecological range. The dry border in the interior has an annual precipitation level of around 800 mm while the southern rim of the Himalayas receives 4000 mm annually. Precipitation reaches its highest level in the summer months, from July through September. On optimal fir sites, temperatures in July average around 10–12 °C. In its coldest location, the frost-free period lasts for only 2–3 months and for 4–6 months the temperatures reach lows of −10 °C, *Abies spectabilis* builds naturally high forests, with 2–6 m high evergreen *Rhododendron barbatum* shrubs. On the upper border of the species, it is possible to find a few firs towering above closed *Juniperus recurva* or *Betula utilis* stands. All stands are predominantly found on acid geological foundations (Dobremez 1976). Today, closed fir forests seldom exist on shady slopes. On the heavily used southern slopes, *Rhododendron arboreum* or *Juniperus recurva* forests have developed as a result of fire clearing, logging, and grazing.

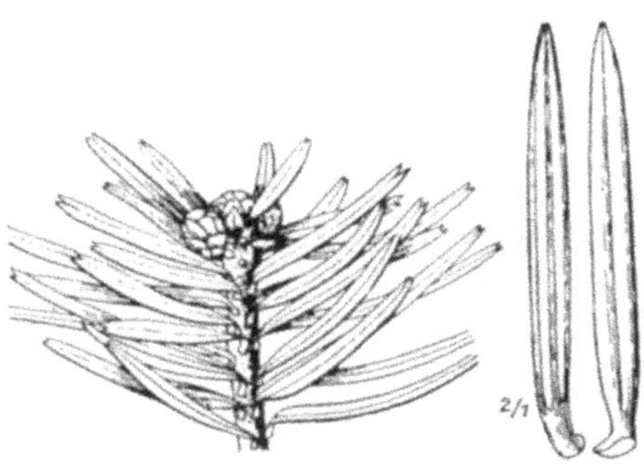

Fig. 83. *Abies spectabilis* (Krüssmann 1972)

Fig. 84. A closed *Abies spectabilis* forest with *Rhododendron barbatum* undergrowth. (Photo Miehe)

Fig. 85. A stand on a dry site at an elevation of 3400 m. The undergrowth is made up of dwarfed, shrub-like *Rhododendron lepidotum*. (Miehe 1982, 1987)

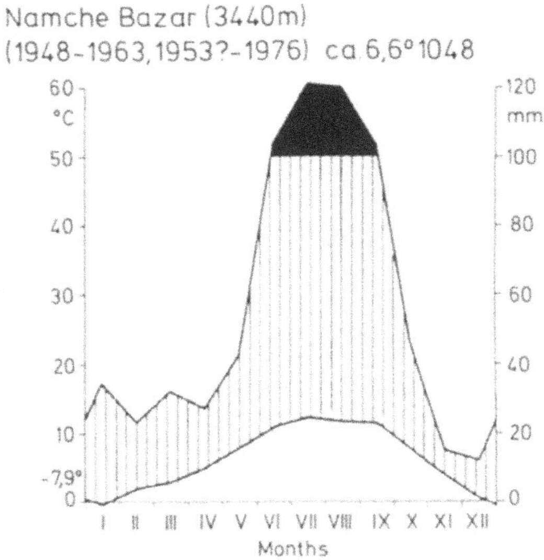

Fig. 86. Climate diagram from a dry valley with fir near Mt. Everest. (Miehe 1987)

Distribution

Horizontal: in the Himalayas south of the main chain from Sikkim to Nepal and Afghanistan. To the East, the area is joined by *Abies densa.*

Vertical: in the Himalayan interior, the lower growth line lies at 3000 m and the upper at 3800 m.

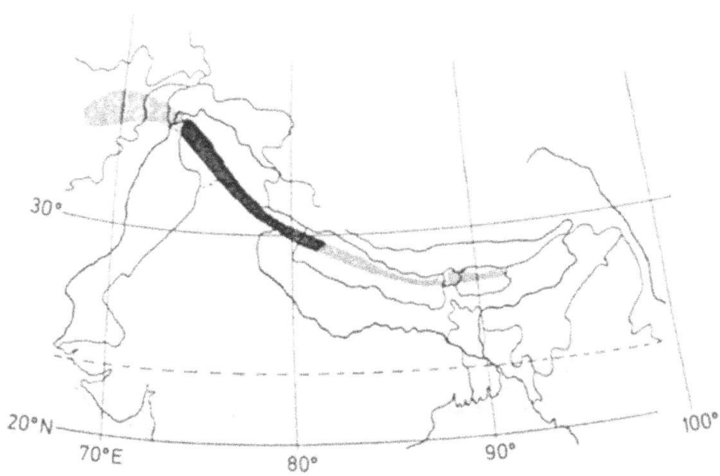

Fig. 87. The geographical distribution of *Abies spectabilis*. In Pakistan the area intersects that of *Abies pindrow* (Dobremez 1976)

Wood

The wood is barely distinguishable from that of *Abies alba*. The earlywood to latewood transition appears to be more gradual than in the latter.

Fig. 88. Radiographs of *Abies spectabilis*

Use in Dendrochronology

Suitability of the Species

Abies spectabilis is best suited for the reconstruction of climate and ecological conditions on timberline sites. The maximal latewood densities are indicators of summer temperatures and multiple, long-lasting growth fluctuations indicate forest dynamic processes. Buildings constructed from *Abies* wood can, presumably, be dendrochronologically dated. Shingle wood is renewed every 7 years.

Previous Uses

There has generally been little dendrochronological work done with *Abies spectabilis* in the Himalayas. Huges (pers. comm.) densitometrically recorded fir chronologies from Kashmir, and Schweingruber (unpubl.) from Nepal. Both studies reveal the restricting effects of summer temperatures on the latewood. Bhattacharyya (1987) made chronologies of additional wood species available. Wu et al. (1988b) provided tree-ring width chronologies of *Abies forestii* from the Hengudan Mountains in China.

Cedrus libani A. Rich.

Cedar of Lebanon

Morphology

This tree has a height of 35 m and a maximum stem diameter of 3 m. The crown is originally narrow, later becoming irregular and broad with spreading, horizontally set branches.

On Cyprus a race can be found with short needles which is identified as *Cedrus libani*, ssp. *brevifolia*.

Fig. 89. *Cedrus libani* var. *brevifolia* (Drawing by Hirzel)

Fig. 90. *Cedrus atlantica* (Drawing by Hirzel)

Cedrus atlantica (Endl.) Carr.

Atlas cedar

Morphology

This tree grows to be 30 m in height with a diameter of 3 m.
The crown resembles that of *Cedrus libani.*

Fig. 91. *Cedrus atlantica* in the Algerian Atlas (natural). (Walter 1968)

Fig. 92. *Cedrus atlantica* near Lake Como, Italy (cultivated)

Site, Site Conditions, Characteristics

Cedrus libani A. Rich.

This light-wood species is native to Mediterranean mountain climates. It can withstand summer drought periods lasting several months, but requires humid winters with heavy snowfall. In Turkey, the species often grows alongside *Juniperus excelsa, J. foetidissima,* and *Abies cilicica.* In Syria, *Cedrus libani* grows in mixed oak-maple-juniper stands.

The range of the species has in historic and recent times been drastically reduced (Phoenicians, Romans, 20th century railway constructors). Today, the famous Cedar of Lebanon has an area reduced to only 300 ha.

In Taurus the trees reach a maximal age of around 500 years; in Cyprus 300 years.

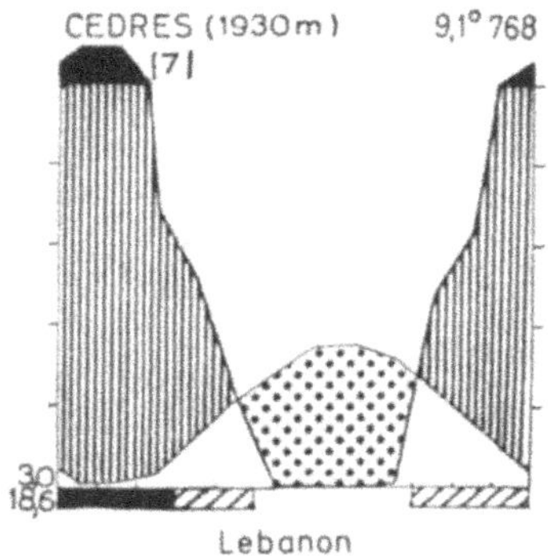

Fig. 93. Climate diagram

Distribution

Horizontal: mainly in southern Turkey (Taurus, Antitaurus, and in Cyprus (var. *brevifolia*).

Vertical: in the mountain and subalpine zones from 1200–2200 m in Turkey and in Cyprus at elevations of around 1000–1500 m.

Cedrus atlantica Carr.

The Atlas cedar has requirements similar to those of *Cedrus libani*. Several different cedar forest societies exist according to variations in precipitation (600–2000 mm/year) and geological foundations (granite, basalt, limestone).

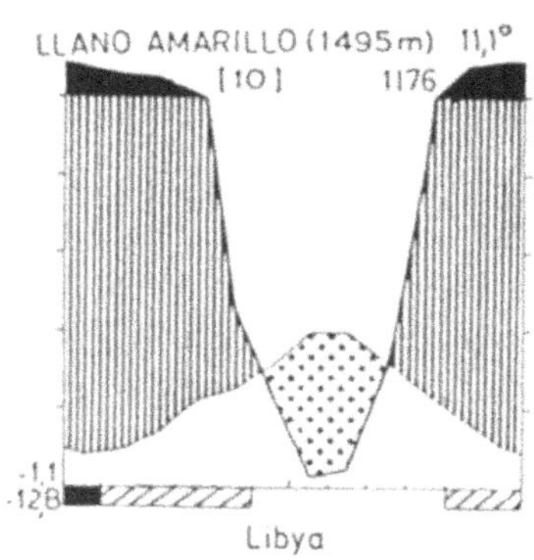

Fig. 94. Climate diagram

They extend from the humid *Quercus ilex-Cedrus atlantica* forest, with rapid-growing trees, monopodial crowns, and heights of 50 m, to semiarid cedar forests near the forest border consisting of short, slow-growing highly branched trees, 10–15 m in height. Most stands suffer from human influence (logging, grazing) (Mayer 1980).

In the Atlas mountains, trees from rapidly growing sites live to be 300 years old. On dry sites they can grow to be 1000 years.

Distribution

Horizontal: in the Atlas mountains. Mainly in the central region, dispersing toward the east.

Vertical: from 1500–2900 m.

Fig. 95. Distribution of *Cedrus atlantica* and *C. libani*

Wood

Both species have heartwood. Their sapwood is pale yellow. *Cedrus libani* has red-brown to dark brown heartwood; *Cedrus atlantica* light brown. Densities average 550 kg/m^3.

Cross-section: the transition from earlywood to latewood is continuous (Fig. 96a, b) or abrupt (Fig. 96c, d). Frequent resin canals in tangential rows (Fig. 96b, d).

Tangential section: the average ray height is 25 cells. The rays are often broad with resin canals (Fig. 96e).

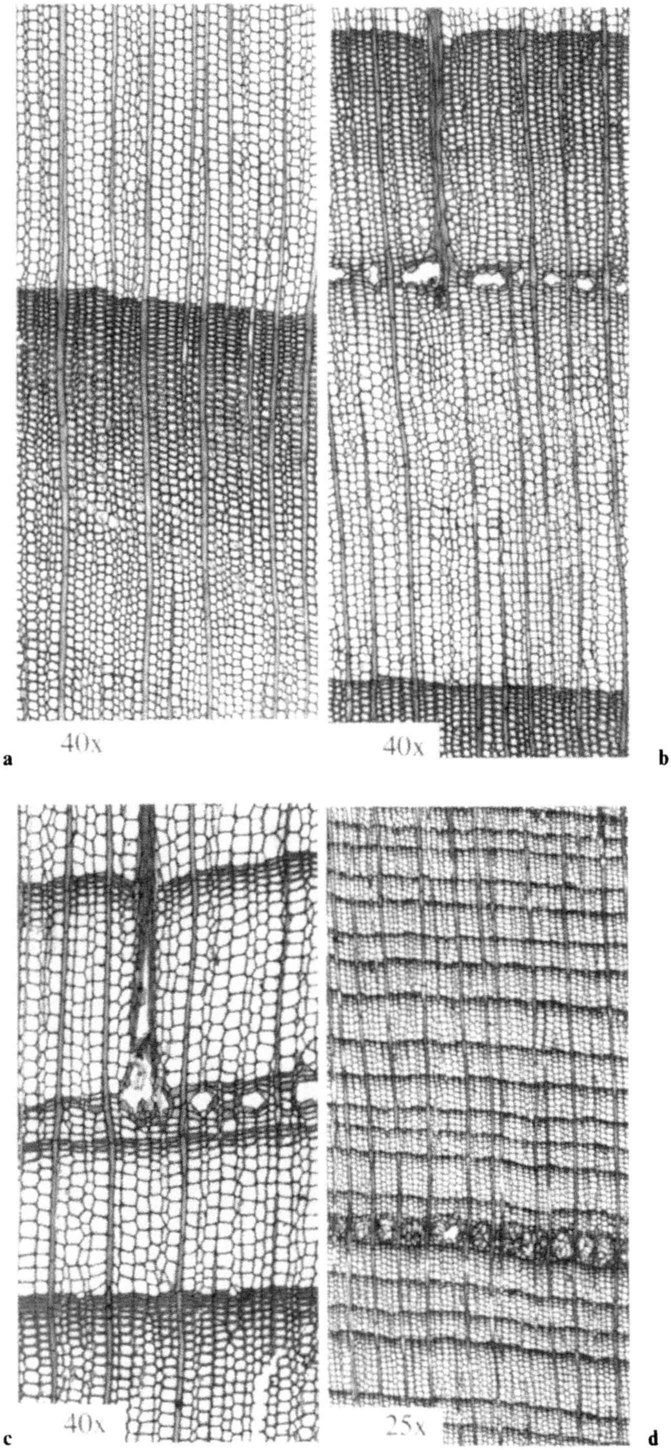

Fig. 96. Cross-sections of *Cedrus libani*

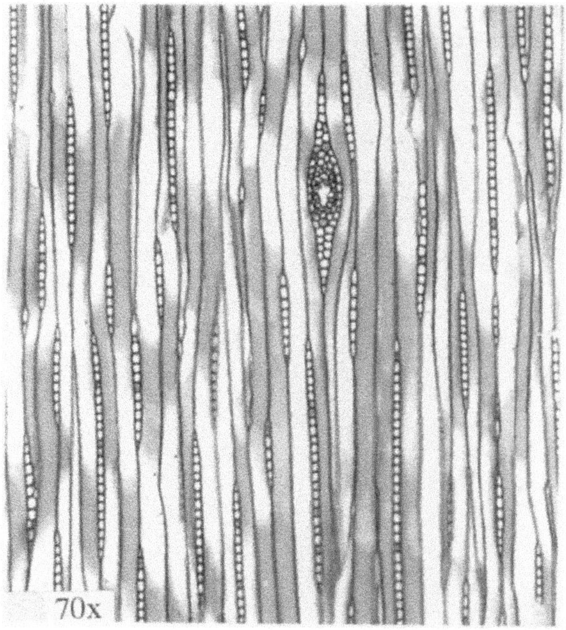

e

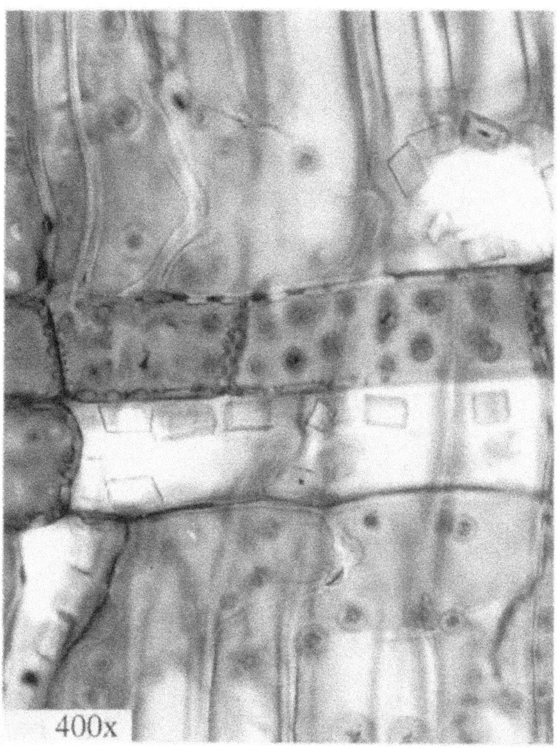

f

Fig. 96. (cont.)

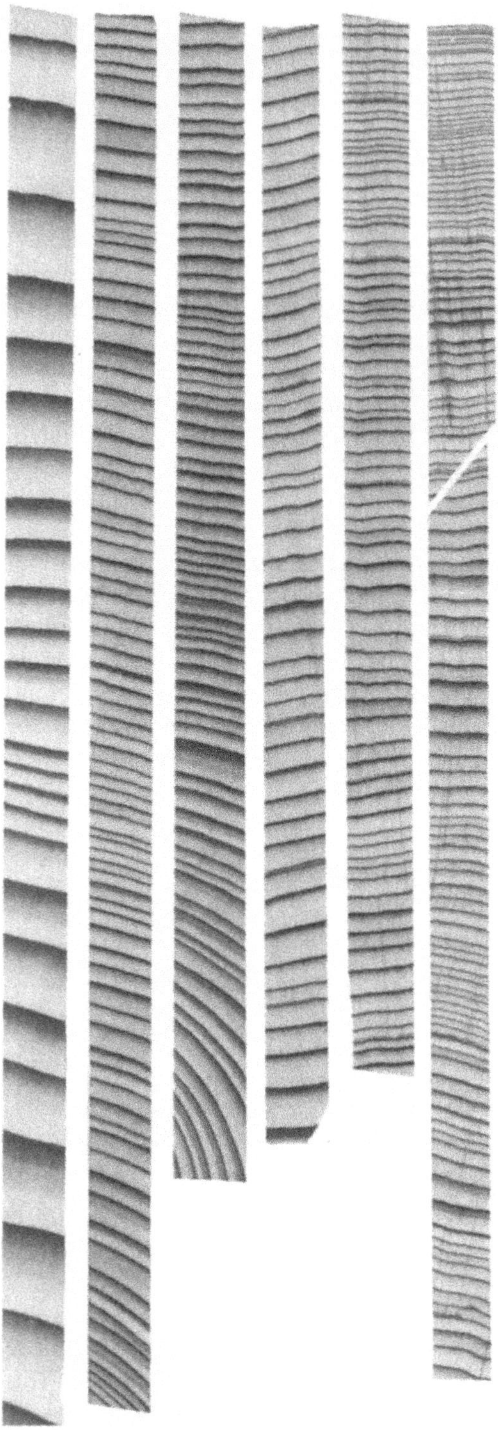

Fig. 97. Radiographs of *Cedrus libani*, var. *brevifolia* (Cyprus)

Radial section (Fig. 96f): tracheids with bordered pits. Rays with ray tracheids. Ray parenchyma cells with broad apertures (taxodioid pits). Thick ray parenchyma cell walls, ray tracheid cell walls are thin. Cells irregularly shaped. Prismatic crystals are common.

Use in Dendrochronology

Suitability of the Species

Both species lend themselves best to dendroclimatological and dendroecological studies, for they exhibit distinct tree-ring boundaries, show few density fluctuations and false rings, have a long life, and are distributed over a wide ecological spectrum. Information on precipitation and temperatures is obtainable. In intermediate altitudes, growth is restricted by many different factors. Radiodensitometric studies hold great promise for climatological research.

Previous Uses

There are presently 213 chronologies from Israel that cover the 3rd and 6th centuries A.D. (Liphschitz 1986). Kienast (1985) studied *Cedrus libani* in Cyprus using dendroecological and radiodensitometric methods. In the Atlas mountains (Morocco), Munaut and Serre-Bachet (1982) and Till (1984) recorded numerous chronologies of recent trees from the year 1287 to the present. Stockton (1985) analyzed chronologies from semiarid sites in Morocco for dendroclimatological reconstruction.

Chamaecyparis obtusa (Sieb. et Zucc.) Endl.

Hinoki cypress, Hinoki

Morphology

A large, 30 to 40 m tall evergreen conifer with furrowed bark and an open, conical crown. The species grows slowly with a maximum stem diameter of 1.5 m.

Site, Site Conditions, Characteristics

Hinoki is found growing in the warm, humid marine climate of Japan. The northernmost tip of its distribution area extends into the cooler climate. This species tolerates shade, is generally slow-growing, and found predominately on

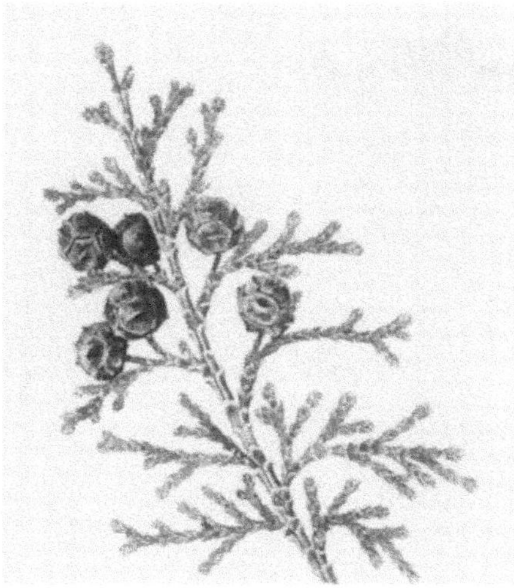

Fig. 98. *Chamaecyparis obtusa* (Krüssmann 1972)

Fig. 99. *Chamaecyparis obtusa* in Japan

granite foundations, in podzol and brown soils. In contrast to *Cryptomeria*, Hinoki is unable to withstand dry sites or sites with heavy snowfall, and avoids the colder regions of northern Japan. The old stands have, for the most part, been destroyed. Hinoki can live longer than 1000 years.

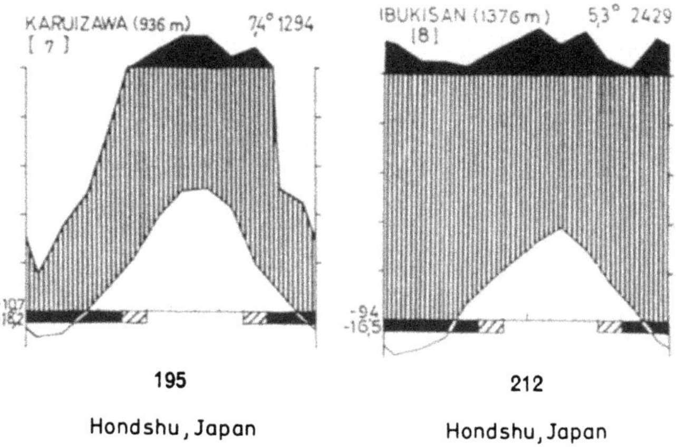

Fig. 100. Climate diagrams

Distribution

Horizontal: endemic to southern Japan, excluding Kiushiu.

Vertical: in Hondshu between 1000–1200 m, in southern regions up to 1500 m.

Fig. 101. Distribution of *Chamaecyparis obtusa.* (Katayama 1968)

Wood

Heartwood is present. The sapwood is yellow-white and its transition into the redish brown heartwood is continuous. The density is 400–500 kg/m^3.

Cross-section: resin canals are absent. There is a clear distinction between the narrow latewood and the earlywood. Parenchyma cells in the heartwood contain colored substances (Fig. 102a, b).

Tangential section: low rays, with an average height of 3–5 cells.

Radial section: the rays are without ray tracheids. The parenchyma cell walls are smooth or with a few nodules, especially in the transverse walls of the axial parenchyma cells. Bordered pits are cupressoid.

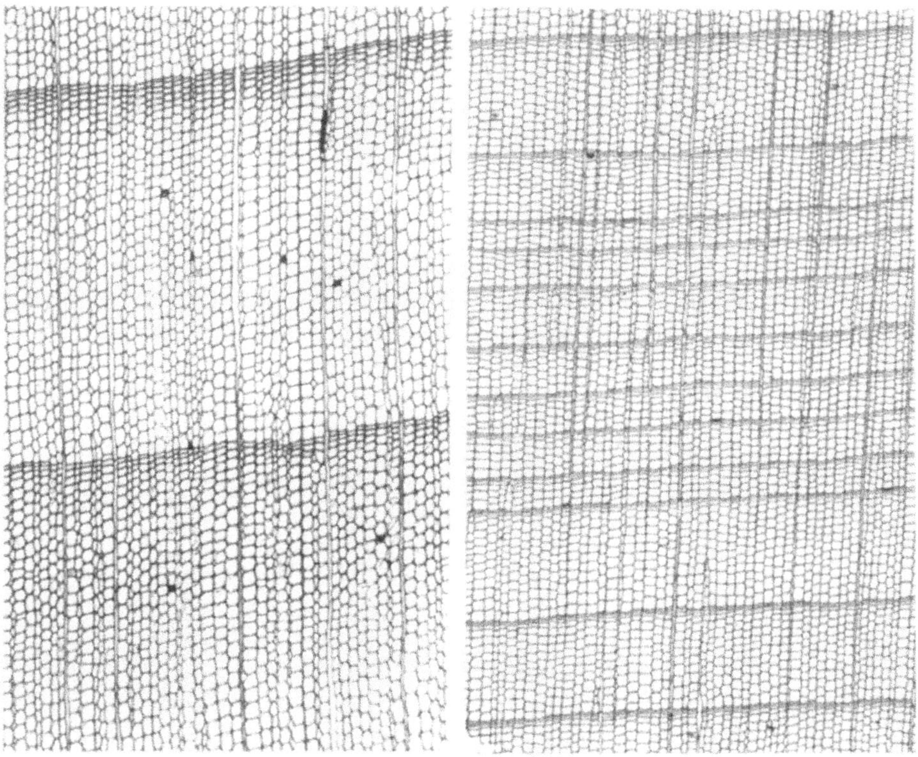

Fig. 102a,b. Cross-sections of *Chamaecyparis obtusa*

Use in Dendrochronology

Suitability of the Species

Because of its durability, the wood was often used to build Shinto temples. Today, the wood is still used as the basis for Japanese laqcer works. Recent (1950) tree-ring sequences from trees growing on reforested stands can be used for various dendrochronological purposes. Density variations in the latewood are minimal, reducing the suitability of this species for densitometric analysis.

Previous Uses

According to Kojo (1987), 1200-year-old, living trees still exist. An unconfirmed report claims that chronologies from recent as well as historical material are available which reach back to the year 38 B.C. Dating is possible within a radius of 200 km.

Many historical and prehistorical examples are given in Mitsutani (1990).

Cryptomeria japonica D. Don.

Japanese cedar, Sugi

Morphology

This very large, up to 60-m-tall evergreen conifer has a deep root system, furrowed bark and a conical crown. The largest trees can, in rare cases, have diameters as great as 7 m. The species can assume many different shapes.

Fig. 103. *Cryptomeria japonica* (Drawing by Hirzel)

Fig. 104. A reforested stand. The individual trees are of equal durability

Fig. 105. An old stand. Each tree has a straight stem (Forestry in Japan)

Site, Site Conditions, Characteristics

The species is found growing in a wide ecological spectrum within warm to temperate climates of Japan. Regional annual temperatures average 8–16 °C,

Fig. 106. An old stand located on a mountain site. Material from such stands is especially suitable for developing long chronologies. (Forestry in Japan)

and annual precipitation measures from 2000 to 4000 mm. *Cryptomeria* grows mainly on granite foundations, in brown soil and podzol. The species tolerates regions with up to 3 m snow cover. It is possible for the trees to live for well over 1000 years, but most old stands have been destroyed.

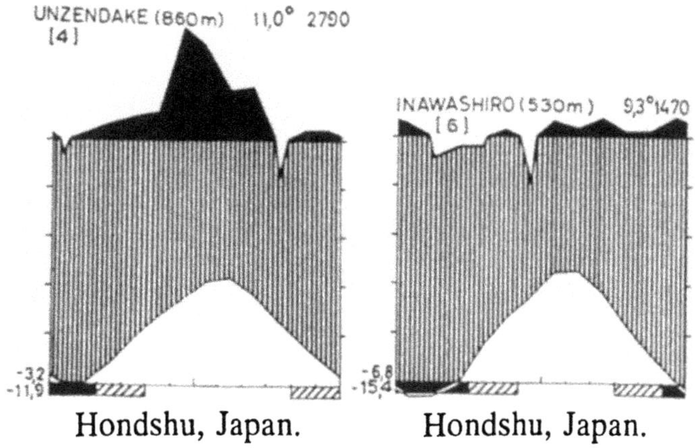

Fig. 107. Climatic diagrams

Distribution

Horizontal: Endemic to Japan. Not found in tropical Kiushiu or cool Hokkaido.

Vertical: from 300 to 1500 m.

Fig. 108. Distribution of *Cryptomeria japonica*. (Katayama 1968)

Wood

Forms heartwood. The sapwood is light yellow to white, while the heartwood is dark brown to red. The density is 400–500 kg/m^3.

Cross-section: resin canals are absent. The latewood is clearly distinguished from the earlywood. The transition from earlywood to latewood is either abrupt or continuous. The parenchyma cells in the heartwood contain colored substances (Fig. 109).

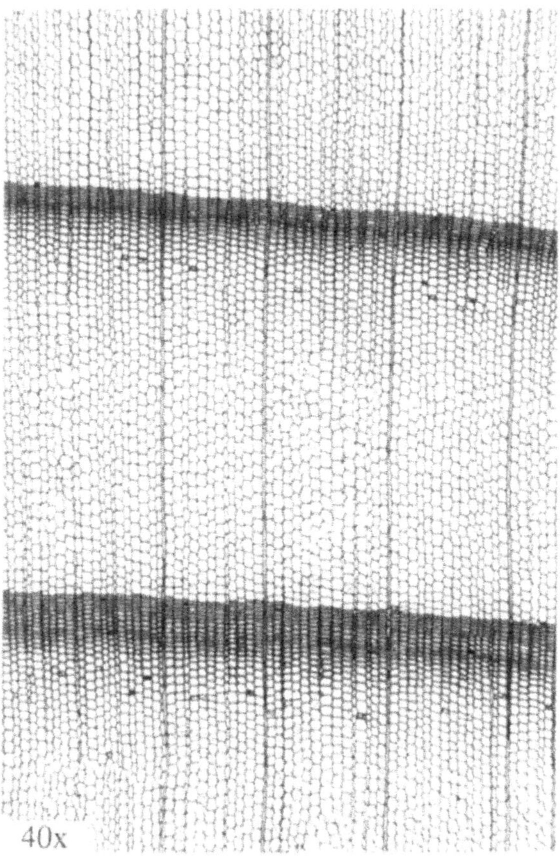

Fig. 109. Cross-section: Resin ducts absent. Distinct latewood zone. Transition from earlywood to latewood abrupt and confluent. Parenchymatic cells with colored contents

Tangential section: the average ray height is six to ten cells.

Radial section: rays without ray tracheids. Parenchyma cell walls are smooth (axial and ray parenchyma) with taxodioid bordered pits.

Use in Dendrochronology

Suitability of the Species

Although the tree rings are distinct, they exhibit only minimal variability. *Cryptomeria* wood has often been used as building material. It grows to be very old and exhibits pointer years. It is therefore possible to use the species for dating purposes. Presumably, the tree-ring sequence of the younger trees can form the basis for growth dynamic studies in managed forests.

Previous Uses

Kojo (1987) has reported dendroclimatological studies of the species taken from three different stands. He found a high autocorrelation. Only 3–4 pointer years are formed per century. The radial growth is mainly restricted by low levels of precipitation in the spring. The effects of forest management are indicated in the tree-ring sequence. Yasuda (1987) determined growth changes in trees growing in highly populated regions. Noda and Higashimura (1990) used *Cryptomeria japonica* ring width to study pollution effects.

Cupressus sempervirens L. var. *pyramidalis*, *Cupressus sempervirens* L. var. *horizontalis*

Mediterranean cypress

Morphology

Aromatic trees and shrubs with narrow to broad crowns. The pyramidal shape is predominately found in populated areas (cultivated form), while in natural stands the broad, spreading shape is more common.

Fig. 110. From Hess et al. (1976)

Fig. 111. The two different Cypress shapes, *pyramidalis* and *horizontalis*, on a site in Greece

Site, Site Conditions, Characteristics

On the Mediterranean islands, the natural range of the species has for the most part been destroyed. Relict areas can be found in the high Atlas mountains and in Libya. Cypress is native to the Mediterranean climate. It withstands long periods of summer drought but not low winter temperatures. It flourishes on limestone, but can be found growing on granite foundations as well. It frequently associates with *Pinus halepensis* var. *brutia*, *Cedrus libani*, *Abies cilicica*, and *Juniperus excelsa*. Today, the species is predominately found on moist sites (gorge areas, northern slopes) which are located on steep hillsides. It is also capable of acting as a pioneer species. (Mayer 1980).

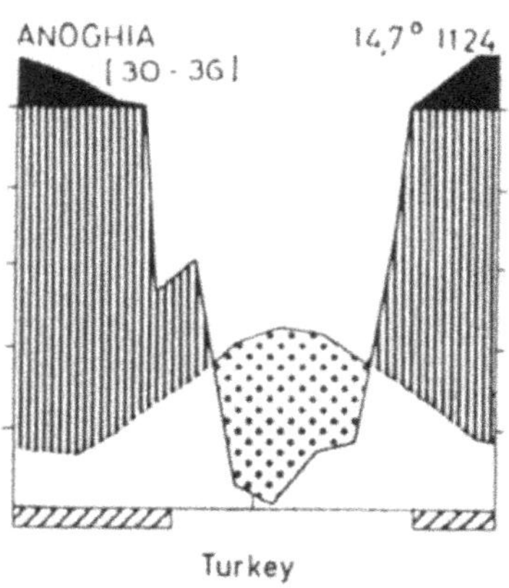

Fig. 112. Climate diagram

Distribution

Horizontal: eastern Mediterranean region: Libya, Crete, Anatolia, Cyprus, Lebanon, Asia Minor. Along the south coast of the Caspian Sea and scattered throughout inland Iran.

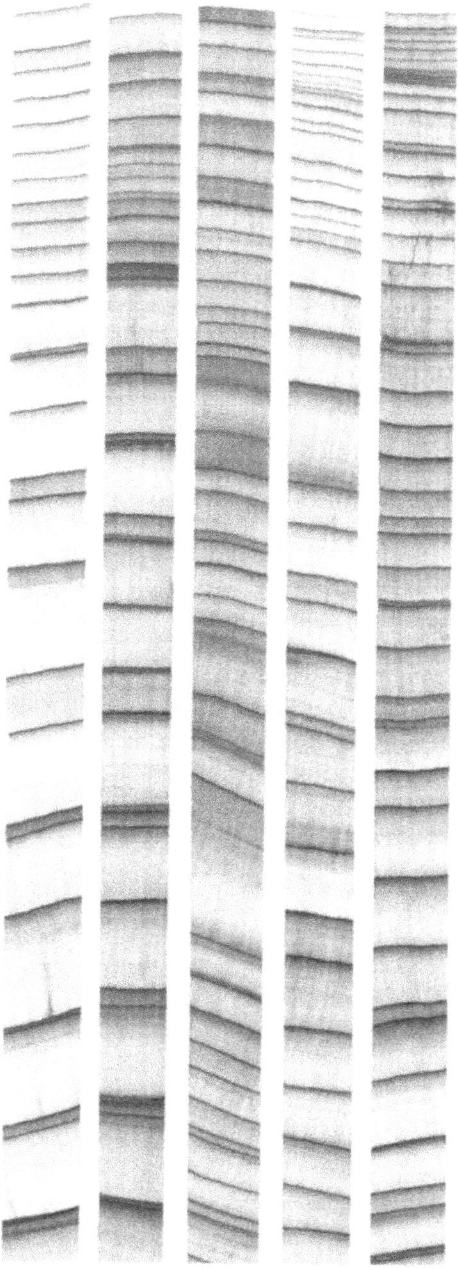

Fig. 113. Radiographs of *Cupressus sempervirens* (Cyprus)

Vertical: on Crete and in Anatolia at 300–1300 (1600) m, in the high Atlas at 1600–1900 m and in Libya at 600–900 m.

Fig. 114. Distribution of *Cupressus sempervirens*

Wood

The wood has a strong odor. The sapwood is yellow, the heartwood brown to red-brown. Density averages 460 kg/m^3.

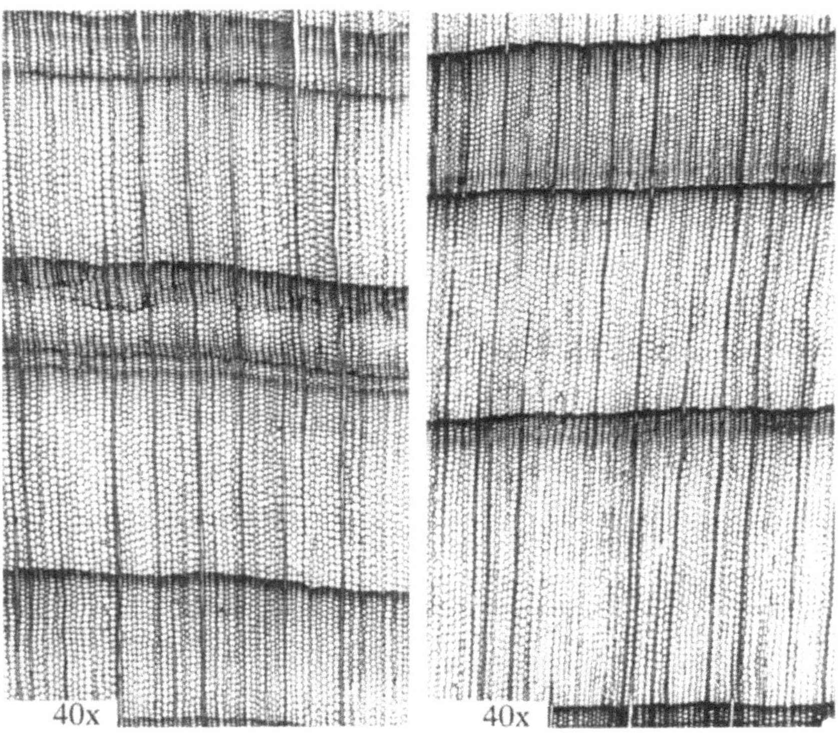

Fig. 115. Cross-sections of *Cupressus sempervirens*

Cross-section: the tree-ring borders are distinct. As a rule, there are many fluctuations in ring width and false rings among the tree rings (Fig. 115a). The axial parenchyma cells contain brown substances.

Tangential section: the average ray height is 10–15 cells.

Radial section: rays consist only of parenchyma cells. The ray cells are all smooth to slightly waved, never with nodules. The bordered pits are cupressoid and therefore have narrow, slit-shaped apertures.

Use in Dendrochronology

Suitability of the Species

Because the species reacts strongly to summer rains, it is not very suitable for dendrochronological purposes. The entire genus *Cupressus* appears to be sensitive to changes in the environment during the vegetation period.

Previous Uses

Dendrochronological work with this species has been minimal. Cambial activity has been studied at the University of Tel Aviv. A historical chronology (1535–1820) from Israel exists. Liphschitz (1986) and Wu et al. (1988b) recorded a chronology using *Cupressus gigantea* (probably identical with *C. torulosa*) from Tibet.

Juniperus excelsa Bieb.

Grecian juniper

Juniperus foetidissima Willd.

Stinking juniper

Morphology

This tree grows to heights of 20 m, with stem diameters of 80 cm, and has a thick, conical-shaped trunk with many partly dead branches.

Fig. 116. *Juniperus excelsa* in the Taurus mountains of Turkey. (Krüssmann 1972)

Site, Site Conditions, Characteristics

Juniperus excelsa forms scattered, pioneer stands in the Mediterranean. In Anatolia and in Iran, the species is a timberline forest. It frequently associates with oak in steppe forests and can withstand low winter temperatures as well as dry summer months.

Fig. 117. *Juniperus excelsa.* (Drawing by R. Hirzel)

Fig. 118. *Juniperus foetidissima* growing near the potential forest border in the Parnassus mountains of Greece. The region is heavily grazed

Juniperus foetidissima

Morphology

This tree grows to heights of 10 m with stem diameters of 60 cm and has a thick, conical-shaped trunk usually with many partly dead branches.

Fig. 119. Climate diagram

Site, Site Conditions, Characteristics

In the Mediterranean, the species generally grows alone or in small groups on the forest border. In Anatolia the main area of distribution lies at intermediate altitudes (1000 m); in Cyprus it mainly grows near the timberline up to 1950 m. For the most part, *Juniperus foetidissima* grows on more temperate sites than *Juniperus excelsa*.

Distribution

Juniperus excelsa

Horizontal: this species is native to eastern Mediterranean countries and Asia Minor. In the east it extends to the eastern border of Iran.

Vertical: north to the coastal region of the Black Sea at an elevation between 300 to 1000 m. In Lebanon from 1700 to 2500 m and in Turkey and Iran up to 3000 to 3400 m.

Fig. 120. *Juniperus excelsa* (Browicz 1982)

Juniperus foetidissima

Horizontal: this species is native to the eastern Mediterranean region. To the east, it reaches the eastern border of Iraq.

Vertical: on Cyprus at 1500 to 1950 m, in Anatolia at 600 to 2000 m. Along the Black Sea the species extends from the mountain zone down to sea level.

Fig. 121. *Juniperus foetidissima* (Browicz 1982)

Wood

It is impossible to distinguish between the two species on the basis of their individual macroscopic or microscopic characteristics. See *Juniperus thurifera* for description.

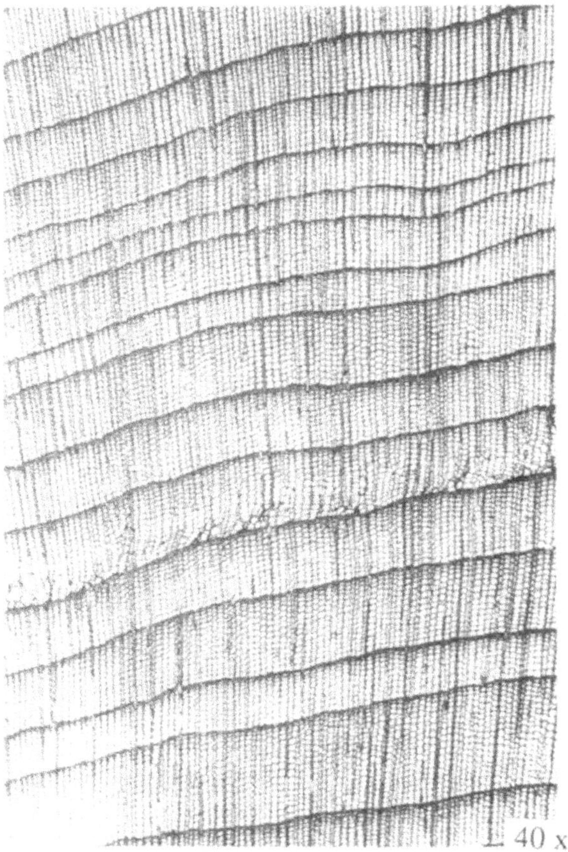

Fig. 122. Tree-ring sequence from a juniper with a frost ring. Callus tissue in the earlywood indicates damage done to young, living cells

Use in Dendrochronology

Suitability of the Species

Because the species often exhibits double rings and false rings, it is of only moderate use for dendrochronological purposes. In the distribution area, with dry summer months and sparse summer rains, the same dating problems are presented as in the American west. It is presumably possible to obtain information, specific to each site, on past winter precipitation conditions.

Previous Uses

Both species were often used in the building of Greek churches and Anatolic royal tombs in Gordian. As Kuniholm and Striker (1983) proved, dating and synchronization, covering wide regions, are possible.

Juniperus thurifera L.

Spanish juniper

Fig. 123. *Juniperus thurifera*

Juniperus phoenicea L.

Phoenicean juniper

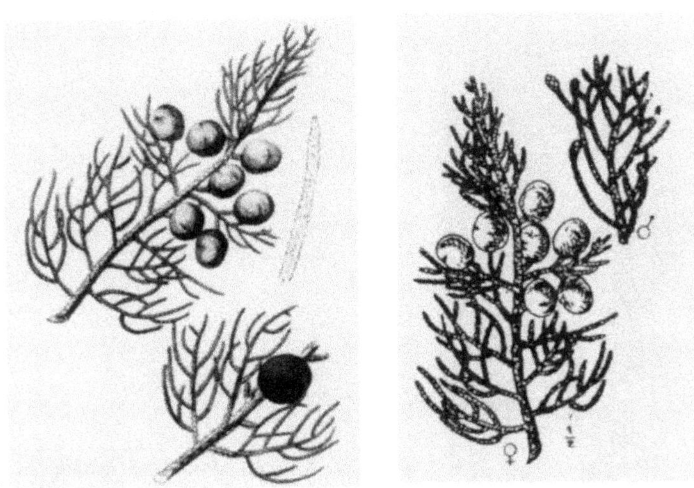

a b

Fig. 124 *Junipera phoenicea*. **a** Pignatti (1982) **b** Dallimore and Jackson (1974)

Morphology

Juniperus thurifera

This small tree with an average height of 10 m, in rare cases 15 m, has a rounded to conical, often irregular crown. The trunks often have branches down to their bases and maximum diameters of 1.5 m. Normally, however, they are significantly thinner and thus the stems are lobed.

Fig. 125. *Juniperus thurifera* on a natural site in St. Crepien by Briançon, southern France. (Krüssmann 1972)

Fig. 126. An old *Juniperus phoenicea* tree from the island of Hierro in the Canaries. (Kunkel 1987)

Juniperus phoenicea

A small tree, rarely growing taller than 5 m. In the chapartal it usually takes on a shrub shape. The stem diameter averages only 10–20 cm, and the stems are flexible.

Juniperus thurifera

Site, Site Conditions, Characteristics

Juniperus thurifera forms scattered stands on limestone bedrocks. The species has adapted to the continental region and withstands summer droughts just as well as low winter temperatures.

Drought-resistant shrubs such as *Juniperus oxycedrus*, *Juniperus sabina*, and *Citrus* species flourish as undergrowth. Regeneration is poor due to grazing. Most trees are less than 100 years old, but in isolated spots it is still possible to find trees that are centuries old.

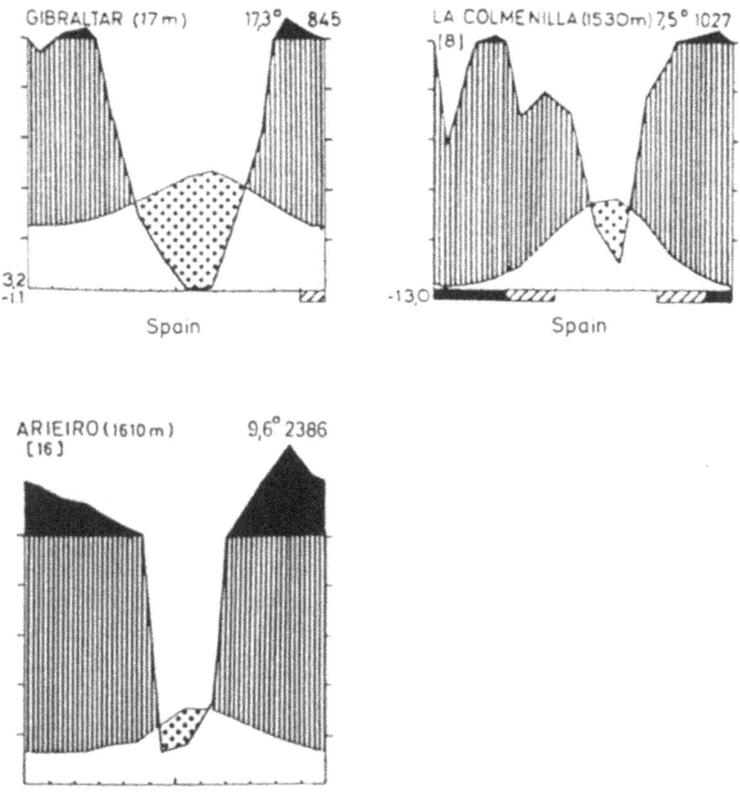

Fig. 127. Climate diagrams

Distribution

Horizontal: north and south of the western Mediterranean.

Vertical: mainly in the mountain zone of central Spain (1400–1900 m) and the subalpine level of the Atlas mountains (1700–3000 m).

Fig. 128. Distribution of *Juniperus thurifera*

Juniperus phoenicea

Site, Site Conditions, Characteristics

Juniperus phoenicea has a broad ecological spectrum within Mediterranean countries. It grows in the semiarid climate of Morocco and in the Canaries as well as in the central Spanish mountain region. The species flourishes in both acid and basic soils. Small trees are found growing in deep soil and bush-like stands are found on beaches. Individual bushes grow on rock foundations. In shrub communities the species frequently grows among *Pistatia*, *Rosmarinus*, and *Cistus* species.

Distribution

Horizontal: mainly in northwestern Mediterranean countries (from Portugal to Greece).

Vertical: from sea level to the upper border of *Quercus ilex* at around 2000 m.

Fig. 129. Distribution of *Juniperus phoenicea*

Wood

The two species cannot be distinguished anatomically from each other.

Cross-section: the tree-ring boundaries are distinct and often exhibit wedging and missing rings (Figs. 130a–c and 131a,b). The parenchyma cells in the heartwood contain brown substances.

Tangential section: low rays, averaging two to five cells. The transverse walls of the axial parenchyma cells have nodules.

Radial section: rays consist of only parenchyma cells. The tangential walls of the ray walls have nodules. Pits in the cross-field are cupressoid.

Juniperus thurifera

Use in Dendrochronology

Suitability of the Species

Older trees from certain individual sites could be suitable for dendrochronological purposes. On normal sites, summer precipitation results in false rings. Trees from dry sites may be used in precipitation reconstruction.

According to Richter (1987), the tree-ring series from Spanish trees exhibit few variations, which makes dating difficult. In regions with a long history, it is possible to find *Juniperus thurifera* used in construction.

Previous Uses

None

Juniperus phoenicea

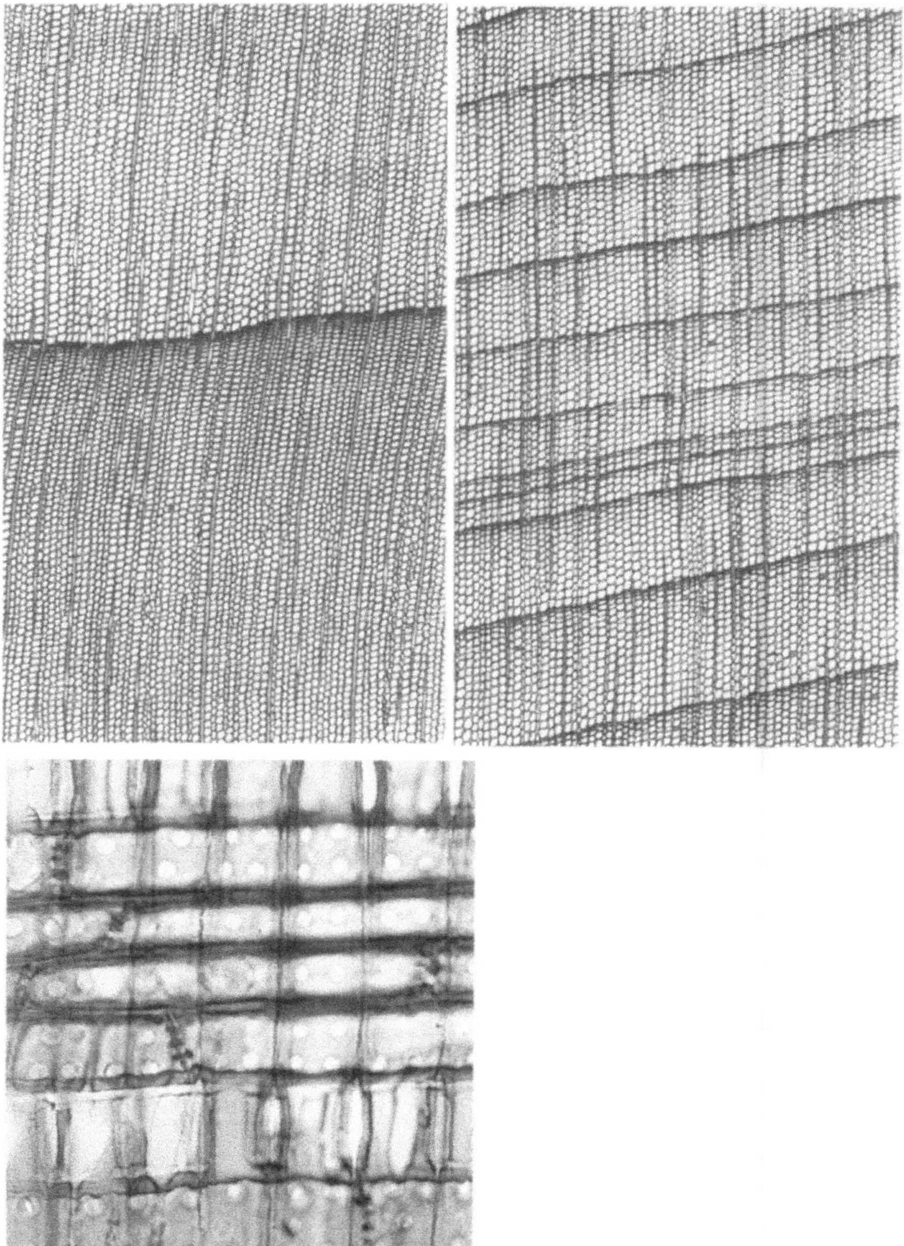

Fig. 130. Cross-sections of *Juniperus thurifera*. **a,b** With heartwood. The sapwood is white to light yellow, the heartwood is dark red-brown. The wood has a pleasant odor and a density of 500 kg/m^3. **c** *Juniperus thurifera* radial 400 x. Note the nodules on the tangential ray walls

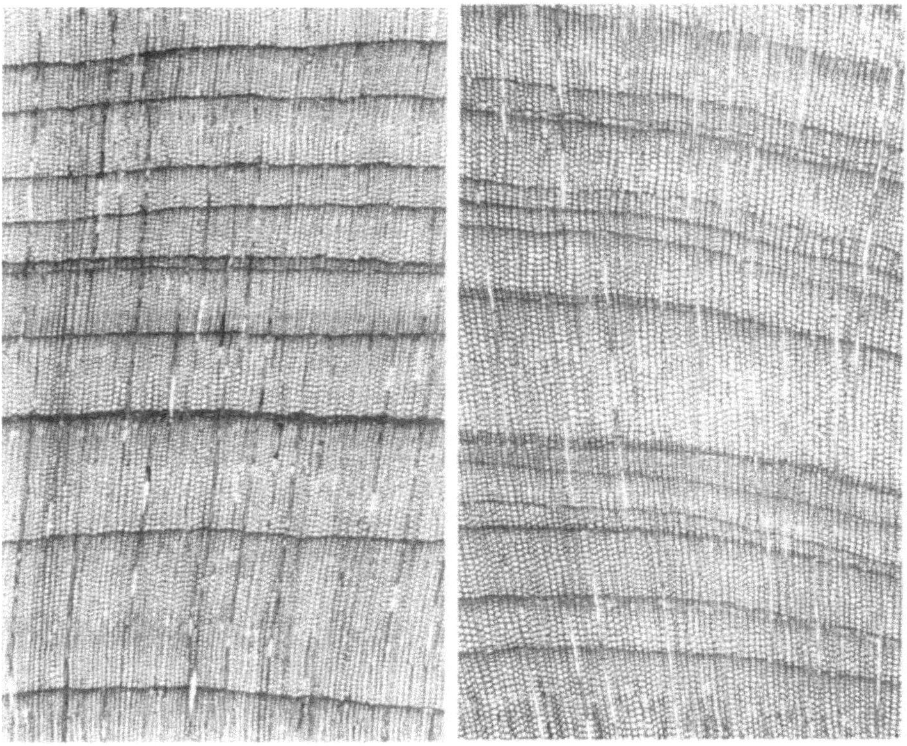

Fig. 131. Cross-sections of *Juniperus phoenicea*. **a** With heartwood. The sapwood is yellowish, the heartwood dark brown. The density is 500 kg/m^3. **b** *Juniperus phoenicea*, radial 400 x. Note the big apertures in the ray piths (cupressoid)

Use in Dendrochronology

Suitability of the Species

Only very few old stands of *Juniperus phoenicea* exist, and these are extremely isolated and varied. Therefore, although the species can live for centuries, it is difficult to use the wood for dendrochronological purposes. The wood is often used for local building purposes.

Previous Uses

Liphschitz (1986) reported on the construction of chronologies with live trees from the Sinai. One chronology reaches back to 1102 A.D., and contains a floating chronology from the time period between the 4th and 1st century B.C.

Juniperus turkestanica Kom.

Turkestan juniper

Morphology

This species is small, with multiple stems and partly dead branches. It grows on rock foundations, at the timberline and at elevations of 3000–3500 m. It has adapted to the heavy precipitation of the continental climate. According to Bitvinskas (1983) and Mukamedsin (1977), average annual precipitation on *Juniperus turkestanica* sites is 500 m and the average annual temperature lies around 2.4 °C. The frost-free period lasts around 100 days.

Juniperus turkestanica is distributed throughout southern Tien–Shan and the northern Altai mountains.

The trees are presumably the oldest growing tree species in Eurasia, living for over 2000 years. According to Zheng et al. (1982) and Bitvinskas (1983), 1500 samples were analyzed to make available a chronology covering 1214 years. The tree ring widths reflect the temperatures of the months of June and July. Graybill (pers. comm.) analyzed recent material.

Fig. 132. *Juniperus turkestanica* at timberline in Tien–Shan mountains (Russia). (Photo Graybill)

Juniperus seravschanica Komarov. Syn. *Juniperus polycarpos*

Juniperus semiglobosa Regel Syn. *J. indica*

Both species have habits similar to *Juniperus turkestanica*. However, these species grow predominately on sites located on the arid forest border of central

Asia (Zheng et al. 1982). The area borders are difficult to determine, because the systematic position of *Juniperus seravschanica* has yet to be clearly established. The western border can, according to Browicze (1982), lie in Iran or Afghanistan. Both species can be used for the reconstruction of summer precipitation. Liphschitz (1986) recorded three chronologies from Iran, which cover the time period from 1659 to the present.

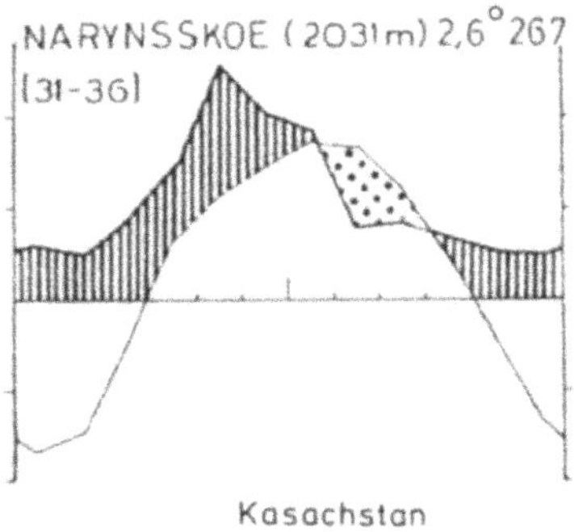

Fig. 133. A climate diagram from the distribution area of *Juniperus seravschanica* Kasachstan

Juniperus przewalskii Kom. (Syn. *Sabina przewalskii*)

This species grows at the timberline in Tibet on the Xizang-Plateau and, according to Zheng et al. (1982), grows to be very old.

The three species may be used for the reconstruction of summer temperatures.

Larix decidua Mill.

European larch

Larix L. russica (Endl.) Sabine ex Trautv. Syn. *L. Sibirica* Ledeb.

Siberian larch

Larix gmelinii. (Rupr.) Kutz. Syn. *L. dahurica* Turcz. ex Trautv.

Dahurian larch

Morphology

These deciduous conifers have generally straight trunks and deep root systems. On optimal sites the trees grow to heights of 40 m, on forest border sites they often develop many stems.

The genus *Larix* has a wide range, covering a broad ecological spectrum. In Europe two high-altitude races exist, (*decidua*, *carpatica*), one low-elevation race (*sudetica*), and one lowland race which grows in the lower part of the Weichsel River (*polonica*). In Russia and China there are two other species *Larix russica* and *L. gmelinii*) with generally intercrossing areas.

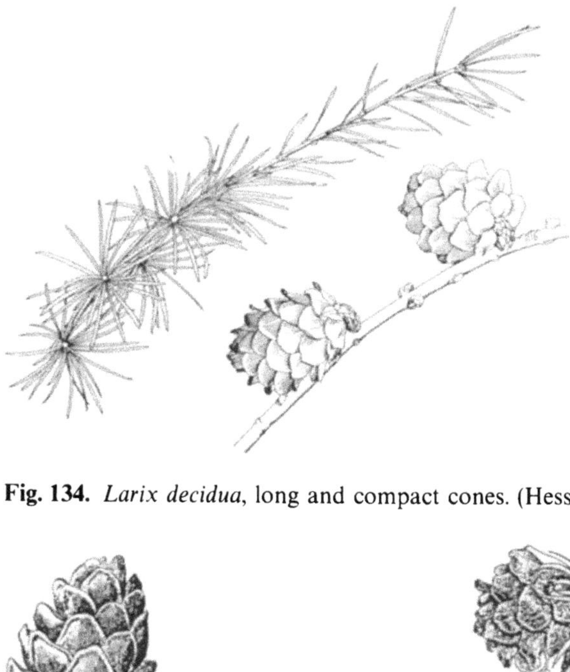

Fig. 134. *Larix decidua*, long and compact cones. (Hess et al. 1976)

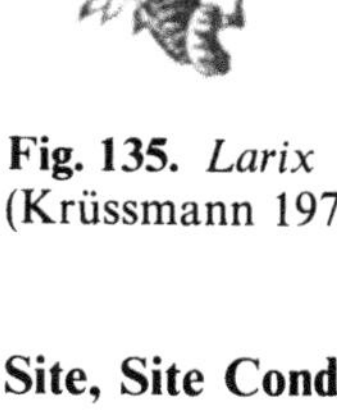

Fig. 135. *Larix russica*, compact cone. (Krüssmann 1972)

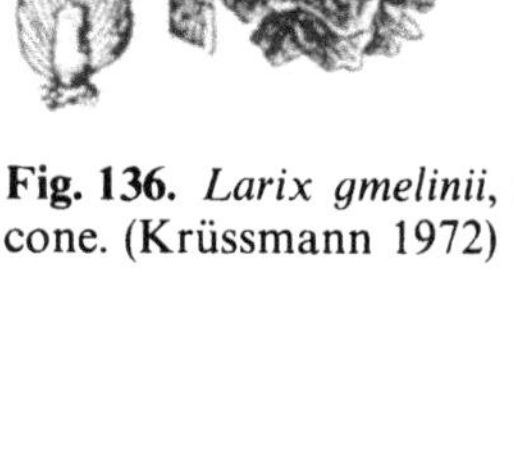

Fig. 136. *Larix gmelinii*, round and loose cone. (Krüssmann 1972)

Site, Site Conditions, Characteristics

Larix decidua

This species is mainly found in European mountain climates. It covers a very wide ecological spectrum. On forest border sites, the species tolerates average

Fig. 137. European larch at timberline. (Fenaroli et al. 1976)

Fig. 138. European larch on an optimal mountain site. From the information service Holz, Germany

Fig. 139. Siberian larch on lake Baikal; second growth after windfall

January temperatures of −8 °C, but it can also maintain stands in regions with average January temperatures of over 0 °C. Temperatures in July reach around 8 °C at high altitudes and at low elevations up to 17 °C. On extreme sites the minimal frost-free period is only 1–2 months. Within the entire range, of the species annual precipitation levels vary between 600 and 2000 mm. At the forest border, *Larix decidua* forms mixed stands with *Pinus cembra*. On lower, drier sites, it associates with *Pinus sylvestris*. At the timberline, *Larix decidua* is the pioneer tree, growing in moraine soil on limestone and silicate foundations. In subalpine sites larch can grow to be over 1000 years old.

Larix russica, L. gmelinii

Both species are indigenous to the continental climate of the European boreal zone. They can withstand long periods of extreme cold and dry winters but

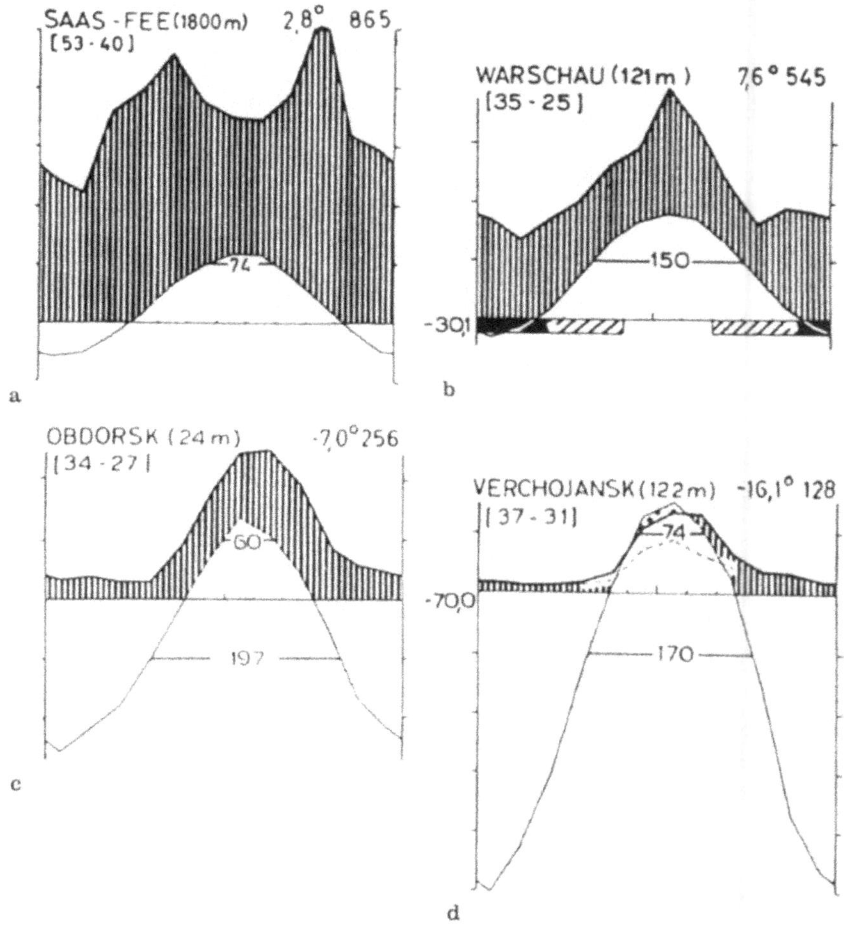

Fig. 140a–d. Climate diagrams

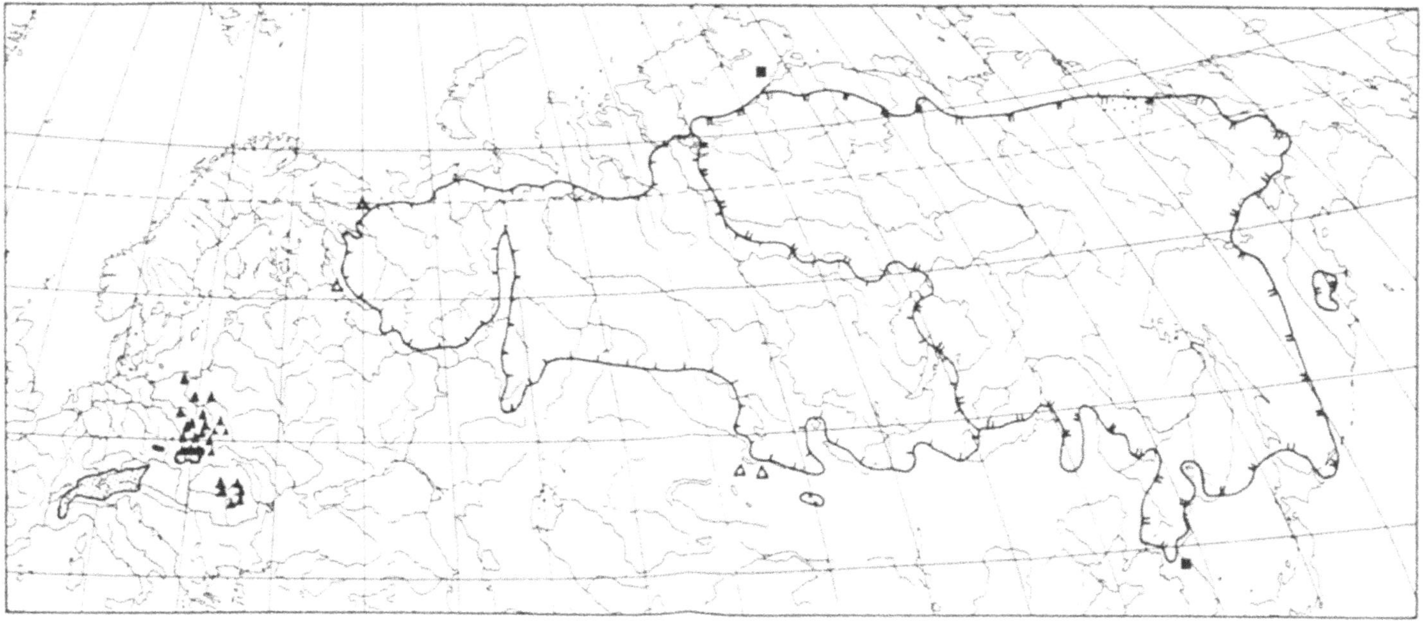

Fig. 141. Distribution of European and Asian larches. (Zoller 1981)

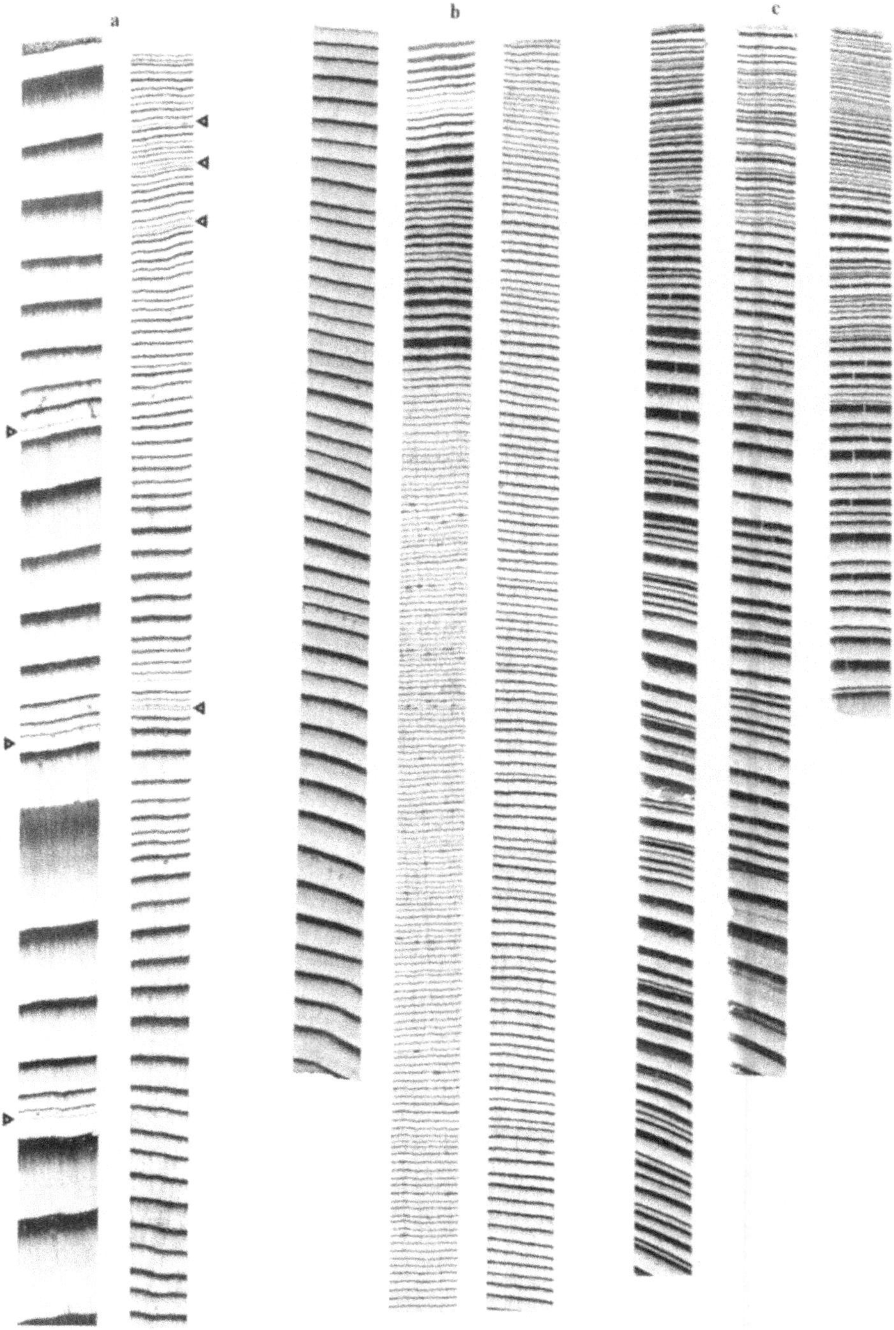

Fig. 142a–c. Radiographs of *Larix* sp. **a** Tree-ring sequence in *Larix decidua* from the Alps, indicating growth reduction caused by larch bud moth infestation. **b** *Larix decidua*, Alps. **c** *Larix russica*, Lake Baikal, Russia

require short, warm summers. Together they form the northern, boreal forest border. Ecologically *Larix russica* and *L. gmelinii*, occupy an intermediate position between *Picea abies* var. *obovata* and *Pinus sylvestris*. They do not advance as commonly into bog regions as *Picea abies* var. *obovata*, and avoid the extremely dry sites of the steppe.

Distribution

Horizontal: *Larix decidua*: main area in the Alps. Relict area in eastern Europe.
Larix russica: western Russia and western Sibiria.
Larix gmelinii: Eastern Siberia and China.
Vertical: *Larix decidua*: from 400 m in the southern Alps, to 2500 m in the central Alps.
Larix russica, L. gmelinii found in the plains and mountains, excluding the Urals.

Wood

With heartwood. The sapwood is yellow, the heartwood red-brown. Distinct tree rings. Depending on the site, the density is extremely variable, ranging between 400 and 860 kg/m^3.

Cross-section (Fig. 143a–c) the resin canals are bordered by 8–12 thick-walled, secretory cells. The transition from earlywood to latewood is normally abrupt (Fig. 143b). In years with *Zeiraphera dinia* infestation, the latewood is less distinct.

Tangential section: the ray height averages 10–25 cells. The resin canals are bordered by 7–12 epithelial cells and are often eccentric in the rays (Fig. 143b).

Radial section: transverse tracheids are present in the rays. Piceoid pits in the cross-fields of the earlywood. Thick transverse and tangential walls of ray cells, occasionally tooth-shaped pits. Biseriate bordered pits frequent in radial walls of the tracheids of the earlywood. The cell walls of the tracheids are lightly thickened between the bordered pits. (crassulae, bars of Sanio).

Use in Dendrochronology

Suitability of the Species

Due to its broad ecological spectrum, larch is fundamentally suitable for dendrochronological and ecological research. For around 2000 years the European larch, most notably in the central Alps, has suffered from periodic (every 6 to 10 years) attack by the larch bud moth (*Zeiraphera diniana*). The defoliating phases are clearly indicated in the tree-ring sequences. Thus, larch from the Alps lends itself better to entomological than to chronological research. Its clearly

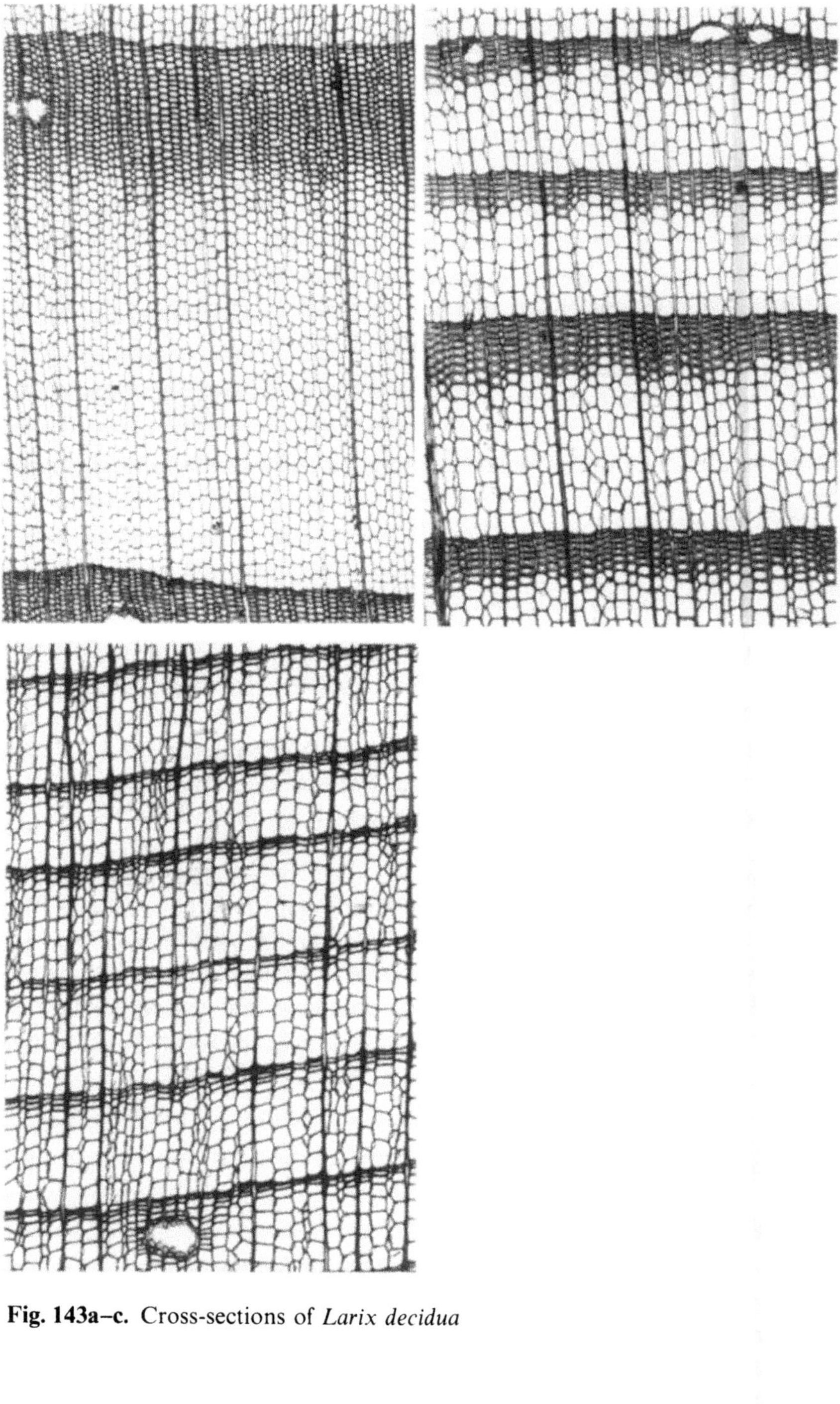

Fig. 143a–c. Cross-sections of *Larix decidua*

delimited latewood zones, longevity, and high sensitivity make larch suitable for tree-ring width as well as radiodensitometric research.

Previous Uses

Despite its extensive distribution, larch species have received relatively little attention. Chronologies from various provinces in Russia exist, which, according to Kairiukstis (1983) reach back to the year 1431 (Baikal). Chronologies from the Alps also exist (Holzhauser 1984, Siebenlist-Kerner 1984). They span 700–1200 years and serve both in the dating of buildings and in the reconstruction of glacial movements. Schär and Schweingruber (1987) employed historic and fossil material from the timberline region to record a 7000-year chronology and, using radiodensitometric methods, reconstructed summer temperatures in central Europe.

In the area around Khabarovsk (eastern Siberia) Malokvasov (1987) carried out dendroclimatologically orientated research on *Larix russica* var. *kurilensis*.

Picea abies (L.) Karst.

Norway spruce

Picea abies var. *acuminata* (Beck) Dall. et Jacks.

Picea abies var. *europaea* Jukev et Parf.

Picea abies var. *obovata* (Ledeb.) Voss. Syn.: *Picea obovata* Ledeb.

Siberian spruce

Dallimore, Jackson, and Harrison consider *Picea obovata* to be a separate spruce species, not to be included under *Picea abies*. Other taxonomists see *Picea obovata* as a race of *Picea abies* in the cold climate (Schmidt-Vogt 1977).

Morphology

An evergreen, shallow-rooted conifer with extremely variable crown and cone shapes. The variability is genetically and ecologically determined. On optimal sites this spruce can grow to heights of over 60 m. In mountain regions the trees have heights between 20 and 30 m. On extreme sites, such as on windy ridges, in avalanche paths, and rockfall slopes or sites with extremely shallow soil, habits and size differ greatly from the norm.

Fig. 144. *Picea abies* var. *acuminata* (Hess et al. 1967). Mature spruce cones hang from the branch

Fig. 145. *Picea abies* var. *obovata*, Lake Baikal

Fig. 146. The root system of a spruce in deep soil. The main roots lie deep under the surface of the ground. Fine roots penetrate even deeper. (Köstler et al. 1968)

Fig. 147. Several examples of crown shapes from free-standing spruces. **a** Columnar. **b** Conical. **c** Candelabrum. **d** Harp. **e** Browsed (grazed) form

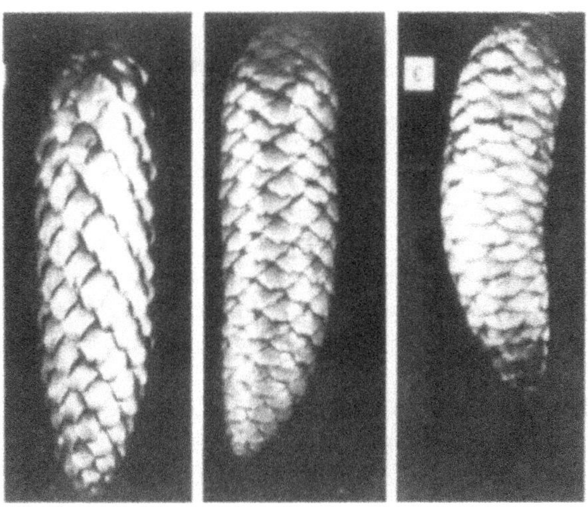

Fig. 148. Cone forms of *Picea abies* ssp. *acuminata* ssp. *europaea* ssp. *obovata*

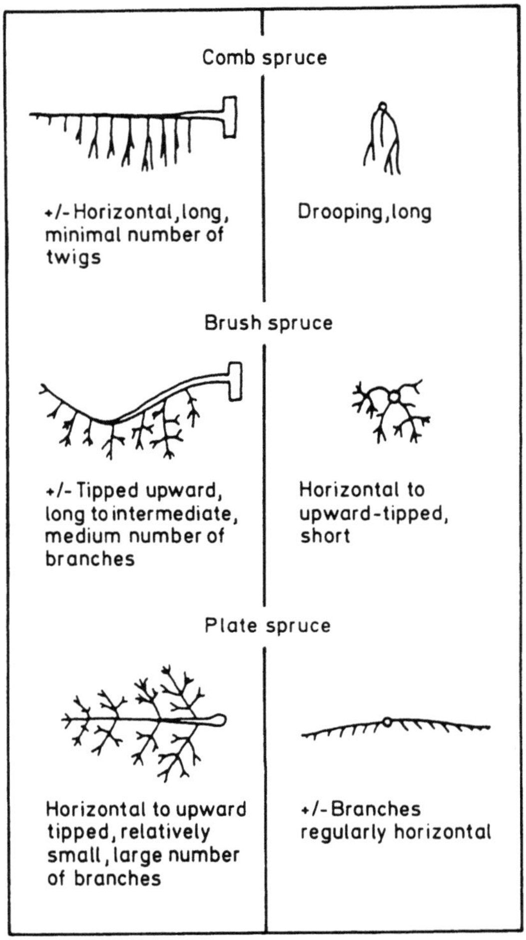

Fig. 149. Three branch types from the Alps. (Mayer 1976; Müller 1986)

Three separate subunits exist, predominately classified by their cone scales.

West of the Eurosiberian area, all three subunits are represented; the *obovata* type is, however, less common. East of the Urals only the *obovata* can be found.

The var. *obovata* is considered by some to be a climate race of *Picea abies*.

The three most important cone-scale types of *Picea abies* are: (1) var. *acuminata*, characterized as long, narrowed, bent wavy, double-tipped, and found exclusively in the east Carpathian mountains; the cones are 12 cm long, (2) var. *europaea*, with rhombic scales, indented tips, found in the Black Forest; the cones are 11 cm long, (3) var. *obovata* with broad, rounded tips, found at high altitudes in the French west Alps; the cones are 9.5 cm long. From Schmidt-Vogt (1977).

Site, Site Conditions, Characteristics

Picea abies is mostly found growing in cold-winter, mountain climates. In mountain regions and especially in the boreal zone, the species forms a clearly marked forest belt.

Picea abies in Europe

The species covers an extremely broad ecological spectrum. In the north, it withstands average January temperatures of −15 °C; in the Alps −4 °C and in lower elevations 0 °C. *Picea abies* can also tolerate minimum short-term cold spells of around −40 °C. In spruce stands in Scandinavia, July temperatures average around 10 °C, and in the Alps climb up to 17 °C. The vegetation period on the upper spruce border in the Alps lasts for around 2 1/2 months. The species avoids low precipitation regions as well as regions with mild winters and strong marine influences. If enough humidity is present, *Picea abies* will grow in very acid and neutral, shallow and deep soils. It forms timberline forests in regions of the Alps with high precipitation. On normal sites, *Picea abies* forms stands in the Taiga. On lower sites, especially in Europe, *Picea abies* associates with beech, and on dry sites with *Pinus sylvestris* and *P. mugo*. Thanks to its excellent technological properties, Norway spruce has found extensive use in industry, and it now grows on many locations outside its natural range. Because the tree is important to forestry, it is difficult to find trees with more than 300 tree rings at lower elevations. In certain regions, the crown is susceptible to sporadic *Lymantria monacha* attacks. The stem is frequently decayed by *Heterobasidion annosum*. In central European stands, the bark beetle is prone to decimate stands of similar age.

Picea abies var. *obovata* (Siberian spruce)

The Siberian variety is much more cold-resistant than the European variety. On its northern border it withstands average January temperatures of −40 °C and minimal temperatures as low as −60 °C. On extreme sites, the vegetation period barely lasts more than a month. On such sites the trees grow in permafrost soils. The varying vegetation periods are indicated in the maximum ring width of the European spruce. In the Alps and Pyrenees, latewood formation takes place from July to September (October), while in Scotland and Scandinavia it occurs between July and August.

Fig. 150. A young spruce stanc on an optimal site. (Information Service Holz, Germany)

Fig. 151. Columnar spruce in the eastern Alps. (Schmidt-Vogt 1977)

Fig. 152. Spruce on a border site in a bog region. (Schmidt-Vogt 1977)

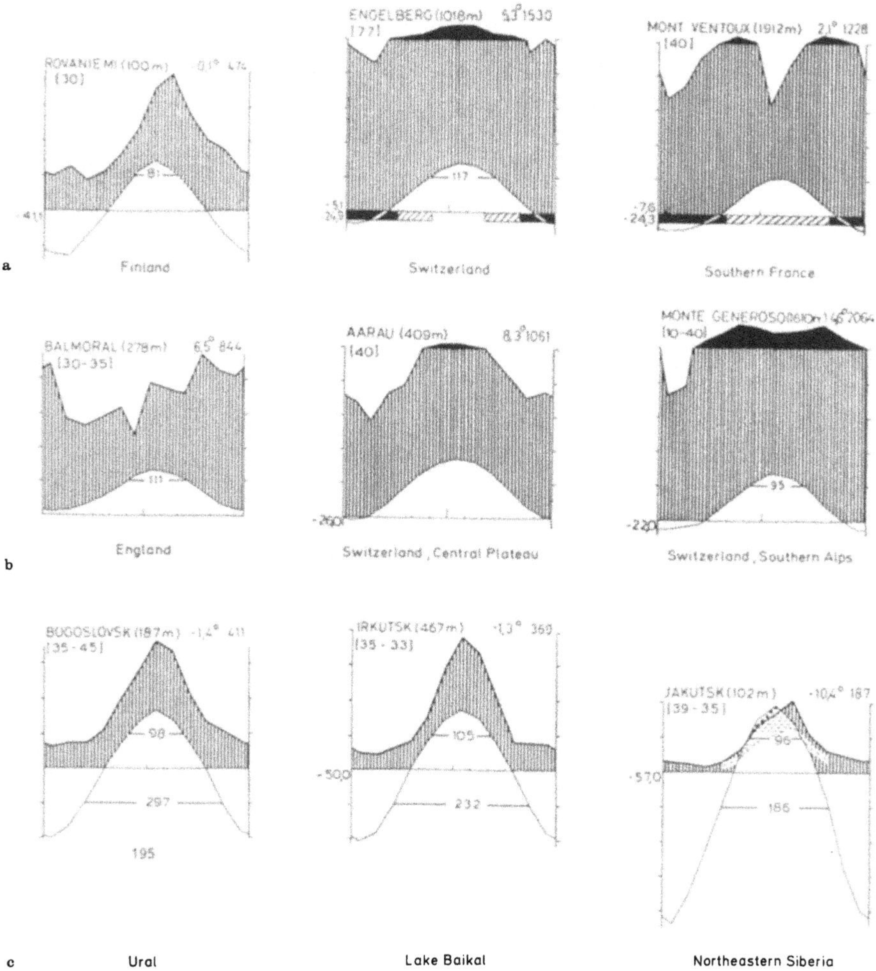

Fig. 153a–c. Distribution area of *Picea abies* var. *obovata*. **a** Climate diagrams from the current area of distribution of Norway spruce (both natural and artificial) Finland, Switzerland, Southern France. **b** Distribution area of *Picea abies* var. *acuminata*, *europaea*, *obovata*, respectively England, Switzerland, Central Plateau, Switzerland, Southern Alps. **c** Outside the natural area of distribution (predominately *Picea abies* var. *acuminata*). Ural, Lake Baikal, Northeastern Siberia

Distribution

Horizontal: transcontinental Eurasia and in European mountain regions. Northernmost point near Murmansk (Kola peninsula at 69° 27′) northernmost point in Siberia, near the river Chatanga (between Jenissei and Lena) at 72° 15′. In western and eastern Scandinavia, the northern border extends much further

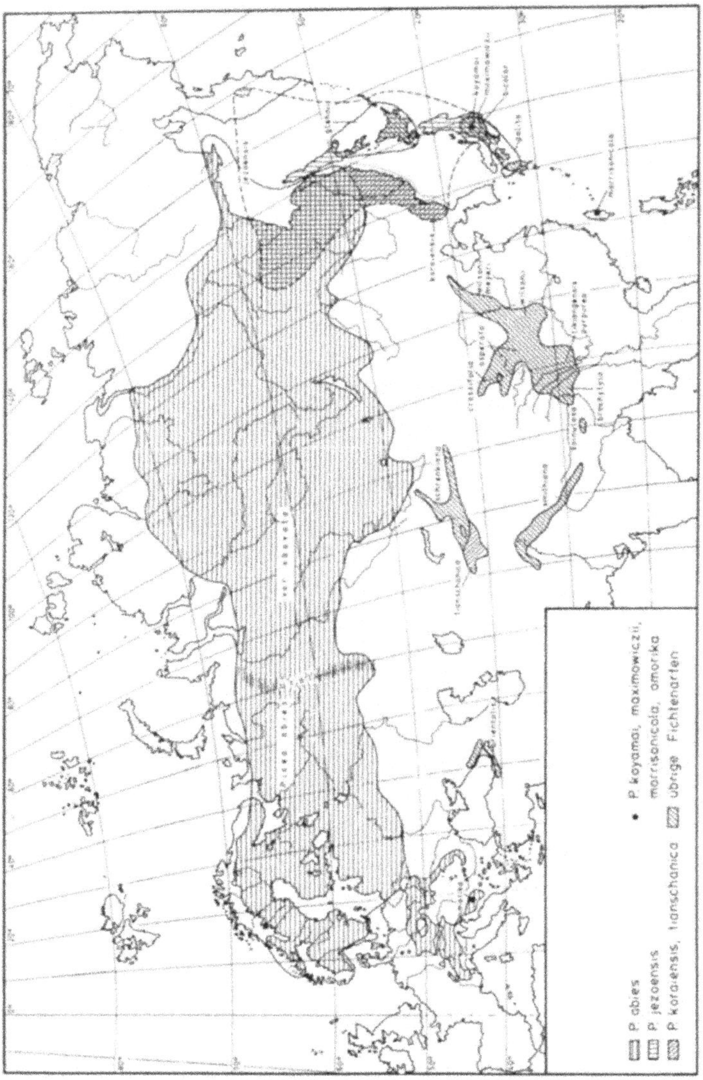

Fig. 154. Distribution of *Picea* sp. in Eurasia

south. The southern area of distribution reaches to the 50th northern latitude (Steppe). Schmidt-Vogt (1977).

Vertical: according to Schmidt-Vogt (1977) the upper *Picea abies* border extends from north to south. At 68 °N the border lies at 400 m (Fig. 155,1) and in southern Tyrol (central Alps) at 2300 m (Fig. 155,7). The lower distribution border lies in the Lüneburgerheide at sea level (Fig. 155,12) and extends southward up to an elevation of 1500 m. (Fig. 155,17, in the Rhodope Mountains, Greece).

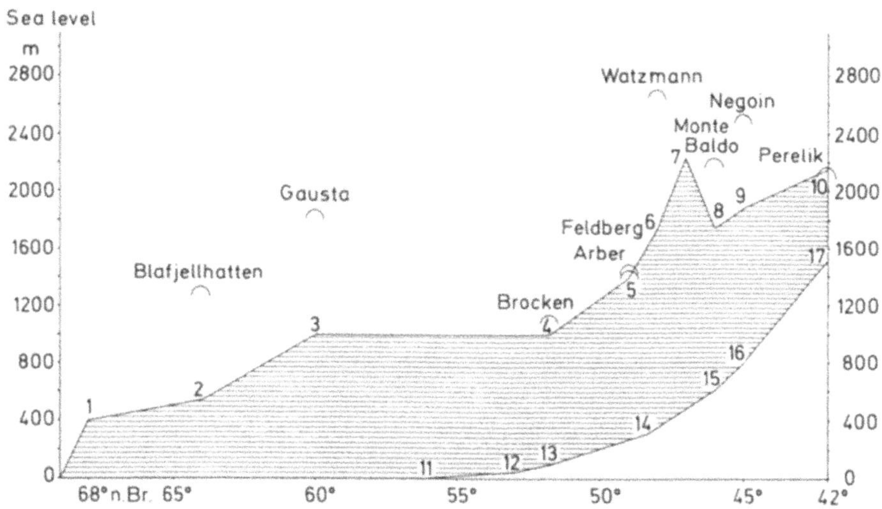

Fig. 155. Vertical distribution of *Picea abies* in Europe

Wood

No (or color-less) heartwood. White to straw-yellow with a silky shine. Moderately distinct tree rings. Frequent resin canals. The density is 300–640 kg/m^3.

Microscopic structure (Fig. 156). See *Larix decidua* and *Picea jezoensis*. It is difficult to distinguish clearly between the wood of spruce and larch. In spruce the transition from earlywood to latewood is continuous and biseriate bordered pits are rare in the radial section of the tracheids.

Use in Dendrochronology

Suitability of the Species

Next to *Pinus sylvestris*, Norway spruce has the highest dendrochonological potential in Eurasia. Spruce grows quite old; 300-year-old trees are common in

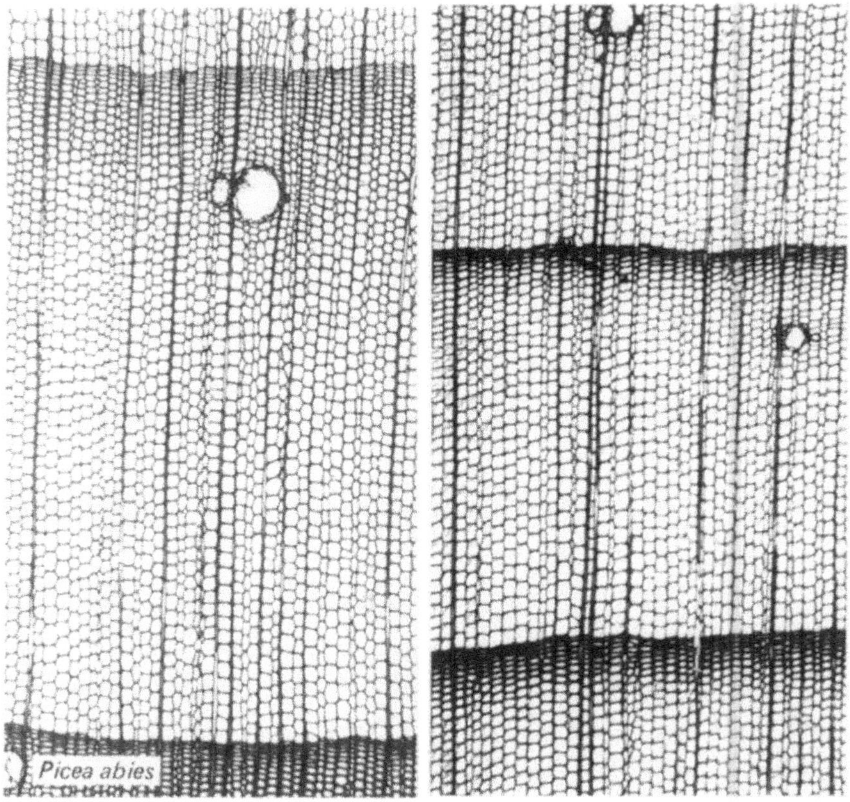

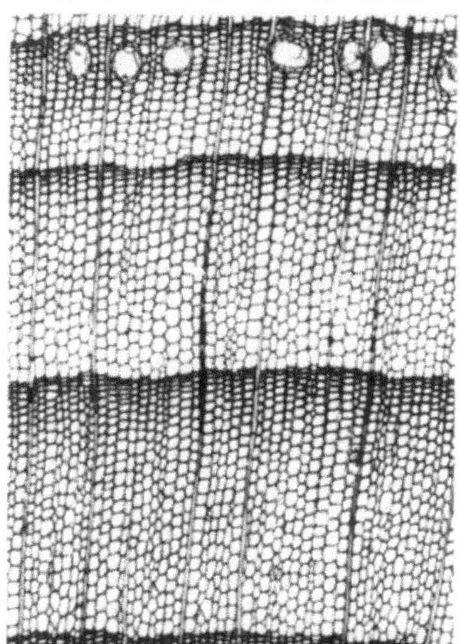

Fig. 156. Cross-sections of *Picea abies*

Fig. 157. Radiographs of *Picea abies*, Switzerland

mountain regions. *Picea abies* wood is often used as lumber and it is a prime source of pulp and paper. Because its ecological spectrum is relatively broad and its tree rings are regularly structured, spruce is suitable for radiodensitometric studies. It can also be used for dating purposes as well as for reconstructing earlier climate conditions.

Previous Use

Spruce has only recently become important in dendrochronology. In the Alps, many buildings have been dated and new standard chronologies are continuously being developed for different regions and elevations.

By radiodensitometric analysis it is possible to reconstruct the summer temperatures in the Alps back to the 10th century (Schweingruber 1988). Spruce wood constitutes an important part of the European sample grid (Schweingruber 1987). Using the maximal densities of trees from high altitudes, it is possible to reconstruct the average temperatures of the months from July to September over the whole of Europe.

Because Norway spruce is the most important forest tree in Europe, it has been the subject of much research. The effects of human emissions on the growth of the tree have been extensively studied (Vins 1962; Eckstein et al. 1983; Schweingruber 1987). Becker et al. (1990) could find no correlation between the extent of emissions and the growth of the tree over large areas.

Dendroecological studies carried out by Kienast (1986) Lingg (1986), and Kontic et al. (1986) showed that, from year to year, ecological site conditions (hydrology, high altitude) have varying degrees of influence on the growth of the tree.

Spruce has also been used for dendrochronologically orientated studies by Russian dendrochronologists. Of special interest is a study from a meteorite crater in the Tunguska region of central Siberia (Kairiukstis 1983).

Picea jezoensis (Sieb. et Zucc.) Carr.

Yezo spruce, Hondo spruce

Picea glehnii. (Fr. Schmidt) Mast.

Sakhalin spruce

Morphology

These two species are very similar. They are both large, shallow-rooted evergreens with generally loose, open crowns, and maximal trunk diameters of 1–1.5 m.

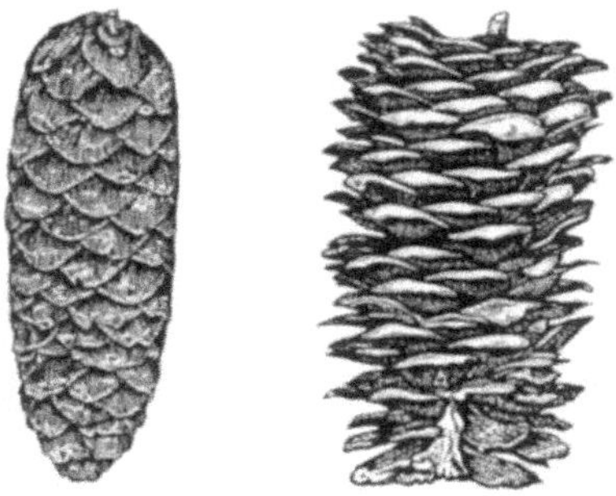

Fig. 158. *Picea glehnii.* (Drawing by Sebek)

Site, Site Conditions, Characteristics

Both species are native to mild to cold marine climates. On northern sites they build pure stands, otherwise they are usually accompanied by *Abies sachalinensis, Quercus crispula, Kalopanax ricinifolia* and *Tilia japonica.*

In cold climate regions the species exhibit broad ecological spectra. They build forests in regions with annual precipitation levels between 1500 to 2000 mm and a temperature range between 30 and −40 °C (with an annual mean temperature of 5 °C). They grow both on limestone-rich and limestone-poor soils.

Fig. 159. *Picea jezoensis* in Hondshu. (Forestry in Japan)

Fig. 160. *Picea glehnii* at a foggy site at the timberline in Hokkaido

Fig. 161. A *Picea glehnii*–*Betula ermanii* forest mountain region of Hokkaido. (Forestry in Japan)

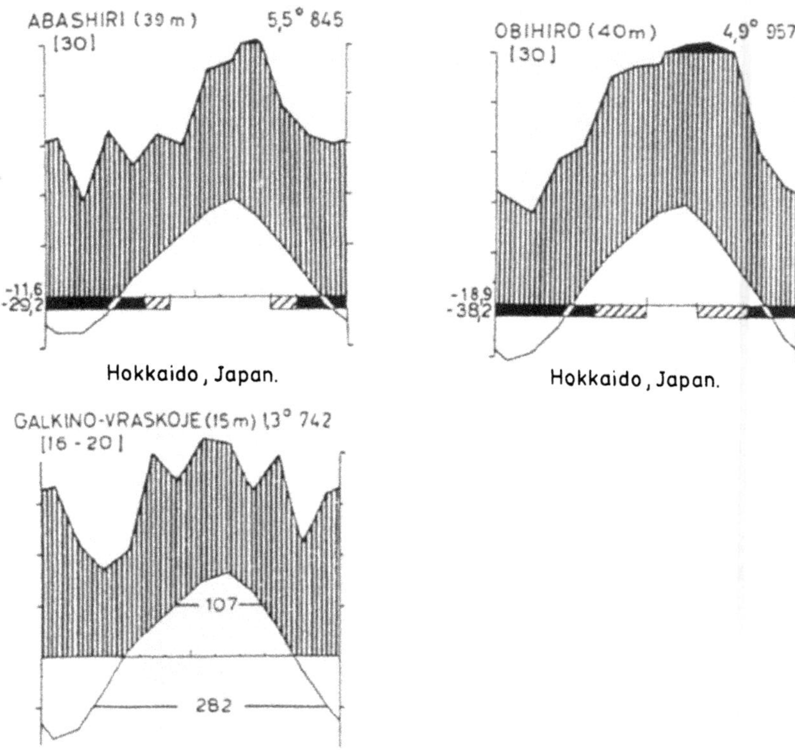

P. jezoensis on Sachalin

Fig. 162. Climate diagrams

Fig. 163. Distribution of *Picea jezoensis.* (Schmidt-Vogt 1977)

Fig. 164. Distribution of *Picea glehnii.* (Katayama 1968)

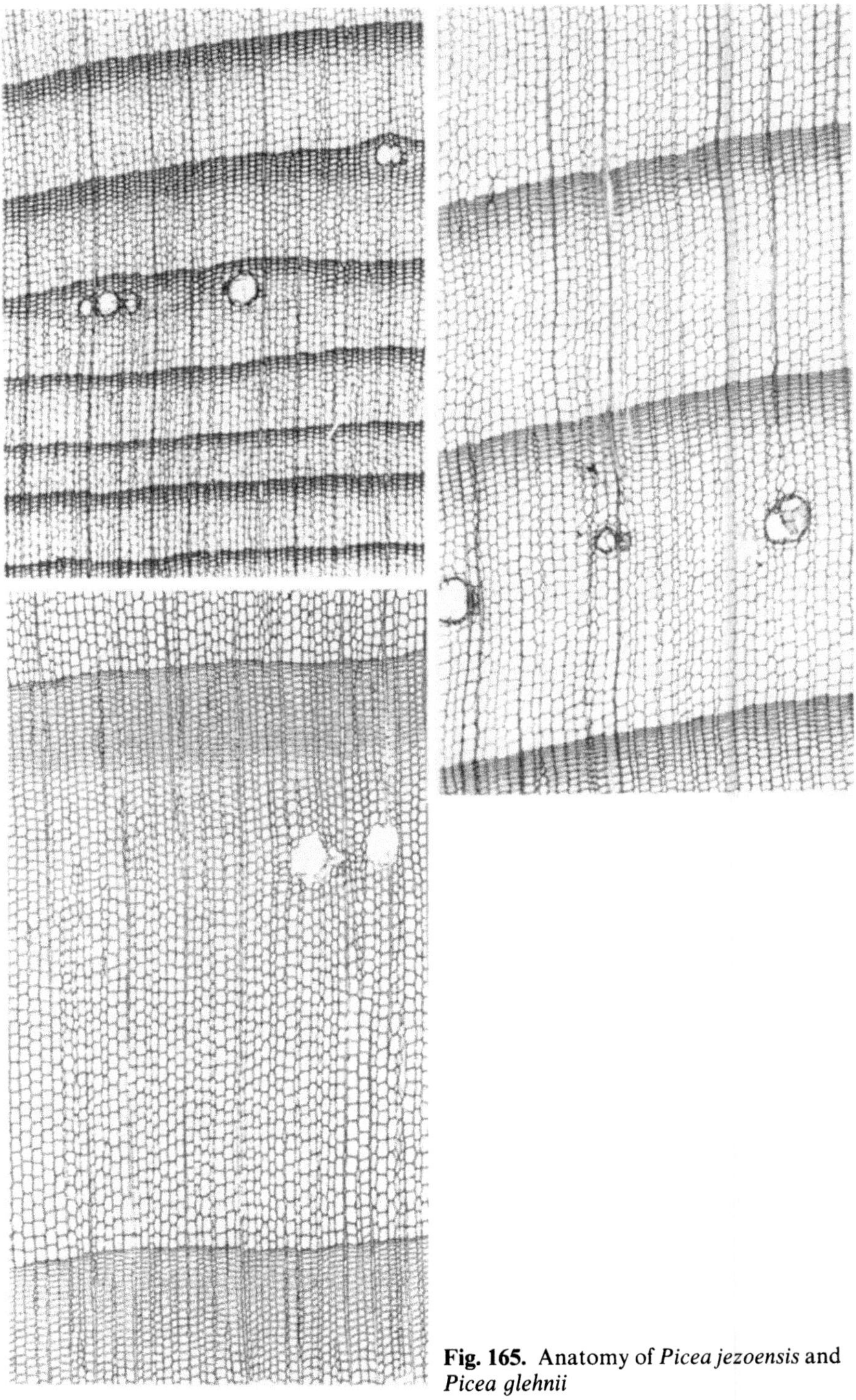

Fig. 165. Anatomy of *Picea jezoensis* and *Picea glehnii*

Fig. 166. Radiographs of *Picea glehnii*

Distribution

Horizontal: the main distribution area of both species lies in Hokkaido. *Picea glehnii* is endemic to Hokkaido, while *P. jezoensis* also grows in eastern Siberia with a special race, ssp. *hondoensis*, occurring in Hondshu.

Vertical: in Hokkaido, both species extend from sea level to the timberline. In Hondshu, *Picea jezoensis* reaches an elevation of 1800 m.

Wood

No heartwood. White to yellow with a slight shine. The density is 400 to 500 kg/m^3.

The two species cannot be distinguished from each other on the basis of their anatomic characteristics.

Cross-section: resin canals are bordered by 8–12 rather thick-walled secretory cells. The transition from earlywood to latewood is generally continuous (Fig. 165).

Tangential section: average ray height of 8 to 20 cells. Resin canals bordered by usually 7–12 secretory cells. The resin canals in the rays are often eccentric.

Radial section: rays with tracheids. Bordered pits in the earlywood with narrow apertures (piceoid). The horizontal and tangential ray walls are thick. The cell walls between the bordered pits in the cell walls of the tracheids are often slightly thickened (crassulae, bars of Sanio).

Use in Dendrochronology

Suitability of the Species

The two species are extremely well suited for dendrochronological research, especially for the reconstruction of summer temperatures using radiodensitometric methods. They can also be used in any other dendrochronological studies.

Previous Uses

Nobori et al. (1990) made radiodensitometric studies from *Picea glehnii* in Hokkaido. They prove the usefulness for dendroclimatological studies.

Picea orientalis (L.) Link

Oriental spruce

Morphology

This large conifer has extremely variable crown shapes. Its habits are similar to *Picea abies*, with the main difference being its very short needles (6–8 mm).

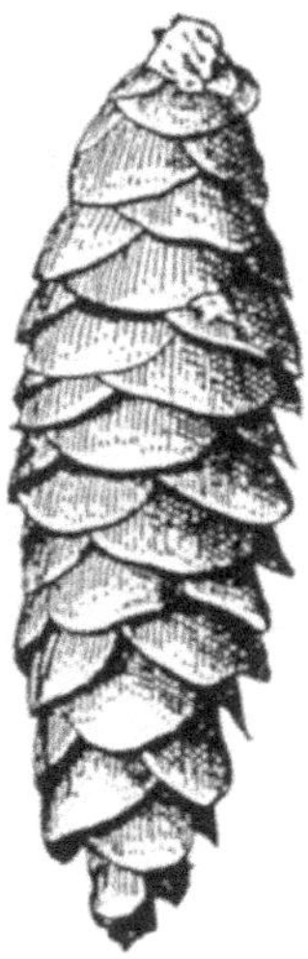

Fig. 167. *Picea jezoensis*. (Krüssmann 1972)

Site, Site Conditions, Characteristics

Picea orientalis is a tree from high, humid, and cool regions with annual precipitation levels of 900–1800 mm (similar to *Picea abies* in the Alps). At lower sites, the species is restricted by lack of precipitation. The upper border climate can be cold. It grows on acid and basic foundations. The maximum age lies between 450 and 550 years. Exploitation of the forest has left only a scattering of

Fig. 168. *Picea orientalis* on the Teberda Preserve, Caucasus/Russia; altitude 1400 m. (Schmidt-Vogt 1977)

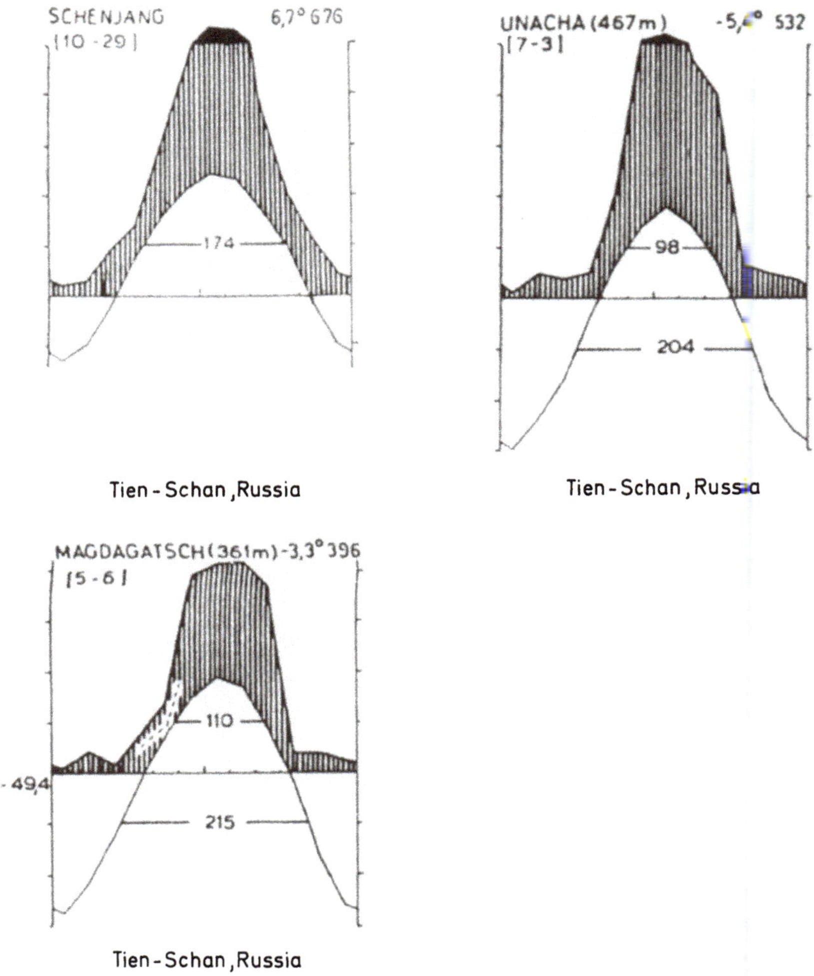

Fig. 169. Climate diagrams from the region just below the actual distribution area where precipitation is higher

older stands which are now restricted to steep, isolated ridges. On lower sites, *Picea orientalis* is accompanied by *Abies nordmanniana*, *Pinus sylvestris*, and *Fagus orientalis*. At higher sites over 1000 m, Oriental spruce is the main tree species and forms pure stands at high altitudes.

Distribution

Horizontal: northern slope of the Pontic Mountains and the lower Caucasus. The area is divided through the dry lowland. In the Caucasus the stands are mixed with *Abies nordmanniana*. In Turkey, *Picea orientalis* covers around 200 000 ha, about 2% of Turkey's forest.

Fig. 170. Distribution of *Picea orientalis* in the Pontic Mountains/Turkey and in the Caucasus/Russia (Schmidt-Vogt 1977)

Vertical: from 600 to a maximum of 2400 m. The lower border lies on the northern slopes of the Caucasus between 1100–1200 m, in Transcaucasia at 1400–1600 m. The upper borders are at 1900–2100 m.

Wood

The wood cannot be anatomically distinguished from that of *Picea abies*; see there for description.

Use in Dendrochronology

Suitability of the Species

Picea orientalis presumably has a dendroclimatological and dendroecological potential similar to that of *Picea abies*. On its lower distribution border, the tree-ring sequences disclose information on precipitation. The maximum latewood densities of trees growing on the upper distribution border offer information on summer temperatures.

Previous Use

Until recently, *Picea orientalis* has only seldom been used for dendrochronological purposes. One available chronology consists of 35 samples from the western Caucasus at an elevation of 2000 m, and reaches back to the year 1808. Two chronologies also exist from the Stravapols region, Ukraine.

Picea schrenkiana Fisch et Mey
Picea tienschanica Rupr.

Schrenk's spruce
Tien–Shan spruce

The two species are taxonomically very similar.

Morphology

A very thin, pyramid and columnar-shaped tree. It reaches heights of 50–60 m and is dense, with many branches.

Site/Distribution

Both species build the conifer forest zone between 1400 and 3000 m, predominately on northern slopes with good water supply (normal precipitation, good soil water supply). The highest sites are found at elevations of around 3600 m. *Picea schrenkiana* usually forms pure stands but is sometimes accompanied by *Abies russica*. On slopes with southern exposure, plant communities exist on dry sites together with *Juniperus pseudosabina*.

In the northern part of the distribution area, *Picea schrenkiana* is predominating while in the southern, dry parts *P. tienschanica* is more common. Both species are restricted to the Tien–Schan mountains (Russia/China).

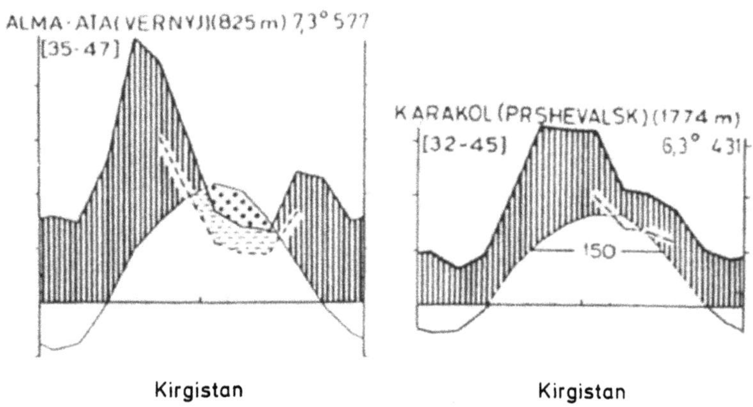

Fig. 171. Climate diagrams from the region. The meteorology station Alma Ata is located outside the natural distribution area

Fig. 172. *Picea schrenkiana* stand in the north of the Tien Schan region, Russia. The columnar shape of the trees and the open stands are typical. (Photo N.M. Borschova)

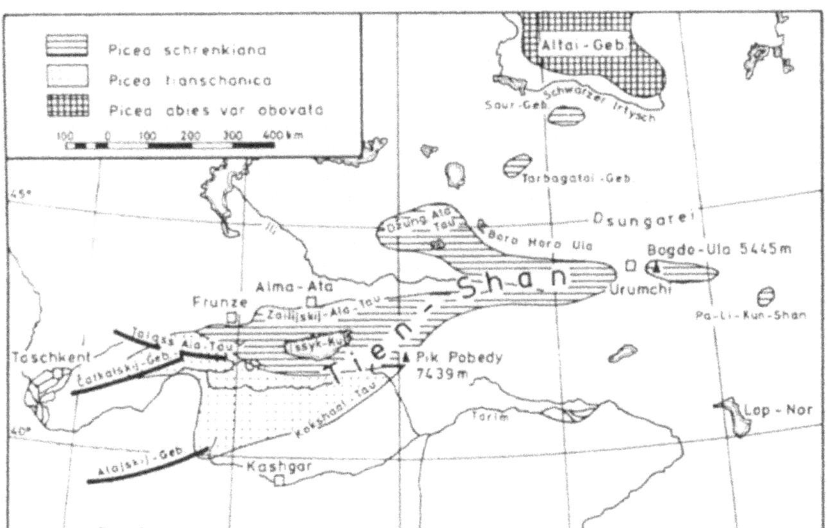

Fig. 173. Distribution for *Picea schrenkiana* and *P. tienschanica* in Tien-Shan and the bordering mountain range (Kasachstan, Kirgistan/ Sinkiang/China). The southernmost extension of *Picea abies* var. *obovata* is in the Altai Mountains. (Schmidt-Vogt 1977)

Use in Dendrochronology

Suitability of the Species

Considering available research, it is probable that summer temperatures can be reconstructed from the maximal ring widths (August to September). Due to the broad ecological spectrum of the species and wide area of distribution, which covers sites of extreme diversity, it is possible to carry out detailed studies of individual micro sites.

Previous Use

Borschova (1983, 1987) (Geographical Institute Alma–Ata, USSR) studied material from various ecological locations in the Tien Shan Mountains. In the western, dry area, growth is mostly influenced by summer precipitation (July and August) on sites at 1400–2900 m. In the eastern, cool moist regions, it is possible to reconstruct temperatures from July and August. Temperature reconstructions go back to 1659. The oldest chronology reaches back to 1590. Wu et al. (1988a) made available chronologies using *Picea likiangensis* from Tibet.

Pinus cembra L. ssp. *cembra*
Pinus cembra ssp. *sibirica* Krylow

Arolla pine

Morphology

This conifer is narrow when young but later becomes wider and heavily branched. It has an irregular crown and a deep, complex root system. On optimal sites in the Alps, the trees grow to heights of 25 m, and in Siberia up to 40 m.

It is difficult to distinguish between the two subspecies. The Siberian subspecies has longer needles and larger cones and seeds with thinner shells than the European subspecies.

Site, Site Conditions, Characteristics

Pinus cembra ssp. *cembra* is native to continental mountain climates with average annual temperatures of 0 °C and low winter temperatures (average January temperature of −9 °C, and a minimal temperature around −40 °C) and low summer temperatures (average July temperature 7–9 °C). The trees do not require much warmth. With a vegetation period of 2 to 3 months and moderate precipitation levels between 1000–2000 mm the species forms forests in

Fig. 174. *Pinus cembra.* (Hess et al. 1967)

Fig. 175. **a** Old stone pines in the central Alps (*Pinus cembra* ssp. *cembra* and **b** near the lake of Baikal (*Pinus cembra* ssp. *sibirica*)

deep soils at the forest border. Arolla pine is frequently accompanied by larch and spruce. In suboceanic foggy regions, *Pinus cembra* competes with spruce. On silicate foundations, Arolla pine reaches 200 m higher than on carbonate rock. Trees growing at the forest border grow to be well over 1000 years old.

Pinus cembra ssp. *sibirica* is common throughout the Siberian Taiga. It grows mainly on sites with normal drainage. In mountain regions it is predominately found on shady slopes.

The seeds of both subspecies are distributed by races similar to *Nucifraga caryocatactes* and *N. C. macrorhynchos*).

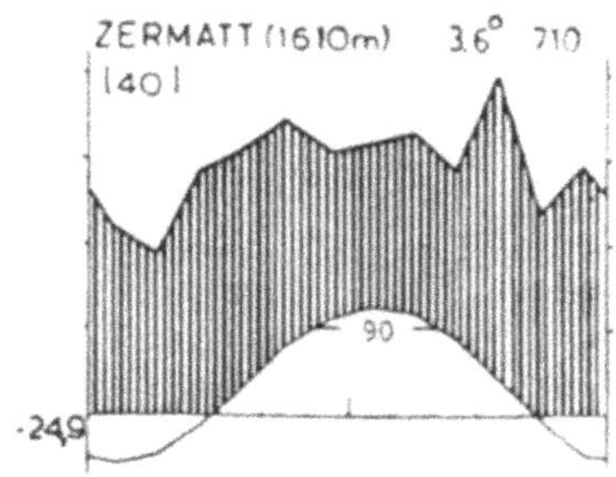

ssp. cembra, Switzerland

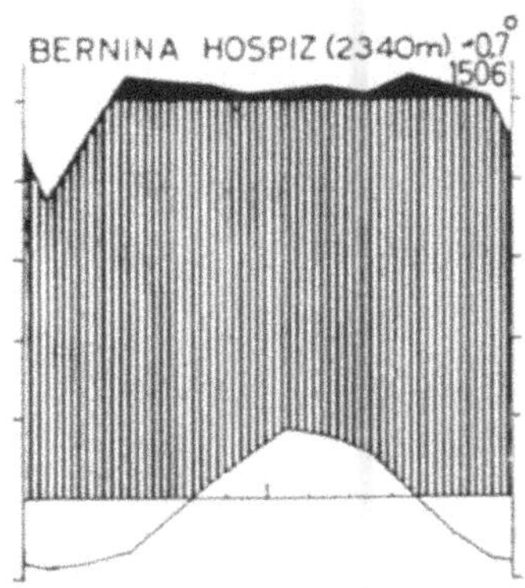

ssp. cembra, Switzerland

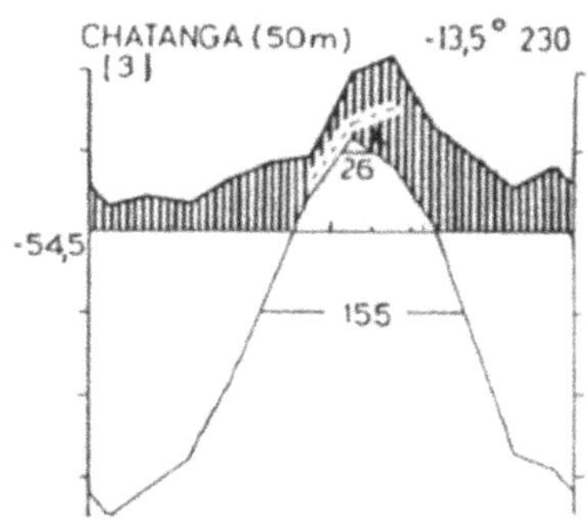

ssp. sibirica, NE-Siberia

Fig. 176. Climate diagrams

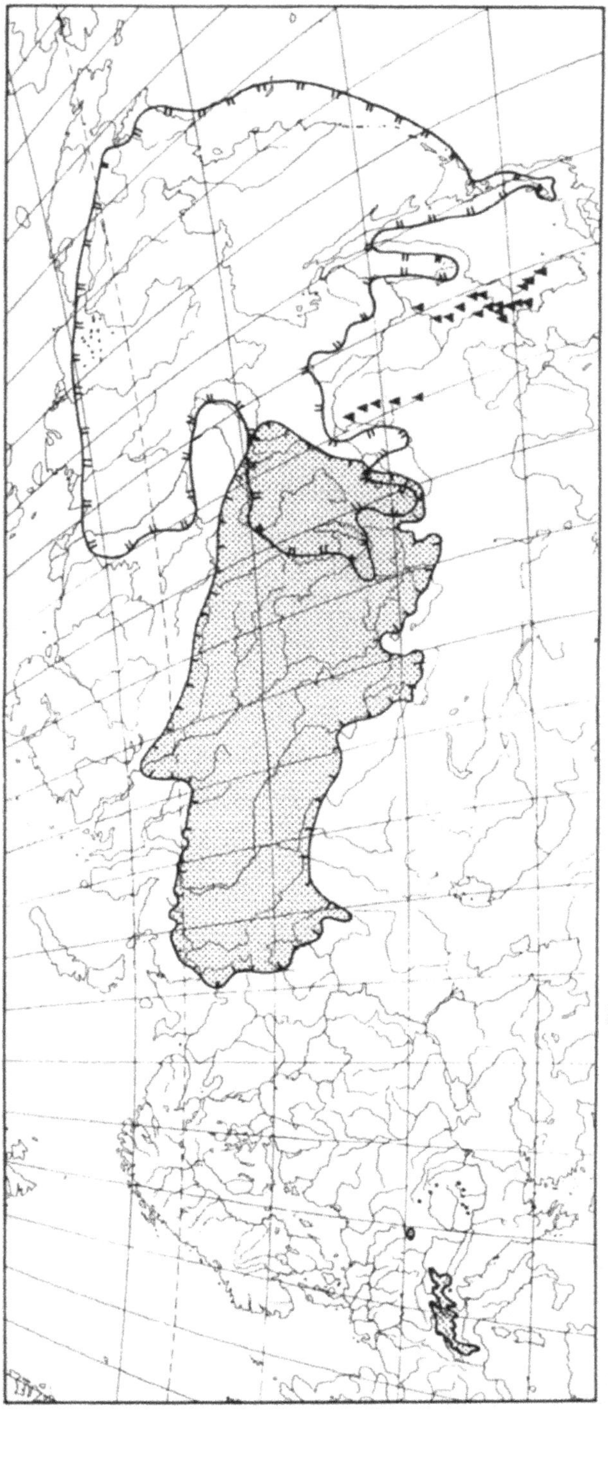

Pinus cembra L. subsp. *sibirica* (RUPR.) KRYL. *P. pumila* (PALL.) REGEL

Fig. 177. *Pinus cembra* L. distribution map. (Zoller 1981)

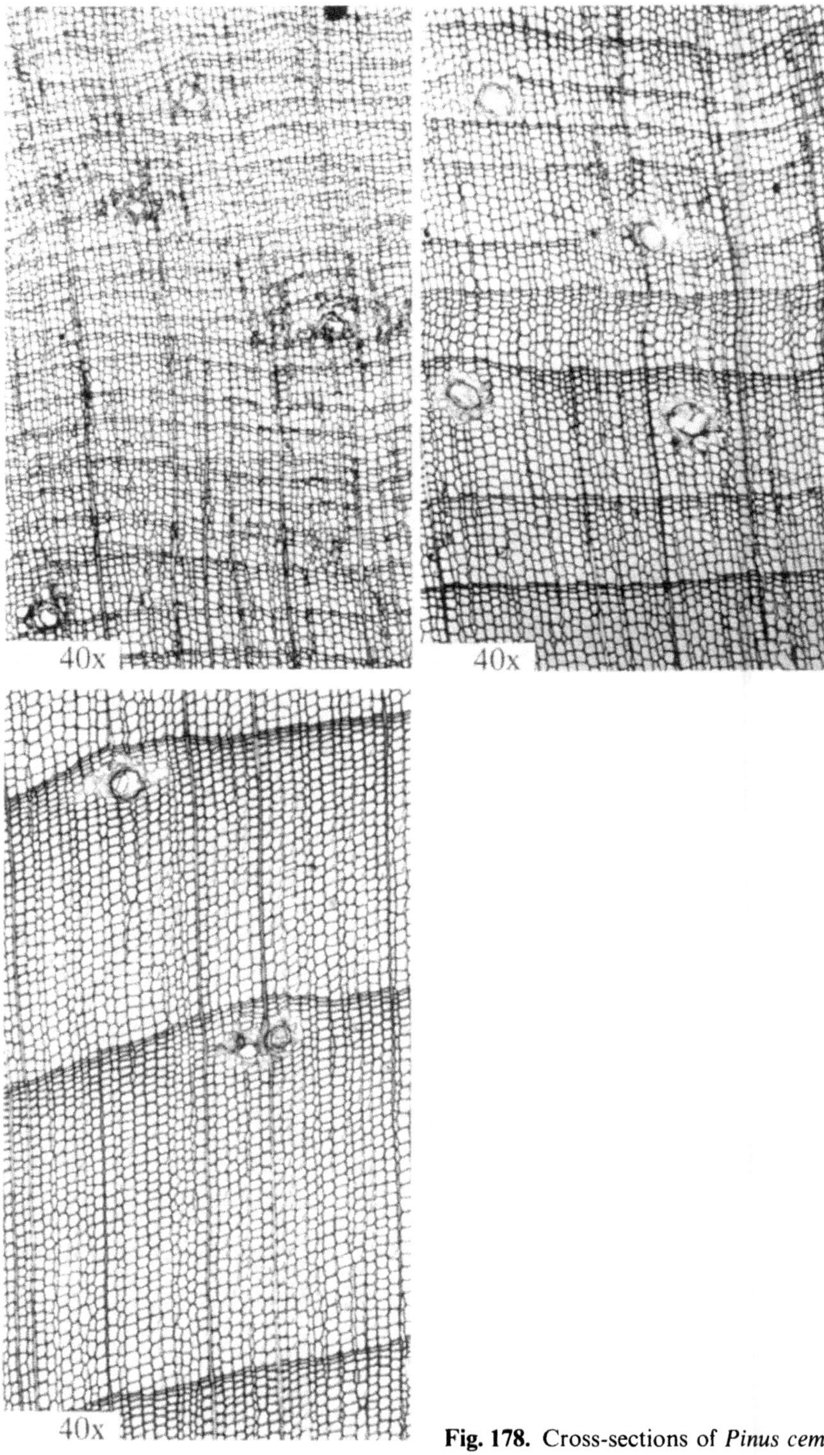

Fig. 178. Cross-sections of *Pinus cembra*

Distribution

Horizontal: *Pinus cembra* ssp. *cembra*: the Alps and the Carpathian Mountains. Distribution in the Carpathian Mountains is very spotted.

Pinus cembra ssp. *sibirica*: North Russian–Siberian Taiga region. In the east Arolla borders on *Pinus pumilo* stands (lower altitude Arolla).

Vertical: *Pinus cembra* ssp. *cembra*: the range extends from under 1400 to 2500 m.

Pinus cembra ssp. *sibirica*: in northern Siberia, the species grows in the lowlands. In Altai, it grows in the mountains.

Wood

Heartwood present. The sapwood is yellow with a silky shine, the heartwood is slightly red. Generally with numerous, dark branches. Strong, pleasant resin odor (Arolla wood odor). The tree rings are often indistinct (Fig. 178a). The density is 360–550 kg/m^3.

Cross-section: the resin canals are bordered by thin-walled epithelial cells. The transition from earlywood to latewood is continuous. (Fig. 178a–c).

Tangential section: ray height averages 5–15 cells.

Radial section: rays with ray tracheids. The tracheid walls are smooth. The ray parenchyma cells have large, open pits. One to two pits per cross-field.

Fig. 179. Stone pine forest in a limestone area of the Swiss northern prealps. Forêt de Lapé

Fig. 180. Radiographs of *Pinus cembra* in Tyrol, Austria

Use in Dendrochronology

Suitability of the Species

For genetic reasons, the latewood densities of Arolla are of minimal variability. Thus, the species is not very well suited for tree-ring width, or even less radiodensitometric studies (Bircher 1982). In Siberia, Arolla could be used in dendroecological and climatological studies because of its extensive area. In Europe, older trees have, in many cases, been restricted to isolated sites.

Previous Uses

Until recently, there has been little research on *Pinus cembra*. For dating purposes, Siebenlist-Kerner (1984) recorded a 700-year Arolla chronology from Tyrol, Austria. Bircher (1982) did densitometric work on fossilized trunks from the Alps. All additional investigations have involved recent material. Artmann (1949) tried to establish a relationship between tree-ring widths and temperatures. LaMarche and Fritts (1971) correlated tree-ring widths with glacier movements in the Alps. Bednarz (1984) dendroclimatologically analyzed Arolla tree-ring widths from Polish, and Kolishchuk (1983) from Ukrainian relict regions. There are reports of four Siberian chronologies, which reach back to 1415 (Kairiukstis 1983).

Pinus densiflora Sieb. et Zucc.

Japanese red pine

Morphology

This large, deep-rooted pine has two needles in a short shoot. It grows to a height of up to 35 m and has a maximum bole diameter of 1.5 m. The bark is deep red-brown. The crown becomes umbrella-shaped and flattened with age.

Site, Site Conditions, Characteristics

Japanese red pine grows in cool to moderately warm marine climates and covers a wide ecological spectrum. With a reduced growth rate it even grows on nutrient-poor, dry as well as moist, soils. Similarly to *Pinus sylvestris*, Japanese red pine grows optimally in normal soils. The species requires sunny sites with an annual precipitation level of 1500 to 2000 mm and avoids regions with low winter temperatures, such as in Hokkaido. Today, large areas on hardwood sites are planted with *Pinus densiflora*. Stands older than 200 years are common.

Fig. 181. *Pinus densiflora.* (Katayama et al. 1968)

Fig. 182. *Pinus densiflora* or *P. thunbergii* on the Japanese 500-yen bill. (excerpt)

Fig. 183. An old site of *Pinus densiflora.* On prime sites, trees reach heights of 35 m. Red-brown stems are typical. (Forestry in Japan)

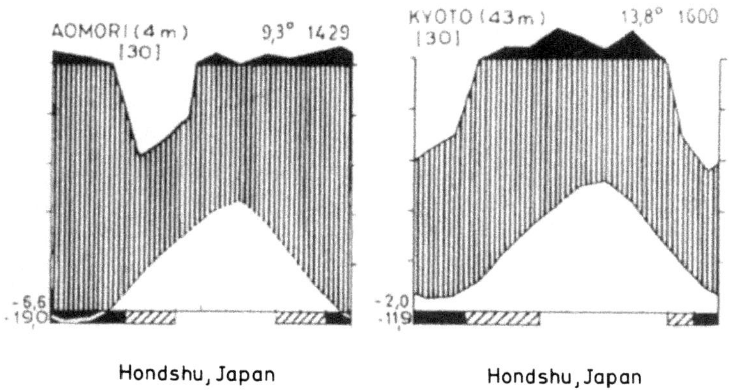

Fig. 184. Climate diagrams

Distribution

Horizontal: Japanese red pine is endemic to Japan, excluding Hokkaido.

Vertical: from sea level to the intermediate mountain region at around

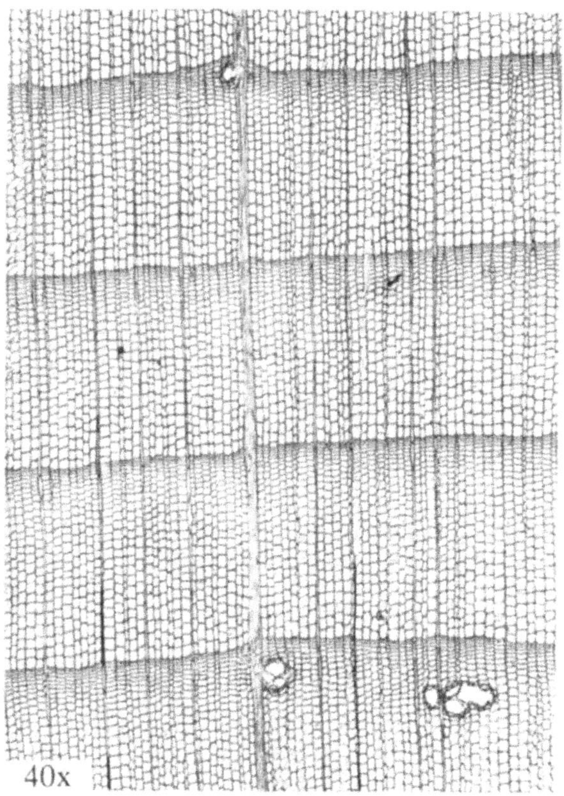

Fig. 185. Cross-section of *Pinus densiflora*

1500 m. The original region has been enlarged by widespread forestation of the species.

Fig. 186. Distribution of *Pinus densiflora*. (Katayama 1968)

Wood

The wood is practically identical with that of *Pinus sylvestris*, see there for description.

Use in Dendrochronology

Suitability of the Species

The tree rings are extremely distinct and exhibit a wide latewood band. The species is especially suitable for ecological purposes. Because only a few old stands exist and the wood was used only locally as lumber, the suitability of the species for climatological and historical research is minimal.

Previous Uses

Otha (1987) and Yokobori and Otha (1983) carried out radiodensitometric studies with trees from industrial sites. They determined reductions in the tree-ring width and the maximal densities as a result of high SO_2 levels.

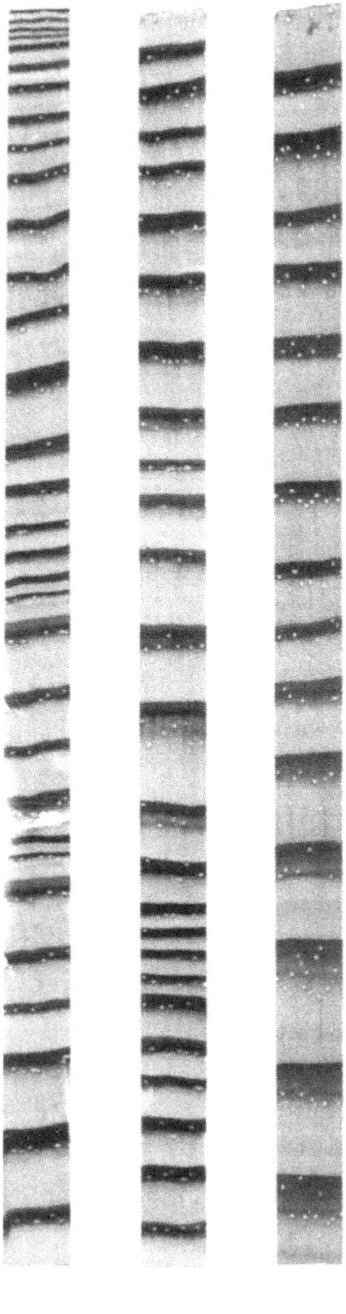

Fig. 187. Radiograph of *Pinus densiflora*. Hondshu, Japan

Pinus halepensis Mill.
Pinus brutia Ten. Syn. *Pinus halepensis* var. *brutia* (Ten.) Elwes et Henry.

Aleppo pine, Jerusalem pine
Calabrian pine

The two species are so closely related that it is often difficult to differentiate between them.

Morphology

The trees grow to heights of up to 20 m and have maximum stem diameters of 90 cm. Their crowns are first pyramid-shaped but later become irregular and open. The cones remain on the tree for several years. The root system is normally complex. On extremely windy sites, the species is often blown prostrate.

Fig. 188. *Pinus halepensis.* Drawing by Hirzel

Site, Site Conditions, Characteristics

Pinus halepensis is native to the summer-dry mediterranean climate. Temperatures in January average no lower than 4 °C, freezing at around −5 °C.

Average temperatures in July are around 22 °C. *Pinus halepensis* requires at least 300 mm of precipitation per year. Due to the well-developed root system, the species can grow in dunes as well as on rock, and is generally indifferent to soil acidity. Together with the evergreen *Quercus* species, *Pinus halepensis* is the main species of the Mediterranean coastal and bush forests. It forms forests together with *Quercus ilex*, *Q. suber*, *Pinus pinaster*, and a variety of shrubs. Pure

Fig. 189. Aleppo pine on a dry, rocky site in the western Mediterranean region

Fig. 190. A typical stand on the Mediterranean coast near Nice. (Rol 1962)

Fig. 191. Aleppo forests are often threatened by fires in the dry summer regions along the Mediterranean

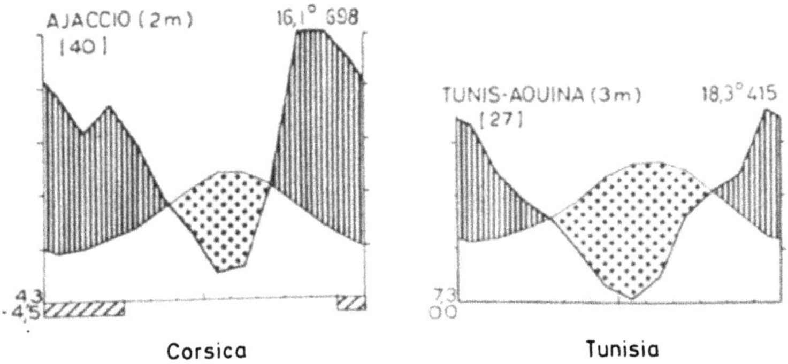

Fig. 192. Climate diagrams

Fig. 193. Distribution of *Pinus halepensis* and *P. brutia*

stands exist, for the most part, only on forested sites. *Pinus halepensis* is the most drought-resistant species in the Mediterranean region. Although it can grow to be quite old, exploitation of the area and forest fires have destroyed most of the old forests. Stands from chaparral are barely 100 years old. The species is very sensitive to fire; it is rarely attacked by *Dendrolimus pini.*

Distribution

Horizontal: in the coastal regions along the Mediterranean. In the western region of the distribution area *Pinus halepensis* predominates. Toward the east, it becomes less common and is eventually replaced by the closely related *Pinus brutia.* Forestation of the species in recent times has greatly increased the size of its area.

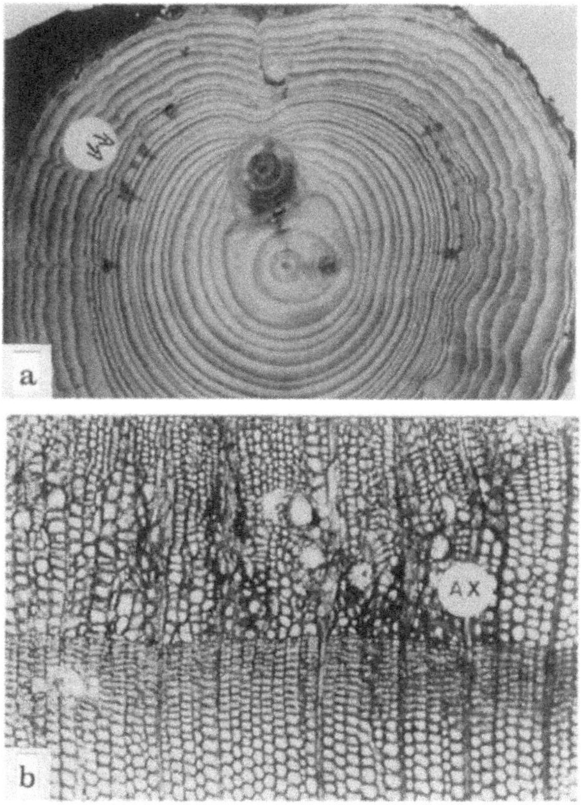

Fig. 194a, b. *Matsucoccus josephii* damage to the tree rings of *Pinus halepensis*. (Liphshitz and Mendel 1987)

Vertical: north of the Mediterranean, *Pinus halepensis* is mainly found growing at low altitudes. However, in northern Africa, it extends to the mountain zone. The upper border lies in Dalmatia at 200 m, in southern Spain at 1000 m and in Algeria and Morocco at 1600 m.

Wood

Heartwood is present. The sapwood is narrow and almost white and the heartwood light yellow. The resin-rich wood contains distinct resin canals. The density is about 650 kg/m^3.

Cross-section: the tree-ring boundaries are very distinct (Fig. 195a–c). Density fluctuations are common. The transition from earlywood to latewood is quite abrupt (Fig. 195b, c). The resin canals are bordered by large, thin-walled epithelial cells.

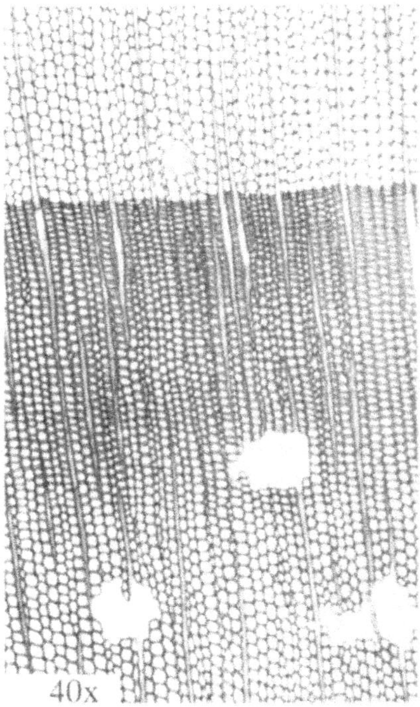

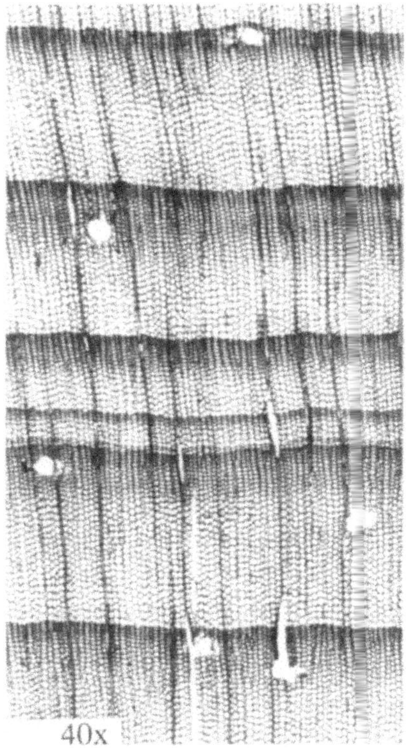

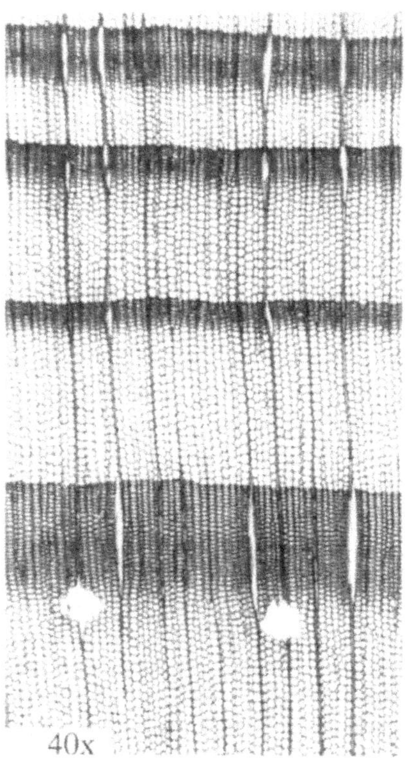

Fig. 195a–c. Cross-sections of *Pinus halepensis*

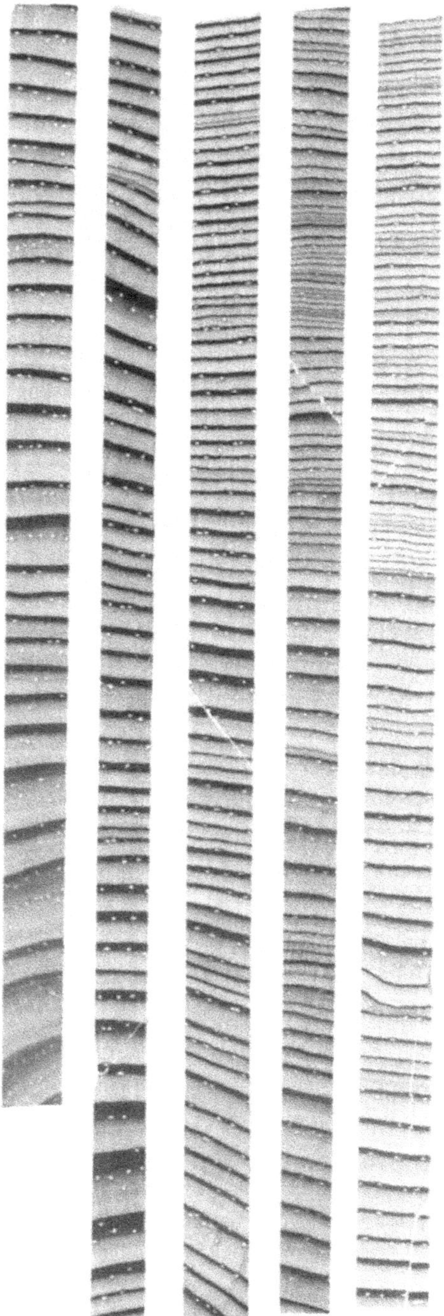

Fig. 196. Radiographs of *Pinus brutia* in Cyprus

Tangential section: the average ray height is ten cells. The cell walls of the resin canals in the rays are very thin.

Radial section: rays with ray tracheids. Parenchyma ray cells with one to three pits in the cross-field. The ray tracheids generally have distinct, notched walls.

Use in Dendrochronology

Suitability of the Species

Aleppo pine wood is suitable for chronological, ecological, as well as climatological purposes. However, early summer rains often result in the development of false rings. Wood found in old buildings or ships can be used for dating purposes.

Previous Uses

As the result of research undertaken by Serre (1973), *Pinus halepensis* is by far the most thoroughly dendrochronologically studied species in the Mediterranean region. Growth rate studies on shoots, needles, and the cambium of both old and young trees from a homogeneous site show that fall and spring precipitation has a decisive effect on the growth when temperatures in February rise above 4 °C. The longest chronology reaches back to 1803. Kienast (1985) used *Pinus brutia* in a radiodensitometric-dendrochronological study in Cyprus. The radial growth, depending on the altitude, is promoted by high temperatures in the months from January through April and by high precipitation levels in the fall. Maximum densities are almost entirely related to high temperatures in March and April. According to Liphschitz and Mendel (1987), damage done by *Matsucoccus josephii* is revealed in the xylem as callus tissue.

Pinus heldreichii Christ
Pinus leucodermis Ant.

Heldreich's pine
Bosnian pine

Morphology

The trees grow to heights of up to 25 m, have dense, flat crowns and a deep root system. Stem diameters average over 1 m. Based on cone types, the genus is divided between *Pinus leucodermis*, which grows in Italy, and *P. heldreichii* from Greece and Bulgaria. Snake-skin-like bark is typical of both species.

Fig. 197. *Pinus leucodermis.* Drawing by Hirzel

Site, Site Conditions, Characteristics

Pinus leucodermis can be found on the forest borders of southern Italy and the Balkans. It withstands long, dry summers, especially at low altitude sites, as well as cold winters. The species forms scattered forests on limestone and its own,

Fig. 198. A 600-year-old *Pinus leucodermis* in southern Italy

Fig. 199. An old stand at the timberline on Mt. Crispo, southern Italy. In the foreground are bush-shaped, pest-eaten beech trees

narrow belt above the beech zone in Italy. The species also forms the forest border in Greece above the *Abies cephalonica* and *Pinus nigra* belt.

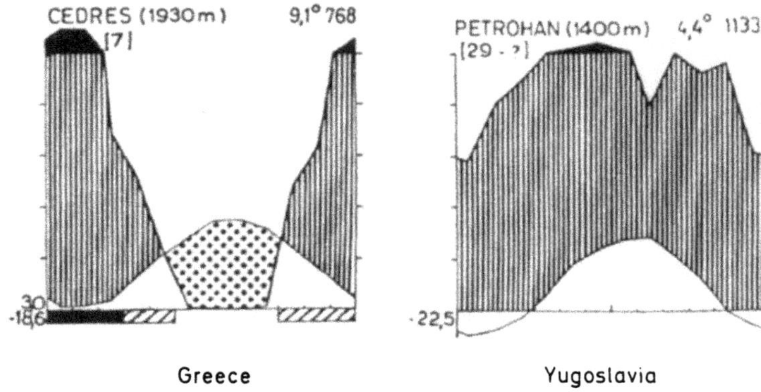

Fig. 200. Climate diagrams

Distribution

Horizontal: Spotted areas in southern Italy and the Balkans.

Vertical: the main area of distribution lies in Italy at an elevation of 1500–1900 m. In Greece, for example on Mt. Olympus, the range extends up to 2500 m.

Fig. 201. Distribution of *Pinus heldreichii* and *P. leucodermis*. (Krüssmann 1972)

Wood

The wood is yellowish and finely structured with a strong resin odor. The density is at 500 kg/m^3. *Pinus leucodermis* cannot be distinguished from *P. pinaster* on the basis of its anatomic characteristics. See *Pinus pinaster* for description.

Fig. 202. An old *Pinus heldreichii* in the Pirin Mountains, Bulgaria

Use in Dendrochronology

Suitability of the Species

The species is limited to a few high-altitude, small sites in the eastern Mediterranean region. Because the trees grow to be quite old and are often used as building material, the species is suitable for chronological and climatological purposes.

Previous Uses

In the eastern Mediterranean region, Kuniholm and Striker (1987) developed a comprehensive chronology spanning several centuries. Serre-Bachet (1985) recorded an 800-year chronology from live trees in southern Italy. In Mediterranean regions poor on trees, the species is the major component of the Mediterranean dendroclimatological sample grid (Schweingruber 1987).

Pinus mugo Turra Syn. *P. montana* Mill.

Mountain pine

Pinus mugo var. *uncinata* Syn. *P. uncinata* Mill. ex Mirb.

Upright mountain pine

Pinus mugo var. *prostrata* Syn. *P. mugo* var. *mugo*

Prostrate mountain pine

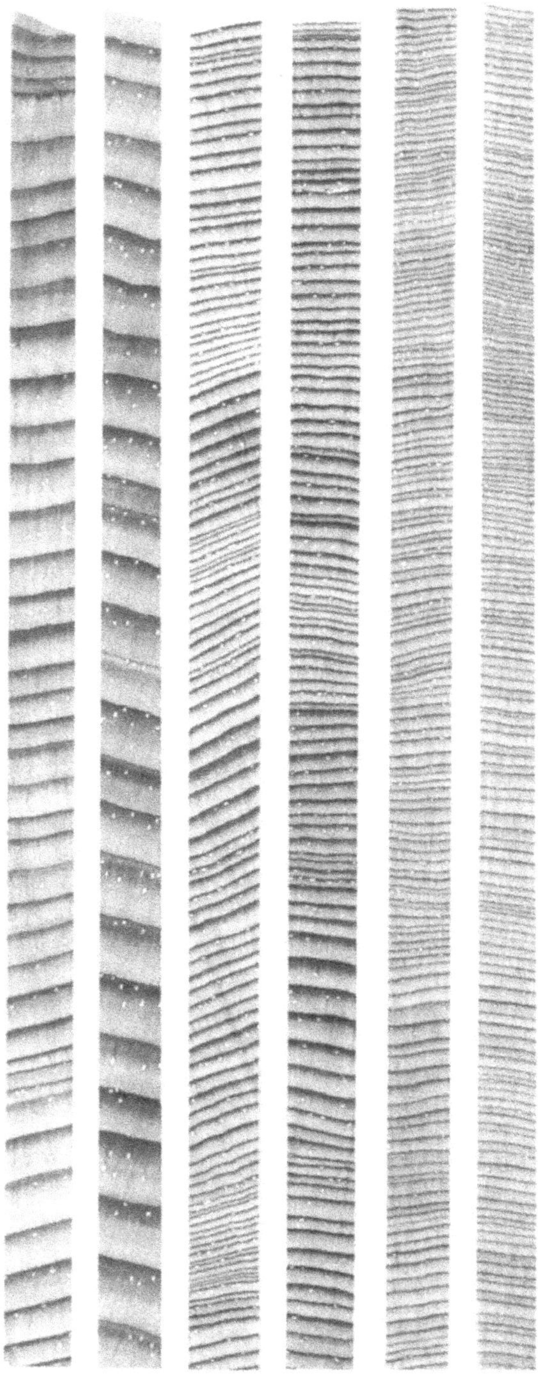

Fig. 203. Radiographs of *Pinus leucodermis*, Mount Olympus, Greece

Morphology

This pine has two needles on a short shoot. It grows in two different shapes. The upright mountain pine is thin, with gray-black bark, reaches a height of up to 20 m, and generally shares the same habits as *Pinus sylvestris*. The prostrate mountain pine, on the other hand, has shrub-like characteristics. The upright species frequently exhibits eccentric cones with hook-shaped scales. The prostrate species has centric cones with flat scales. Between the two forms and the different cone characteristics, subsequent, transitional forms exist.

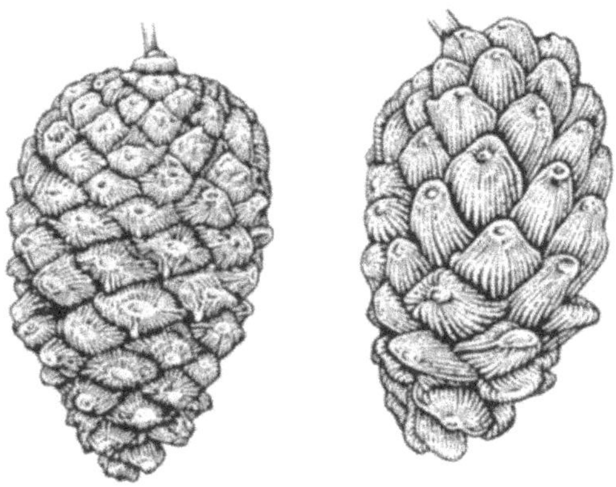

Fig. 204. a Centric cones. **b** Eccentric mountain pine cones. (Drawing by Sebek)

Site, Site Conditions, Characteristics

Pinus mugo predominately grows on subalpine sites in shallow, rocky soils or bogs. It flourishes on limestone foundations and crystalline rocks. In regions with limestone or on moist sites, for example in Flysch, in the northern lower Alps, the upright species forms the forest border. The prostrate species forms a belt above the spruce-larch-stone pine zone and borders on the alpine level.

Fig. 205. Prostrate mountain pine from the eastern Alps. (Fenaroli 1976)

Especially in the eastern Alps and in the Carpathian Mountains, *Pinus mugo* covers large, almost impenetrable areas (see *Pinus cembra* for climate diagram).

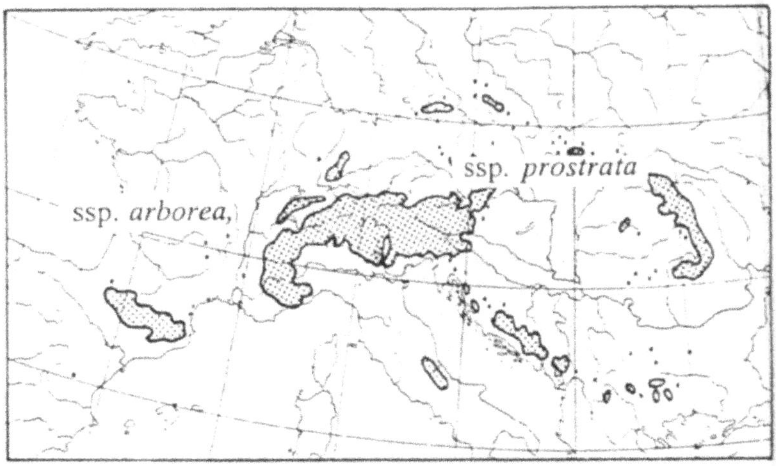

Fig. 206. *Pinus mugo* Turra, distribution map. (Zoller 1981)

Fig. 207. A stem cross-section with healing and compression wood from an avalanche slope. (Bitterli 1987)

Distribution

Horizontal: upright mountain pine: above all in the western half of the distribution area: Pyrenees, Jura, Black Forest, southern Bohemia, and the eastern Alps to lower Engadin (Switzerland).

Prostrate mountain pine: predominately in the eastern half of the distribution area: Carpathian Sudeten, and Dinaric mountains, the maritime Alps, the Appenines and the western Alps to the Swiss province of Tessin.

Fig. 208. Upright mountain pine near the forest border in the Swiss Alps

Vertical: the upper borders of both growth shapes lie in lower Engadin, Switzerland, at an elevation of 2300 m. They form a belt between 1800 and 2200 m. In the swamps (bogs) they extend to an elevation of 450 m in the Swiss central plateau.

Wood

It is impossible to differentiate between the wood of *Pinus mugo* and that of *P. sylvestris* on the basis of their anatomic characteristics.

Use in Dendrochronology

Suitability of the Species

Upright mountain pine, is extremly well suited for dendroclimatological and dendroecological research. It lives long, up to 500 years, can be found on many different sites, including extreme sites, and it has very distinct tree rings.

In contrast, prostrate mountain pine is only suited for dendroecological purposes because the prostrate stems are subjected less to climatic than to local influences.

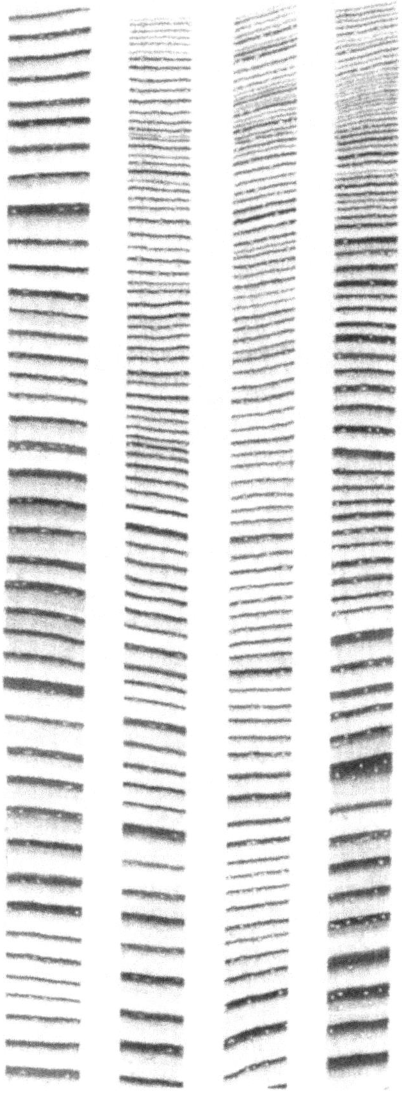

Fig. 209. Radiographs of *Pinus mugo* ssp. *uncinata*, Pyrenees

Previous Uses

Upright mountain pine has until recently been used only rarely. It is a component of the European sample grid from Bräker and Schweingruber (1984) as well as Genova (1986). Ruiz-Flaño (1988) undertook dendrochronological studies on *Pinus mugo* from three different sites.

Tree-ring sequences from prostrate mountain pine were studied by Bednarz (1986), Bitterli (1987), and Kolishchuk (1990). By examining the abrupt growth changes and wound overgrowths of the irregular, eccentric stems, it is possible to obtain information on mechanical soil alterations, snow movements, and wind effects.

Pinus nigra Arnold

Black pine

Morphology

On optimal sites, this tree grows to a height of 30 m and has a broad, flat crown. On dry sites and windy summits, the trees are stout with thick stems.

Pinus nigra has an unusually large number of polymorphic and geographic races. Some of them are: *Pinus nigra* var. *calabrica*, var. *caramanica* (Crimean pine), var. *dalmatica*, var. *laricio*, var. *maritima* (Corsican pine), var. *nigra* (Austrian pine), var. *pallasiana*, and var. *salzmannii*. *Pinus nigra* var. *laricio* is taller and has a larger stem diameter than the others.

Fig. 210. Drawing by Hirzel

Site, Site Conditions, Characteristics

This light tree builds pure stands on southern slopes in the Mediterranean region at elevations of around 1000–1400 m. In the lower zones the stands are mixed with *Pinus pinaster* and *Quercus ilex*, in the highest zones with *Fagus sylvatica* and *Abies* sp. as well as *Pinus sylvestris*. *Pinus nigra* requires more than 700 mm of precipitation per year and can tolerate as much as 1800 mm. The average annual temperature must not fall below 7.5 °C. *Pinus nigra* is, for the most part, indifferent to soil acidity and grows on limestone and silicate foundations. The species is relatively sensitive toward all types of pests. Outside

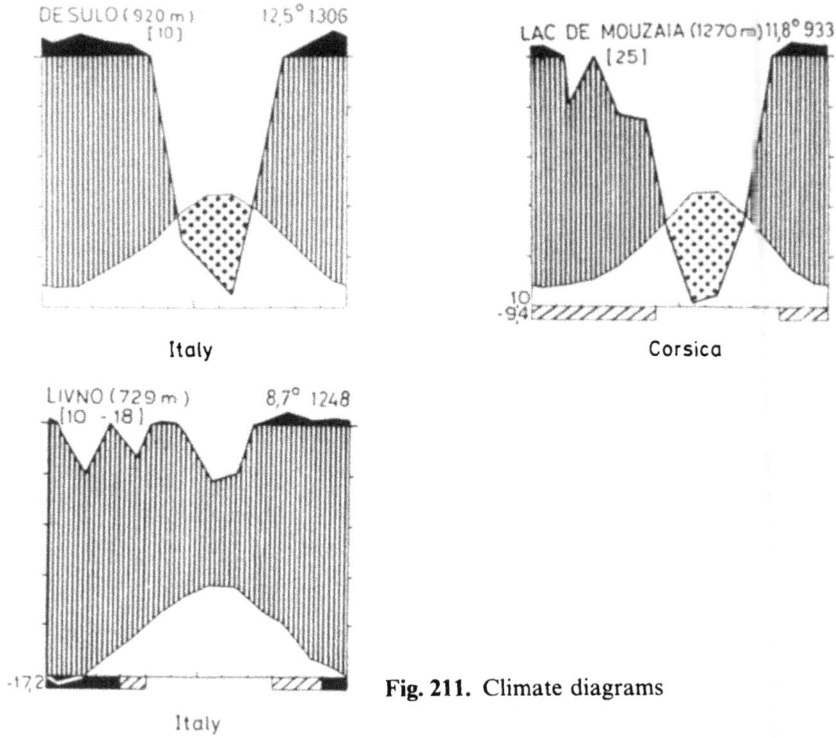

Fig. 211. Climate diagrams

Fig. 212. *Pinus nigra* ssp. *calabrica* on Mt. Etna in Sicily

its natural area of distribution, *Pinus nigra* is mainly attacked by the fungus *Gremmeniella abietina* (shoot decline).

Fig. 213. *Pinus nigra* ssp. *laricio* in central Corsica

Fig. 214. *Pinus nigra* ssp. *salzmannii* in central Spain

Distribution

Horizontal: predominately in eastern Mediterranean countries.

Vertical: the main area of this polymorphic species lies in the mountain and subalpine zones. Its optimal elevation is at 800–1200 m. In Corsica the upper border lies at 1800 m and in Turkey at 2000 m.

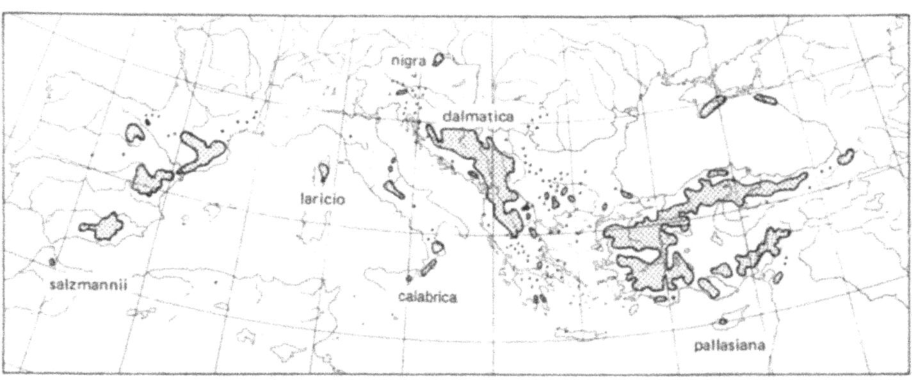

Fig. 215. Distribution of *Pinus nigra*. (Zoller 1981)

Wood

With heartwood. The sapwood is almost white; the heartwood is red-brown to bright red. In extremely resin-rich wood, the resin canals are very distinct. The density is 550–680 kg/m^3.

The wood cannot be distinguished from that of *Pinus sylvestris* and *Pinus mugo* on the basis of its anatomic characteristics. See *Pinus sylvestris* for description (Fig. 216).

Use in Dendrochronology

Suitability of the Species

The tree rings are distinct. Radiographic research disrupts the large resin canals in the narrow tree rings. It is to be presumed, on the basis of the broad ecological amplitude, that on the lower border of the distribution area, tree growth is influenced by low precipitation. On the upper border, in the beech and fir region, cell wall growth (maximum density) is restricted by late summer temperatures. Dendrochronological research, especially from eastern Mediterranean countries, promises to be successful (Cyprus and Turkey).

On natural stands in the Wienerwald, 600-year-old trees can be found, while in Corsica the trees can grow to be 1000 years old.

In every region, resin is harvested from *Pinus nigra*. Stem damage caused by the resin harvest results in irregular tree-ring growth.

The straight-growing trunks make excellent lumber. In ship building, the longest stems are used for the masts. In Austria, *Pinus nigra* serves in ground and water construction.

Previous Uses

The wood has mainly been used for dating historic and prehistoric complexes in Turkey and Greece by Kuniholm and Striker (1987), who succeeded in formulating an almost 1000-year, absolute chronology and a floating chronology which spans many centuries. Kienast (1985) performed radiodensitometric studies on *Pinus nigra* from Cyprus and determined that the tree-ring growth is usually restricted by precipitation. The maximal densities of trees from forest border sites allow for the reconstruction of summer temperatures (Schweingruber 1985).

Richter (1987) investigated recent and historic samples taken from various sites in the mountain ranges of Spain. He came to the conclusion that species from these regions are best suited for chronological and climatological purposes.

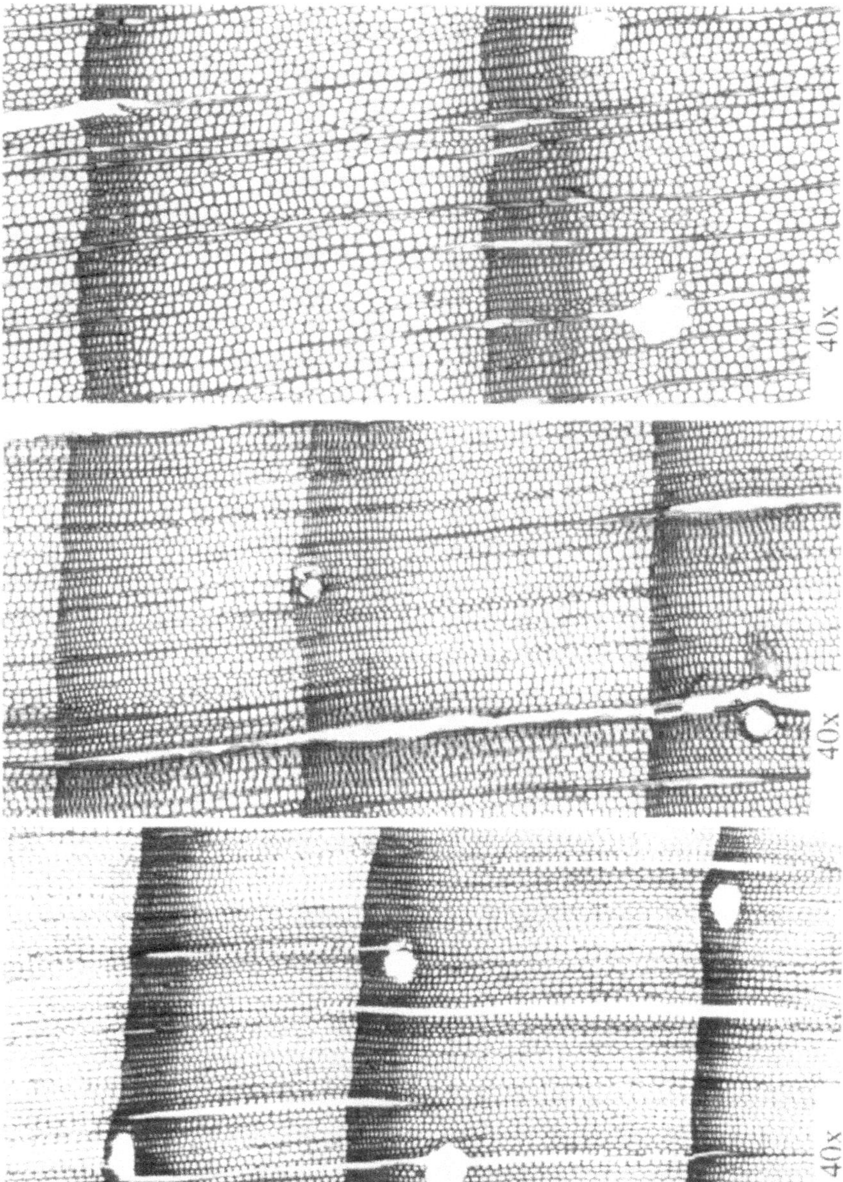

Fig. 216a–c. Cross-sections of *Pinus nigra*

Fig. 217. Radiographs of *Pinus nigra* from a natural site in Cyprus (six samples at the *left*) and plantings in Switzerland (two samples at the *right*)

Pinus pinaster Ait. Syn. *P. maritima* Poiret

Maritime pine, cluster pine, seaside pine

Morphology

This 30-m-tall tree has a full, broad crown and a stem diameter frequently greater than 100 cm.

Fig. 218. Cones from *Pinus pinaster*

Site, Site Conditions, Characteristics

Pinus pinaster is a conifer from the temperate Mediterranean climate. It forms pure stands on lower mountain sites with average annual temperatures of 2–4 °C, July temperatures of 20–22 °C and an annual precipitation of 600–800 mm. Pure stands grow on silicate foundations. At lower altitudes, *Pinus pinaster* is accompanied by *Quercus ilex*, and at higher sites by *Pinus nigra* ssp. *laricio*. It grows quickly and is therefore valuable in the forest industry. Trees over 100 years old can only be found on protected sites. *Pinus pinaster* is often damaged by forest fires. Damage done to the crowns by *Matsucoccus feitandi* or *Dendrolimus pini* (pine spider), can be detected in the tree rings. The trees are frequently injured by resin harvest.

Fig. 219. Rapid-growing *Pinus pinaster* in western France. (Badré 1983)

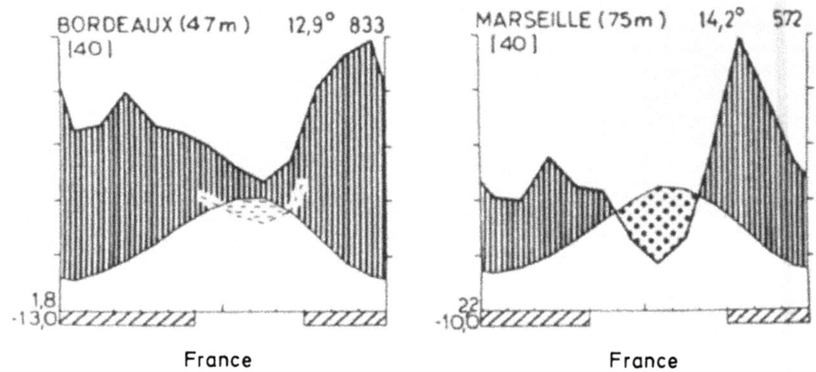

Fig. 220. Climate diagrams

Distribution

Horizontal: in the coastal regions of western Mediterranean countries.

Vertical: at low altitudes to the mountain zone, forming forests at 400–900 m. The upper borders lie in Corsica at 1500 m and in the Moroccan Atlas at 2000 m.

Fig. 221. Distribution of *Pinus pinaster*

Fig. 222. Resin harvesting and resin wounds on *Pinus pinaster* trunks in Spain

Wood

With heartwood. The sapwood is wide and almost white, the heartwood is red to dark red. In resin-rich wood, the resin canals are very distinct. The density is around 550 kg/m^3. The description is the same as for *Pinus pinaster*, *P. leucodermis*, and *P. halepensis*.

Cross-section: the transition from earlywood to latewood is gradual to sharp. There are usually frequent density fluctuations (Fig. 223a, b). Large resin canals with thin-walled epithelial cells.

Tangential section: the average ray height is ten cells. The walls of the ray parenchyma cells are generally thin.

Radial section: the rays have ray tracheids with few distinct notched walls. Pits in the cross-field are rather large and almost taxodioid. There are three to five pits per cross-field in the earlywood. *Pinus pinaster* is anatomically indistinguishable from *Pinus leucodermis.*

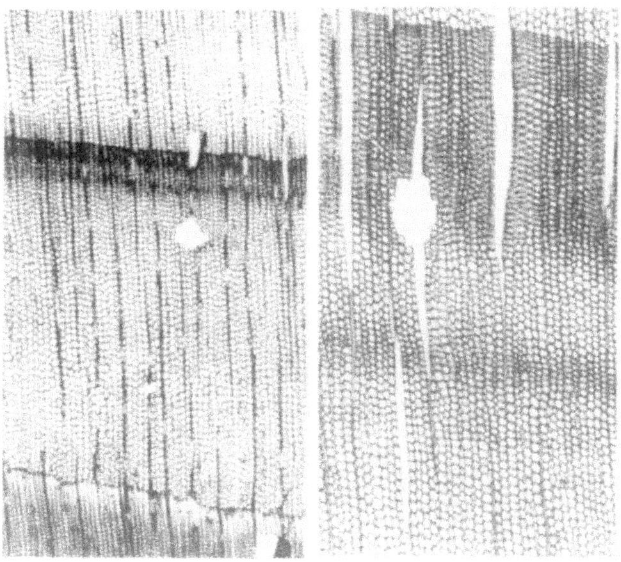

Fig. 223a, b. Cross-sections of *Pinus pinaster*

Use in Dendrochronology

Suitability of the Species

Pinus pinaster is only of limited use in dendrochronology and climatology because very few old stands still exist, and trees from the lower mountain level often have false rings. Most trees have been damaged by resin harvest. *Pinus pinaster* wood found in old buildings can be used for the constructing of longer chronologies (Richter 1987).

Previous Uses

A few younger, recent trees have been studied in Tunisia (Serre-Bachet 1982) and in the Sierra Albaracin, Spain (Richter 1987).

Pinus pinea L.

Stone pine, umbrella Pine

Morphology

This conifer has two needles in short shoots and a decorative, umbrella-shaped crown. On optimal sites it grows to a height of 25 m and has a trunk diameter of 100 cm. In coastal regions, on dunes, *Pinus pinea* develops extremely deep tap roots that often extend down to the ground water surface.

Fig. 224. *Pinus pinea.* (Drawing by Hirzel)

Site, Site Conditions, Characteristics

Pinus pinea is native to the dry-summer, marine climate of the Mediterranean. It grows mainly on dunes, but does not avoid granite foundations. These sites have average January temperatures of over 6–8 °C, July temperatures higher than 22 °C, and annual preciptitation levels of 500–800 mm. On moist ground, it builds individual, widespreading beach forests. *Pinus pinea* requires more moisture than *Pinus halepensis.* In relict stands, trees can still be found that are several centuries old.

Fig. 225. Typical umbrella-like tree form

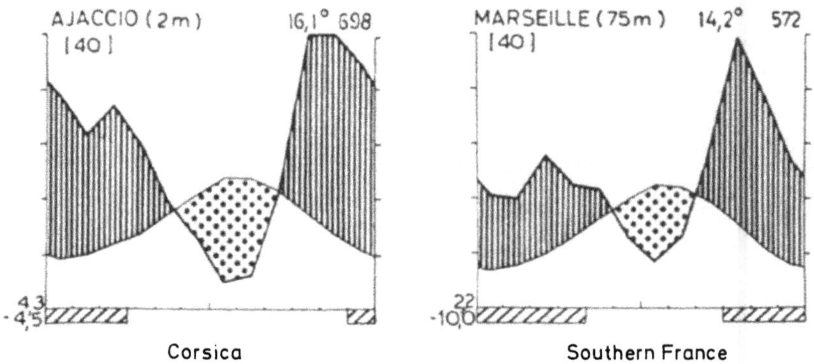

Fig. 226. Climate diagrams

Distribution

Horizontal: mainly in northwestern Mediterranean countries. In contrast to that of *Pinus halepensis*, the range extends to Portugal. The species is not found in northern Africa.

Vertical: at low elevations near the coast. In southern Spain it extends to a maximum elevation of 1000 m.

Fig. 227. Distribution of *Pinus pinea*

Wood

With heartwood. The sapwood is generally narrow and light yellow-red. The heartwood is yellow-red. The density is 600 kg/m^3.

Fig. 228a, b. Cross-sections of *Pinus pinea*

Cross-section: distinct to missing tree rings. Large resin canals with thin-walled secretory cells (Fig. 228a–b).

Tangential section: the ray height is normally 8–15 cells.

Radial section: the rays have thin and smooth-walled ray tracheids. Pinoid to taxodioid pits in the cross-field. There are two to three pits per cross-field.

Use in Dendrochronology

Suitability of the Species

Large, old stems are of limited dendrochronological use because the trees on deep soil along the coast develop false rings. For constructing longer chronologies, wood from mines and ships can be used. Earlier, *Pinus pinea* was often used for railway ties.

Previous Uses

De la Hera et al. (1981) analyzed three trees from a dry site in Spain. Fluctuations in the ground water surface could be reconstructed from trees growing near the coast. Pividori and Cherubini (1990) studied the stand dynamics in an area around Florence. In the 16th century, Leonardo da Vinci described the formation of tree rings and their relation to precipitation.

Pinus sylvestris L.

Scots (Scotch) pine

Morphology

This conifer has both well-developed surface roots and a deep tap root. It takes many different forms. In northern Europe, and in the high and intermediate mountain sites, it has a narrow crown. In marine and semiarid climates the

Fig. 229. *Pinus sylvestris*

crowns are bushy, sometimes umbrella-shaped. On optimal sites, the trees grow to heights of 40 m but on adverse sites only 5–15 m. In rare cases, the stem diameters are larger than 100 cm. The species can be classified into many ecological types corresponding to its vast distribution area.

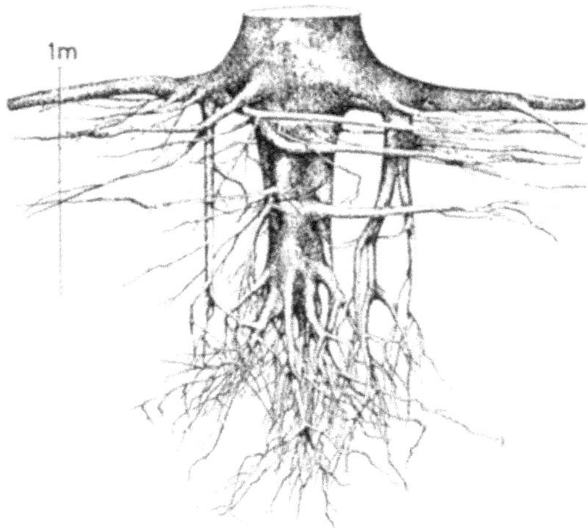

Fig. 230. The root system is very well developed. (Köster et al. 1968)

Site, Site Conditions, Characteristics

Pinus sylvestris covers a very wide ecological spectrum. It grows in the cold continental climate of northern Asia as well as in the semiarid climate of southern Europe. On northern European sites, temperatures in January average

Fig. 231. Scots pine at the timberline at an elevation of 1900 m in central Spain near Madrid

around $-15\,°C$ (the minimal temperature is around $-40\,°C$), in southern Europe around 8 °C. On the northern forest border, temperatures in July are about 10 °C, while in the Mediterranean region they can reach 22 °C. In western Europe, annual precipitation is 2500 mm but in southern Europe only 400 mm. The vegetation period lasts about 2 months. *Pinus sylvestris* is indifferent to soil acidity. It withstands loose sand soils on limestone and crystalline rocks and forms pure stands on extreme sites. On favorable sites, Scots pine is often

Fig. 232. Scots pine at the timberline at an elevation of 500 m in Scotland

Fig. 233. Scots pine at the northern timberline on a windy site at Karelien, Russia

Fig. 234. Scots pine at the timberline in Scotland (Caledonian pine)

Fig. 235. A Scots pine forest with identically aged trees, a result of a previous fire. (Bern Highland, Switzerland)

suppressed by more demanding species such as oak and beech. On all natural sites, *Pinus sylvestris* grows to be well over 300 years old, and it reaches its maximum age of 600 years on dry sites.

Pinus sylvestris crowns are periodically attacked by Lepidoptera caterpillars, especially pure stands in dry regions of northern Germany, eastern Europe, and Russia. *Bupalus piniarius*, *Panolis flammea*, *Lymantria monacha*, and *Dendrolimus pini* are the main pests. The reduction of the needle surface is reflected in the tree rings. Growth is often stunted by the fungus *Heterobasidion annosum*.

Distribution

Horizontal: transcontinental Eurasia

Vertical: predominately a plains tree. Its maximum elevation lies in Norway at 2000 m. The trees are commonly found growing up the sides of southern slopes.

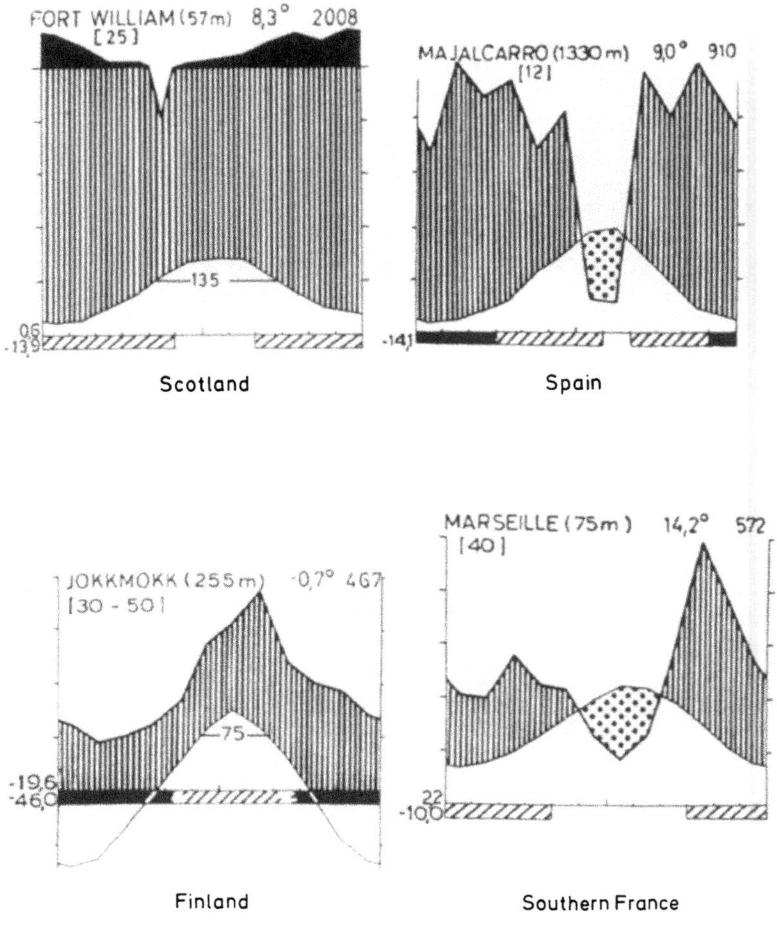

Fig. 236. Climate diagrams

Wood

With heartwood. The sapwood is yellow and the heartwood is red. There is a strong resin odor. The tree rings are distinct. The density is 310–860 kg/m^3.

Cross-section: resin canals with large, thin-walled epithelial cells. Sharp transition from earlywood to latewood.

Tangential section: average ray height: 5–15 cells.

Radial section: rays with ray tracheids. Transverse tracheid walls, tooth-shaped. Ray parenchyma cells with large, open pits. Mostly one large pinoid pit per cross-field.

This description is applicable also to *P. mugo* and *P. nigra*.

Fig. 237. Distribution of *Pinus sylvestris*. (Zoller 1981)

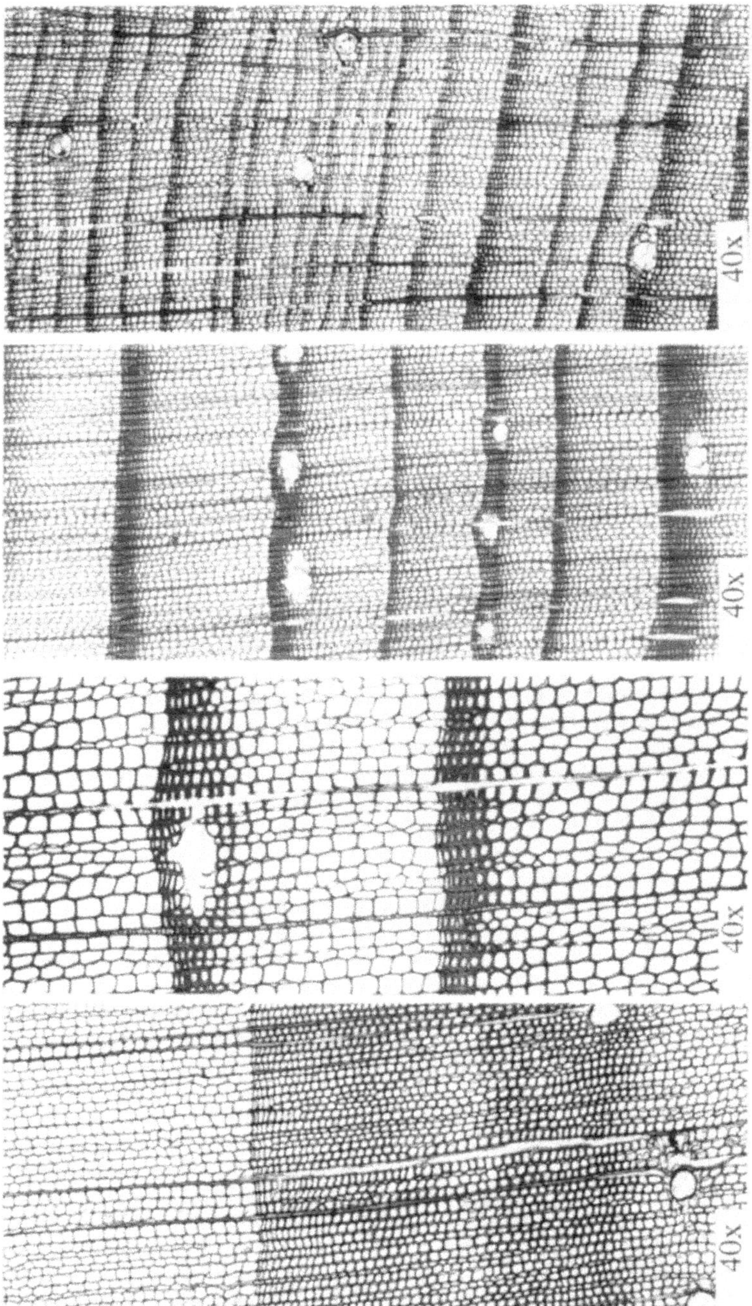

Fig. 238a–d. Cross-sections of *Pinus sylvestris*

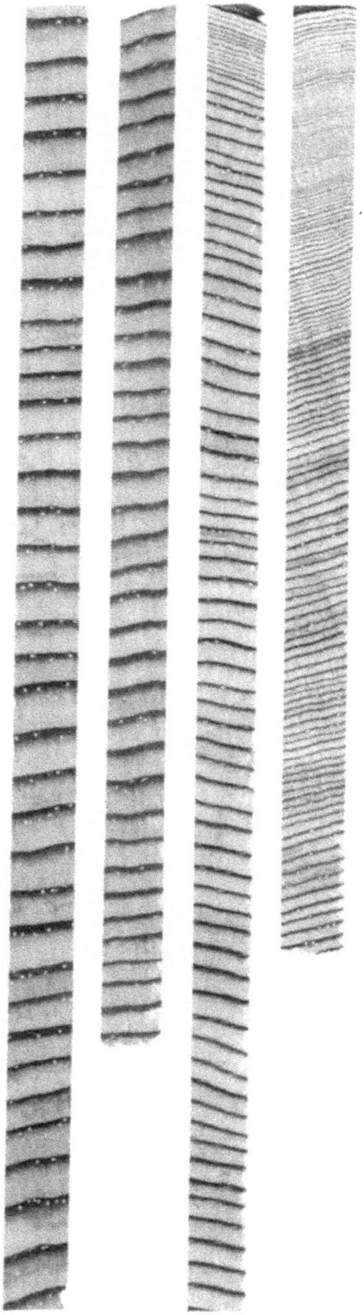

Fig. 239. Radiographs of *Pinus sylvestris*, Lake Baikal (Russia)

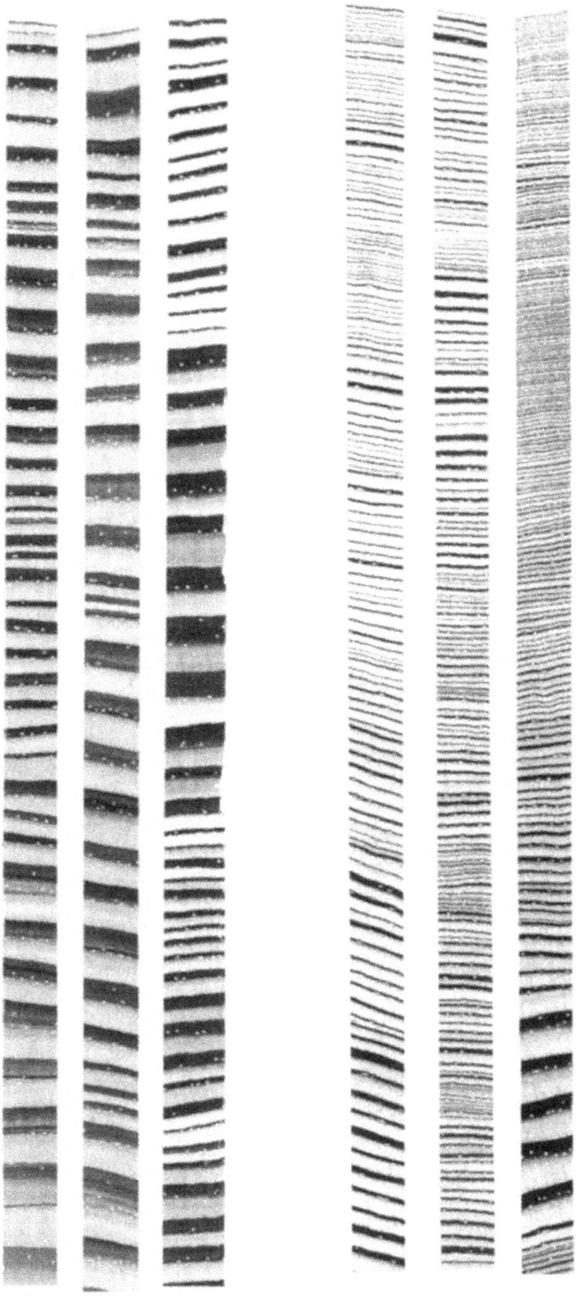

Fig. 240. *Left* Radiographs of *Pinus sylvestris* from dunes on the North Sea (Germany). *Right* *P. sylvestris* from a dry slope in Wallis (Switzerland)

Use in Dendrochronology

Suitability of the Species

Next to Norway spruce, Scots pine possesses the highest dendrochronological potential in Europe because it grows on so many different sites. It is also suited for radiodensitometric studies. Regional and site homogeneous chronologies need to be constructed. Ancient stems found in northern bog regions offer a tremendous reservoir for historic and dendroclimatic studies. Using radiodensitometric analysis it has been possible to construct several year-long climatologically interpreted rows from the Holocene period. In historic times, pine wood was only used for building material when no other coniferous species were available.

Previous Uses

Corresponding to the wide geographical and ecological distribution of Scots pine, the species has been used in many studies. *Pinus sylvestris* is a western component of the sample grid in Europe, Russia, and Siberia (Bitviskas 1974). Bitviskas and coworkers constructed a several-year chronology along a transect from Murmansk to the Black Sea and from Karlia to Kamchatka. Briffa et al. (1988) and Schweingruber (1988) used the species to reconstruct summer temperatures in Europe. Sirén (1963) and Bartholin (1983) synchronized a 1500-year, climatologically interpretable chronology from Scandinavia. Ecological studies with recent material taken from bogs as well as from normal and dry sites have above all been performed in Russia. The observations deal mainly with problems of forest development, hydrology, and geomorphologic processes. In western Europe, Tessier (1982) completed methodic statistical experiments on material from southern France. Kienast (1985, 1987) analyzed samples obtained at high altitudes in Switzerland, and Richter and Eckstein (1986) made available a 305-year recent chronology from Spain.

Taxus baccata L.

Common yew, English yew

Morphology

This slow-growing evergreen reaches a height of 10–15 m. It generally has several trunks and red, meaty fruits.

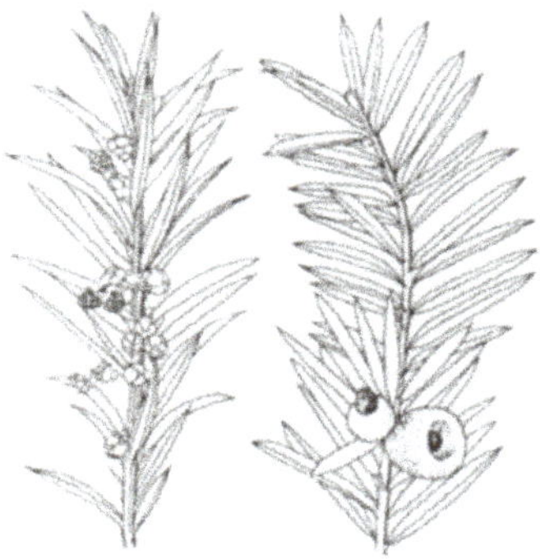

Fig. 241. *Taxus baccata.* (Hess et al. 1962)

Site, Site Conditions, Characteristics

Yew trees are found in sub-oceanic climates. The trees reach their largest dimensions in regions with even temperatures, for example in southern England and on the southern coast of the Caspian Sea. To avoid the threat of frost, the trees often grow under the protective canopy of deciduous trees. The species is able to withstand century-long periods of shade. It flourishes in deep, limestone soils. The trees can live for as long as 800 years.

Fig. 242. Yew as undergrowth in a beech forest, Zürich, Switzerland

Fig. 243. An ancient yew in Bavaria. (Goerss 1981)

Fig. 244. Garden yew

Distribution

Horizontal: the present area has been broken up by human exploitation. (Manufacture and commerce with arrows, bows, and crossbows). Yew trees are missing in the continental climate regions of Europe, i.e., in the central Alpine valleys of eastern Europe. Relict species can be found on shady sites in valleys of the Alps with continental climates (Via Mala). Despite its poisonous leaves, the tree is frequently planted in gardens. The main distribution area occurs in western Europe and along the Caspian Sea. The western border is the same as that of beech.

Vertical: from the plains up to a maximal elevation of 2000 m in southern Spain. In the Alps the upper border is at around 1400 m.

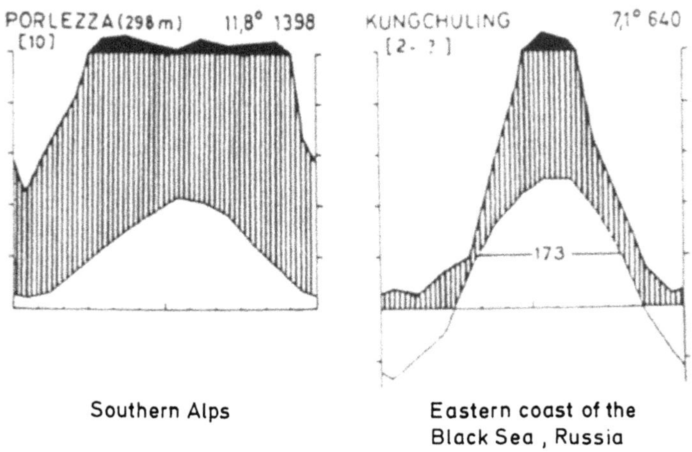

Fig. 245. Climate diagrams

Fig. 246. Distribution of *Taxus baccata* (Zoller 1981)

Wood

Heartwood is present. The sapwood is yellow, the heartwood is red-brown. The wood is extremely dense. The tree rings are distinct and often wave-shaped. The density is 700–800 kg/m^3.

Cross-section: no resin canals. The transition from earlywood to latewood is continuous. The tracheid walls are generally thick-walled.

Transverse section: the ray height is 5–12 cells.

Radial section: the tracheids are with distinct spirals. Rays without ray tracheids. The pits are cupressoid.

Use in Dendrochronology

Suitability of the Species

The trees grow to be quite old and are widely distributed. The wood has frequently been used for building material, and its growth ring structure is simple and distinct. For these reasons yew wood is fundamentally suited for dendrochronological purposes. However, the trees, which often grow in the shade of deciduous species, form irregularly shaped trunks (spiral grain, lobed stems) which limit the usefulness of the species.

Previous Uses

Taxus baccata has been dendrochronologically used only by Parsapajouh et al. (1986). The material originated from the warm climate of northern Iran. Despite the many individual growth reactions, it is possible to use radiodensitometric means to gather data on the correlation between temperatures in February and

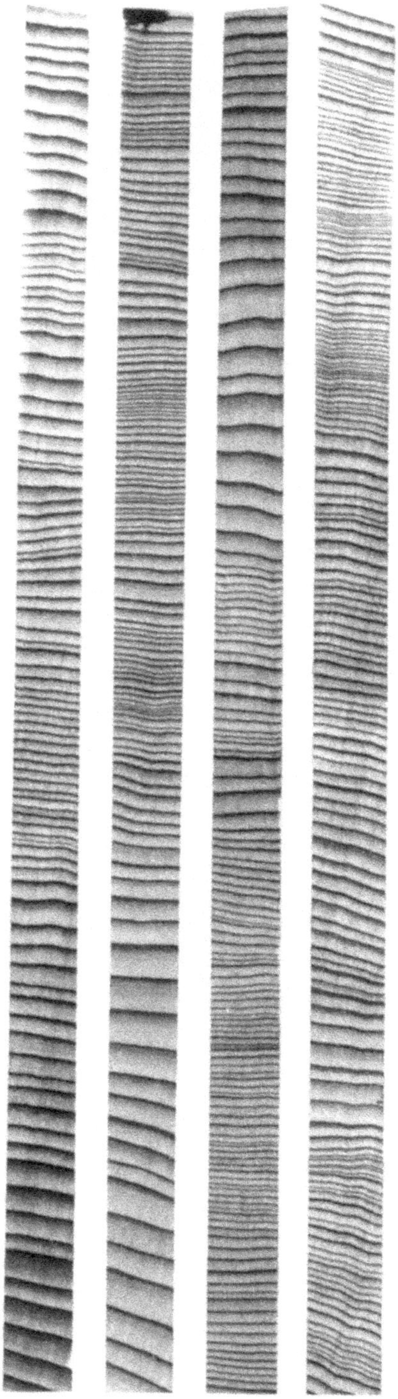

Fig. 247. Radiographs of *Taxus baccata*, Iran

March, earlywood, and ring widths and maximal densities. Zheng et al. (1982) recorded chronologies from recent *Taxus cuspidata* in eastern Asia.

Tetraclinis articulata (Vahl) Mast.

Alerce, Arar Tree, African thuja

Morphology

This small, cypress-like tree, with a height of 5–8 (15) m, has a loose crown and an exceptionally deep root system. The maximum stem diameter is 70 cm.

Site, Site Conditions, Characteristics

Tetraclinis is native to the warm Mediterranean climate with its several months of summer drought. It forms loose stands on sites with average July temperatures of over 23 °C and January temperatures of over 6 °C. Annual precipitation in the area averages 300–700 mm. The stands often develop into bush-like forests as a result of grazing and forest fires. The species can be found on silicate and carbonate soils. It is accompanied by bushy trees such as *Juniperus phoenicea*, *J. oxycedrus*, and *Quercus roundifolia*, as well as by bushes such as *Ephedra fragilis*, *Rosmarinus officinalis*, and the palm *Chamaerops humilis*. In principle, the tree has a long life expectancy. However, due to the excellent technological qualities of the wood, and its resistance to fungal growth and insect damage, it has been heavily exploited throughout history. Today, only a few stands exist that can be used for dendrochronology.

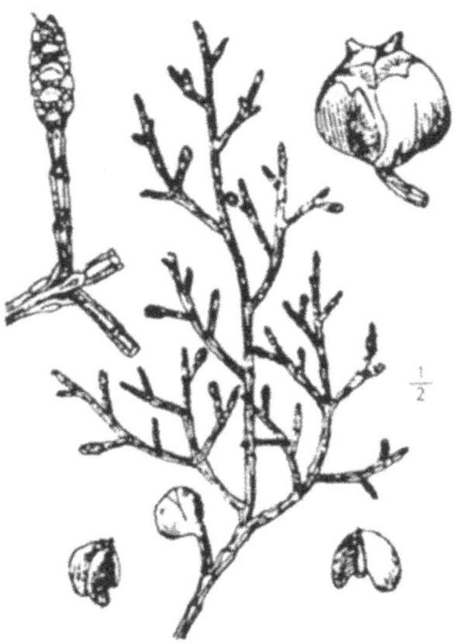

Fig. 248. *Tetraclinis articulata*. (Pignatti 1982) 1:2

Since Roman times, exquisite wood carvings have been produced and exported throughout the Mediterranean countries. Today, root tubers are collected for veneer production. The high flammability of the wood makes it useful for incense and as an embalming material.

Fig. 249. *Tetraclinis* at the timberline in the high Atlas, Morocco. The forests have been destroyed by grazing and logging

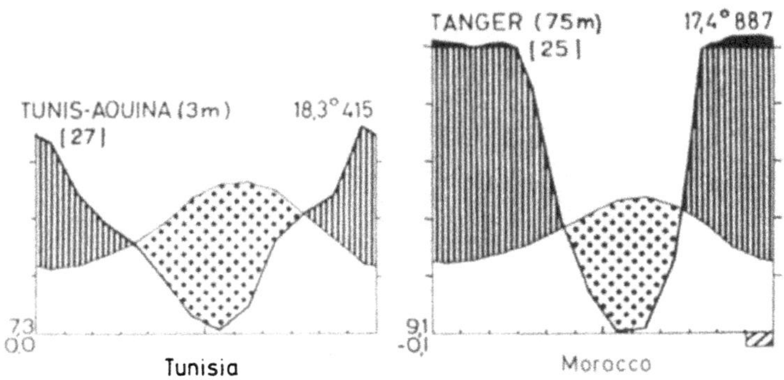

Fig. 250. Climate diagrams

Distribution

Horizontal: in southern Mediterranean countries. Mainly in the Atlas Mountains in Morocco. The distribution in Spain is minimal.

Vertical: from sea level up to an elevation of 1800 m. Optimal elevation is between 400 and 800 m.

Fig. 251. Distribution of *Tetraclinis articulata*

Wood

With heartwood. The sapwood is yellow; the heartwood is brown to brown-red. The density is around 600 kg/m^3.

Cross-section: the tree-ring border is rarely distinct (Fig. 252a, b). The transition from earlywood to latewood is continuous. The growth ring borders are sometimes missing (Fig. 252a). There are no resin canals. The axial parenchyma cells contain dark substances.

Tangential section: average ray height: 8–12 cells. The horizontal walls of the axial parenchyma cells are thin and smooth.

Radial section: bordered pits with thick cell walls, and two pit openings (Fig. 252d). The tracheid walls have small warts (Fig. 252e). Cupressoid pits in the cross-field (Fig. 252e).

Use in Dendrochronology

Suitability of the Species

Because its tree rings cannot easily be recognized, and false rings caused by summer rains are often present, the species has only limited dendrochronological value. It is possible to derive a continuous tree-ring sequence by synchronizing several radii. In principle, however, the tree rings can reflect past precipitation levels.

Previous Uses

A study by Serre (1969) on material from Tunisia illustrates the difficulties that can occur in synchronization.

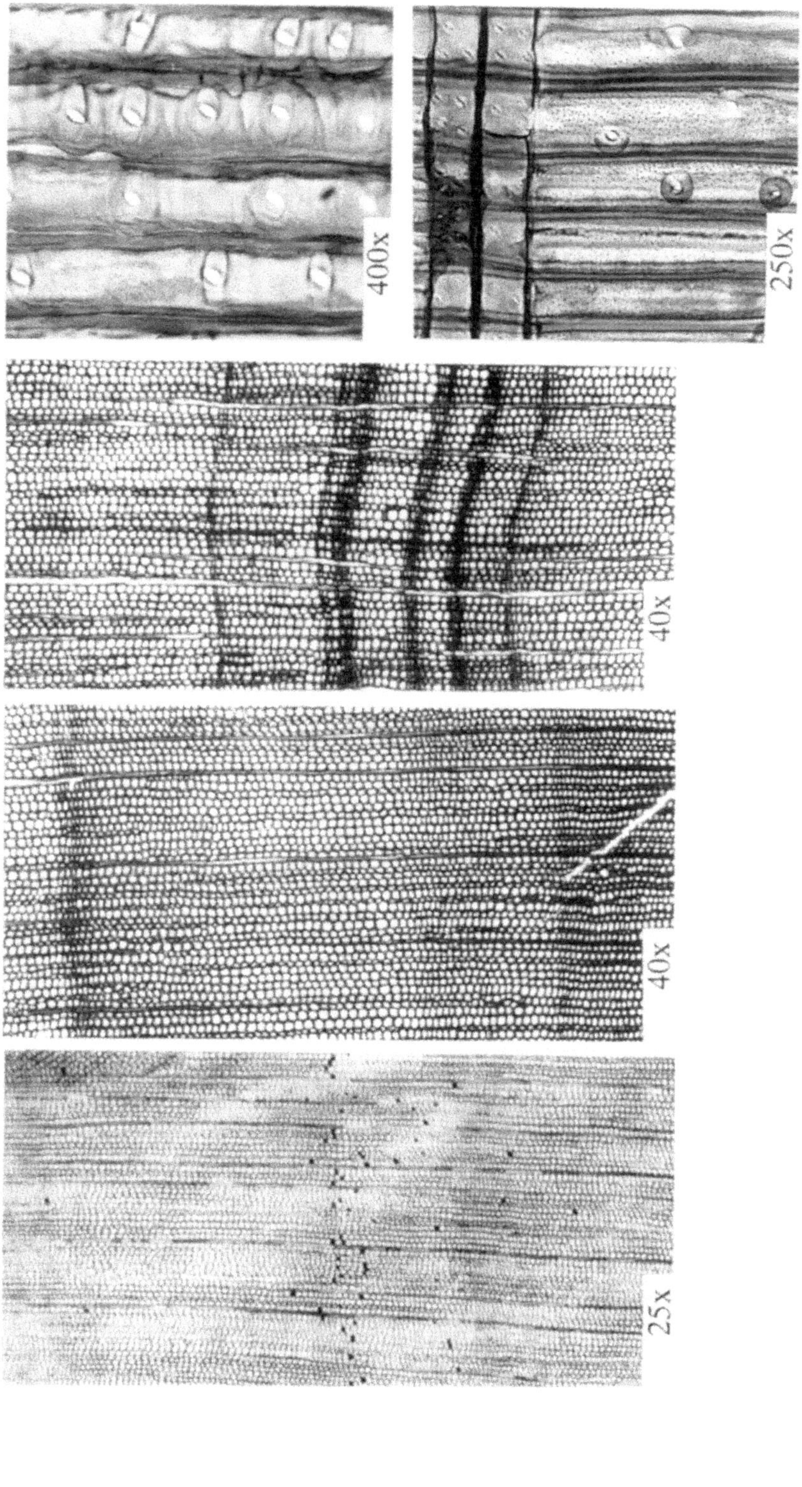

Fig. 252 a–e. Anatomy of *Tetraclinis articulata*

Alnus glutinosa Gaertn.

Common alder, black alder

Alnus incana Willd.

Gray alder, speckled alder

Morphology

This deciduous tree grows to a height of 30 m and has a monopodial trunk with a diameter of 50 cm. Its crown is narrow and its root system extends deep into the ground.

Fig. 253. *Alnus glutinosa.* (Hess et al. 1967)

Fig. 254. *Alnus incana* on a river bed in Switzerland

Fig. 255. *Alnus glutinosa* in a riparian forest, Poland

Site, Site Conditions, Characteristics

Black alder grows in many different climates. In Europe it avoids northern Scandinavia with its relatively short summers, as well as the very dry summer region of southern Spain. The species flourishes in humid, warm summer

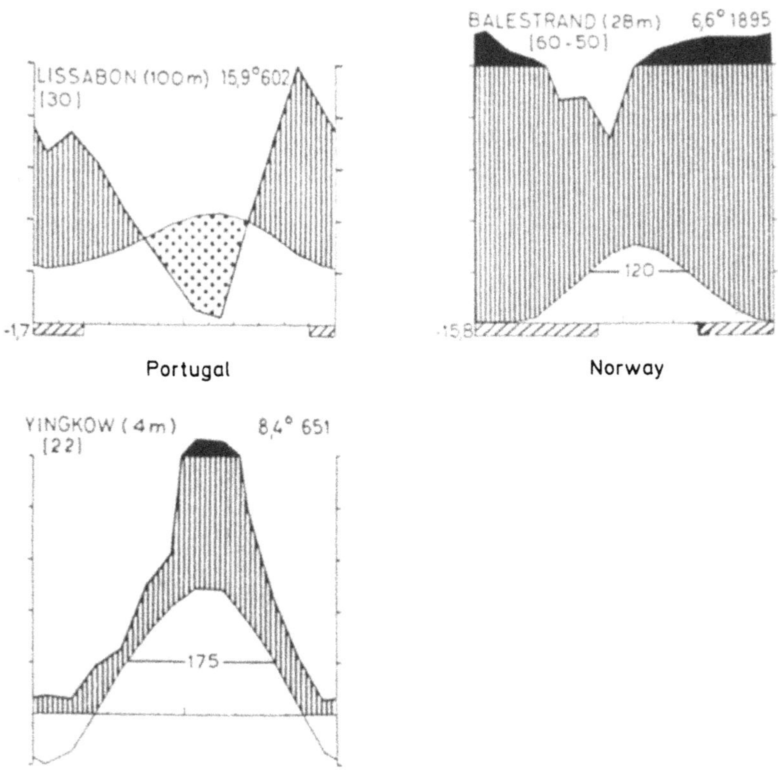

Fig. 256. Climate diagrams

Pleistocene dunes in lowlands of eastern Europe. Together with *Betula pubescens*, black alder forms tall forests on annually flooded areas. It grows optimally on constantly fresh to moist loam soils with a thick cover of humus and is able to grow in oxygen-poor soils. Under such conditions, the trees usually grow to be between 100 to 120 years old. Black alder forms stands along the banks of streams, where it is often accompanied by ash and elm. Capable of binding gaseous nitrogen, (root symbiosis with *Actinomyces alni*), black alder often improves the soil in which it grows.

Distribution

Horizontal: practically throughout Europe, absent in northern Scandinavia. Predominately found in northeastern Europe.

Vertical: mainly in the lowlands. In the Alps, the species reaches an elevation of 1800 m in Engadin (Switzerland).

Fig. 257. Distribution of *Alnus glutinosa*

Wood

Without heartwood. Freshly cut alder wood has a deep orange color. The density is around 480 kg/m^3.

Cross section: diffuse-porous. Vessels in radial rows and groups. Aggregate rays are usually, but not always, present and very distinct. (Fig. 258).

Tangential section: uniseriate rays, in close proximity to aggregate rays often small seriate rays (2–3 cells wide). Average ray height: 10 to 25 cells.

Radial section: scalariform perforation plates. Rays generally homogeneous.

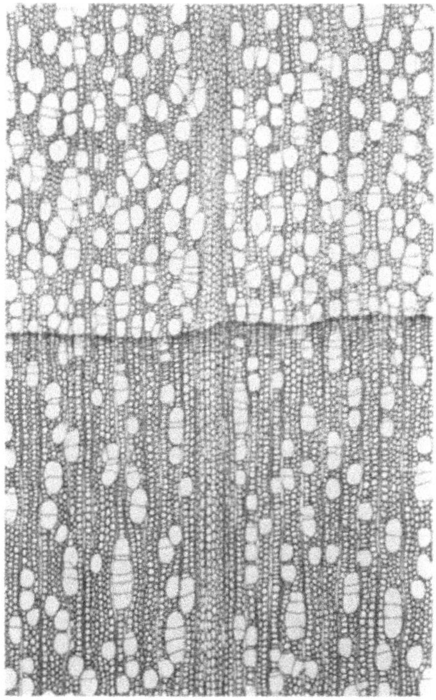

Fig. 258. Cross-section of *Alnus glutinosa* (40 x)

Use in Dendrochronology

Suitability of the Species

Black alder has a relative short life. It is above all suited for reconstructing forest dynamics in large, annually flooded riparean forests of eastern Europe. In many cases, the tree rings reflect yearly fluctuations in the ground water surface. The tree-ring borders, especially in narrow rings are difficult to recognize.

Gray alder, which grows along riverbanks, is especially suited for the reconstruction of river dynamic processes. The trees, however, only grow to be around 60 years old.

Previous Uses

The species has only seldom been used. According to Kairiukstis (1983), there are around 30- to a maximum of 110-year-old chronologies from recent material taken from the Baltic region (Lithuania). Elling (1966) incorporated black alder from southern Germany in dendrochronological studies. He observed many missing rings and recognized the promoting effects of warm summer months and the restricting effects of cold winters as well as precipitation in summer on the growth.

Betula verrucosa Ehrl. Syn. *B. pendula* Roth.

Silver birch, European white birch

Betula pubescens Ehrh. Syn. *B. alba* L.

Silver birch, European White birch

The two species are taxonomically very similar with frequent hybridization. The differentiation between them is not always easy.

Morphology

Mid-sized, deciduous, rapid-growing, shallow-rooted trees with maximal heights of 30 m. Their bark is at first white and smooth, becoming black and fissured with age.

Site, Site Conditions, Characteristics

Both species occupy a broad ecological spectrum. They both require sunny sites and are pioneering species. They are widely distributed in cool-humid to cool-continental climates and form forest especially in the boreal zone. The trees can live for 200 years, but are usually younger than 100 years.

They differ in ecological characteristics.

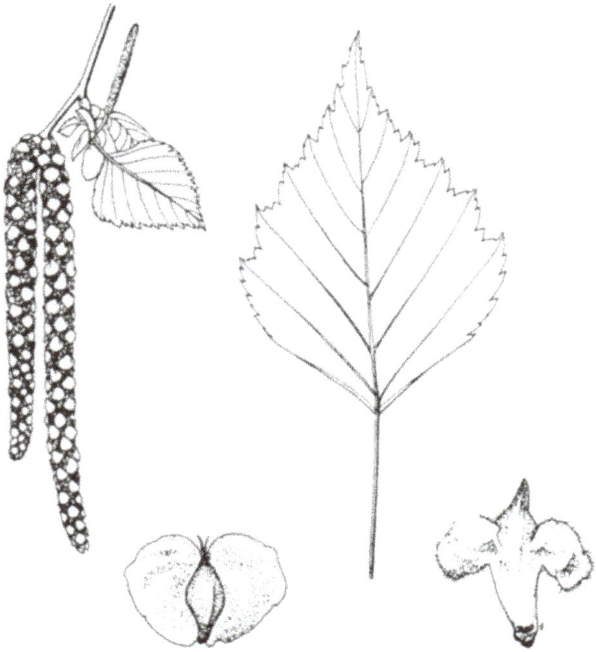

Fig. 259. *Betula verrucosa* (Hess et al. 1967)

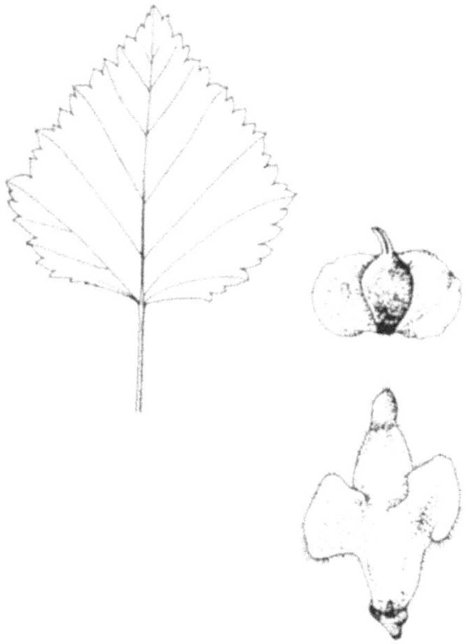

Fig. 260. *Betula pubescens.* (Hess et al. 1967)

Fig. 261. *Betula verrucosa.* In Emmental, Switzerland, the trees are cultivated for making brooms

Fig. 262. *Betula pubescens* on the bank of lake Mauen, Switzerland

Betula verrucosa

Predominately found on dry sites with rocky, sandy, loam, or humus foundations; they cannot tolerate periods of summer drought, but can withstand cold winters. *Betula verrucosa* is commonly found in light deciduous and conifer forests and forms forests in, Scandinavia, the Baltic countries, and Russia.

Betula pubescens

Predominately found in limestone-poor, moist, swampy regions; tolerates the extreme winter temperatures of Siberia but not summer droughts; predominately in bogs, often accompanied by *Alnus glutinosa*. Forms forests along the Polar Sea.

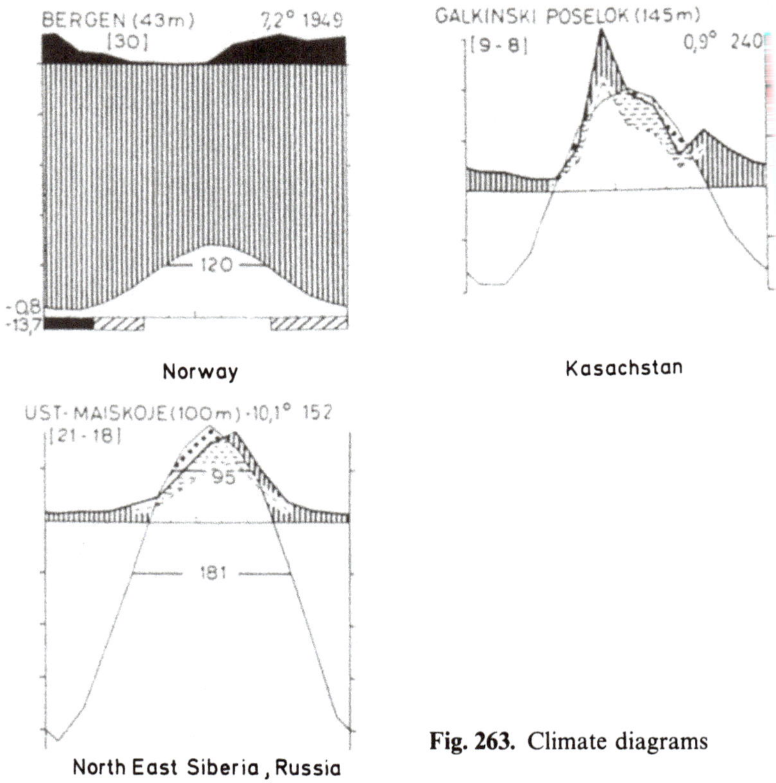

Fig. 263. Climate diagrams

Distribution

Betula verrucosa

Horizontal: western Europe to the Baltic Sea. North to the estuary of Jenissei, south to the central Asian steppes. Not found in the marine region of northern Norway.

Vertical: from the lowlands to the forest border.

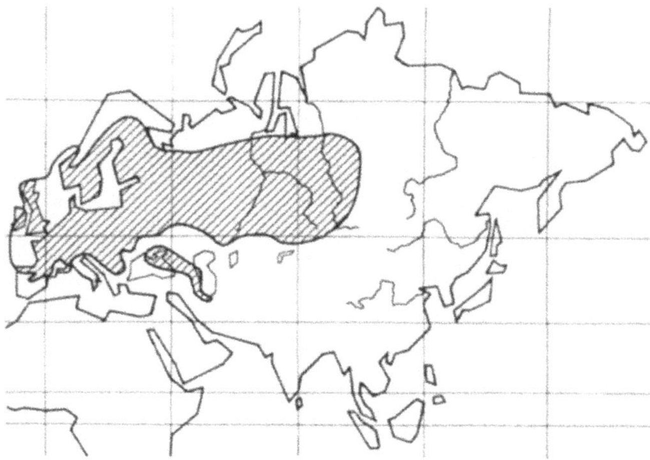

Fig. 264. Distribution of *Betula verrucosa*

Betula pubescens

Horizontal: similar to *B. verrucosa*. In western Europe to Iceland and in eastern Europe to Lena.

Vertical: as *B. verrucosa*.

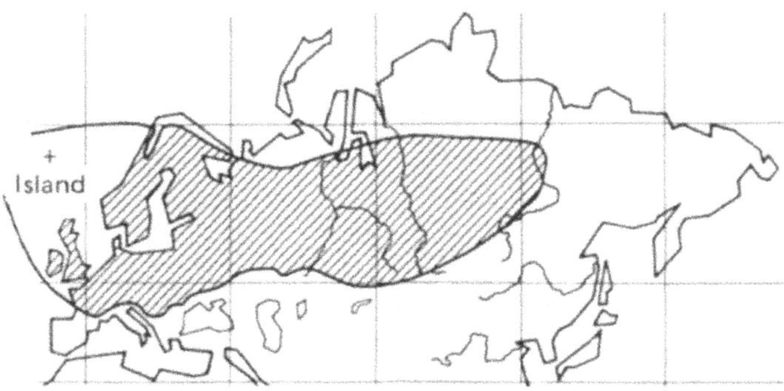

Fig. 265. Distribution of *Betula pubescens*

Wood

Without heartwood, light red, density around 550 kg/m^3.

Cross-section: diffuse- to semi-ring porous. Vessels rather sparsely scattered in regular radial files of two to four cells and vessel clusters. Diffuse apotracheal parenchyma.

Tangential section: rays mostly two to four cells wide, also often uniseriate. Height of uniseriate rays: 1–10 cells, height of multiseriate rays: 10 to 25 cells.

Radial section: scalariform perforation plates with 10–25 bars. Very numerous and extremely small pits at the ray-vessel intersections. Rays homogeneous to heterogeneous (type 1). All fibers are libriform.

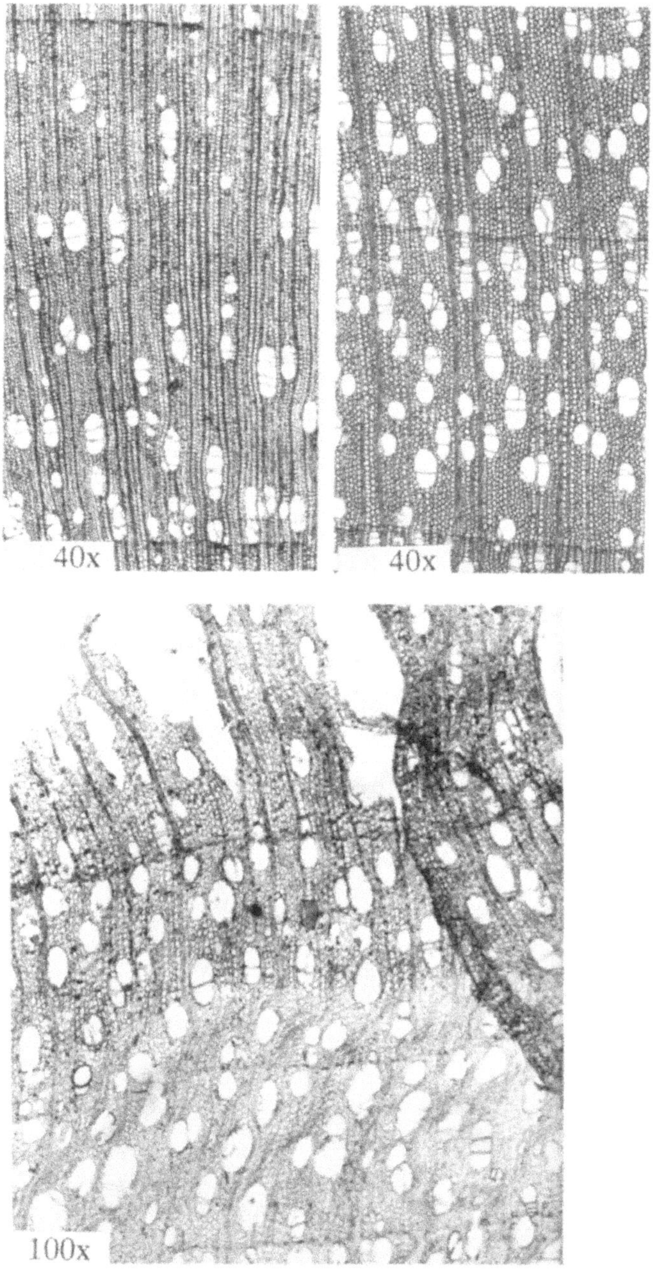

Fig. 266a, b. Cross-sections of a heavily decomposed, subfossilized birch, Switzerland

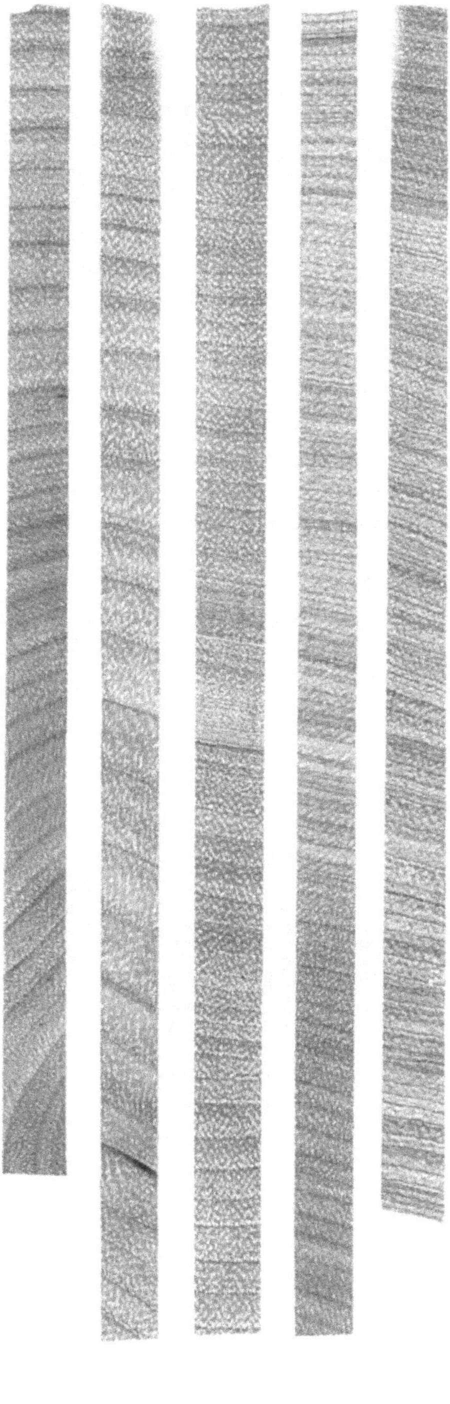

Fig. 267. Radiographs of *Betula verrucosa*, Scotland

Use in Dendrochronology

Suitability of the Species

Both species lend themselves well to dendroecological analysis. The trees seldom live longer than 100 years and were rarely used as lumber in historic or prehistoric times. Thus, birch tree-ring sequences are, for the most part, only suited for studies on short-term ecological events, i.e., in the glacier perimeter in bogs, on newly settled sites, and in storm-damaged areas.

The indistinct tree rings can be made recognizable by coloring a transversal section with a brown stain and then rubbing over it with white chalk. The pores will appear white, the meristem dark.

Previous Uses

Although the species are widely distributed, they have seldom been used in dendrochronology. In Greenland and Scandinavia, Kuivinen and Lawson (1962), Aas (1969), Kuhlmann (1979), and Treter (1984) analyzed fluctuations of the northern forest border by determining the age of local birch trees. Staschel (1987) worked in the perimeter of a Norwegian glacier and determined a high sensitivity in the tree rings. With the indexing of indicator years, it is possible to determine the intensity of growth-restricting factors. Iseli et al. (1990), dated the beginning of the reforestation of a cultivated area in southern Switzerland.

Castanea sativa Mill.

Sweet chestnut, Spanish chestnut, European chestnut

Morphology

This large, deciduous tree is deeply rooted, has a broad spreading crown and an exceptionally thick stem (more than 2 m in diameter). The largest trees have circumferences of 23 m.

Fig. 268. Chestnut fruit

Site, Site Conditions, Characteristics

Castanea sativa is native to the Submediterranean climate. It grows poorly in the periphery regions with average January temperatures of 2.5 °C and mid July temperatures of 22 °C. It has a wide range in regions with mild winters and an annual precipitation between 500 and 1500 mm. It prefers acid soils. The main area in Europe lies in northern Greece, where the species is accompanied by *Tilia tomentosa*, *Fraxinus ornus*, *Ostrya carpinifolia*, and *Carpinus orientalis*. In the northern periphery, along the southern Alps, the stands are mixed with *Quercus petraea* and sometimes with *Fagus sylvatica* and *Betula* sp. The species reaches a maximum age of 1000 years. Because the stems are frequently decayed at the center, it is often difficult to determine the exact age of the tree. As a result of human exploitation (firewood, fruit and leaf harvesting, grazing), so-called Selven or Palina forests have emerged. Since the 1940s *Castanea sativa* has suffered heavy damage from the east Asian fungus, *Endothia parasitica* (the fungus was transported from America to Genua, Italy, in 1938 in the wood from munition boxes and spread throughout Europe). Sprout forests now grow on many former Selven forest sites.

Fig. 269. A chestnut tree with bark cancer. (Müller and Stierlin 1990)

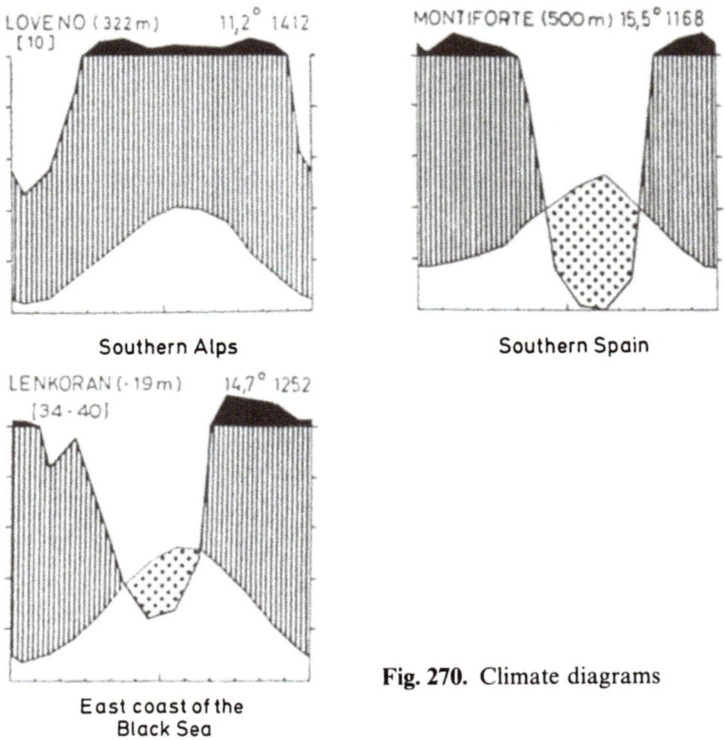

Fig. 270. Climate diagrams

Distribution

Horizontal: the wild form is native to southeastern Europe. The fruit tree has been widely dispersed through human cultivation. Since the time of the Romans,

Fig. 271. Main distribution of *Castanea sativa*

Castanea sativa has propagated itself spontaneously. The main area of distribution occurs in Macedonia and Thessalia.

Vertical: *Castanea sativa* is found in Mediterranean countries at intermediate elevations. The upper borders lie in the southern Alps at 1700 m, in southern Italy at 1500 m, and in Anatolia at 1700 m.

Wood

Very similar to that of *Quercus robur*. Anatomically, *Castanea* differs from *Quercus robur* mainly by the absence of large rays.

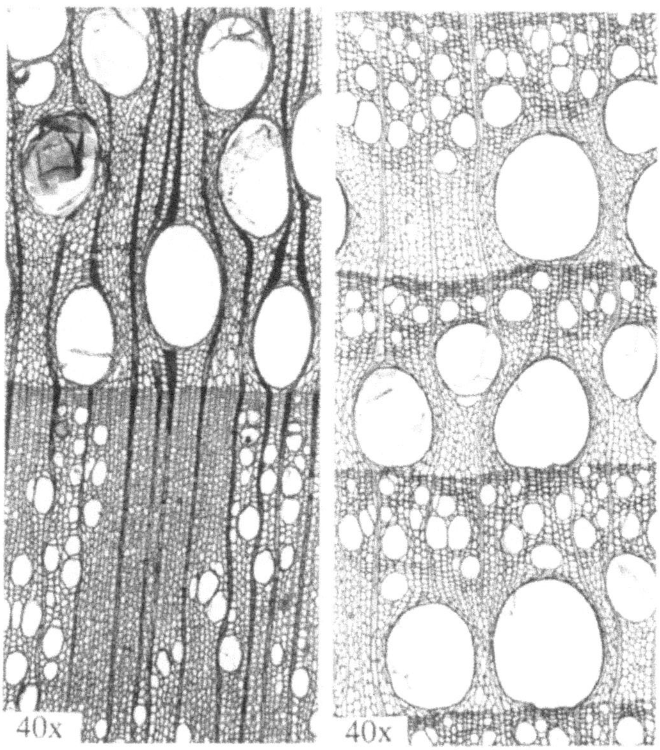

Fig. 272a, b. Cross-sections of *Castanea sativa*

Use in Dendrochronology

Suitability of the Species

Castanea sativa is generally poorly suited for dendrochronological purposes. As a cultivated tree, it frequently grows near populated regions and is thus subjected to many human influences. These influences are reflected in the tree rings (growth change, healing, decay). Because the stems often are split parallel to the rings, it is difficult to take samples by boring. Dendrochronological studies related to the fungal attack of *Endothia parasitica* could prove valuable.

Previous Uses

Donati and Orcel (1988) constructed a chronology (1551–1984) from southern Switzerland (Tessin), using recent to 230-year-old trees and samples taken from buildings. The chronology serves to date historic objects found in the southern Alps. According to Genova and Garcia (1984), *Castanea sativa* in the Montseny Mountains near Barcelona, Spain, reacts to the varying duration of sunlight.

Fagus sylvatica L.

European beech, Common beech

Morphology

A large deciduous tree with a broad crown. It grows to a height of over 30 m and has a trunk diameter of over 150 cm and a deep root system.

Fig. 273. *Fagus sylvatica*

Site, Site Conditions, Characteristics

European beech is native to the subatlantic climate. It avoids the dry, continental regions of Europe. Only in the eastern regions do the average January temperatures fall below 0 °C. The species also avoids regions with late frost. July temperatures in its main area of distribution average more than 16 °C. The annual precipitation level in these regions exceeds 800 mm. Beech flourishes on fresh, deep limestone soils. It avoids extremely dry sites and sites with very wet soil. Pure beech stands are found in central Europe at elevations between 500 and 900 m. *Abies alba* or *Picea abies* accompanies *Fagus sylvatica* on its upper border. On dry sites, the upper border stands are mixed with *Carpinus betulus*, *Quercus* sp., and *Pinus sylvestris*. The trees reach an average age of 120–160 years and, in rare cases, 300 years. The species is prone to attack by the insect *Melolontha vulgaris*. The damage is reflected in the tree rings.

Fig. 274. Large beech trees build a canopy over various deciduous trees

Fig. 275. Beech from forest-regenerated sprouts

Distribution

Horizontal: in Europe, in the north, to southern Sweden; does not occur in the most southern areas. In the east, the distribution area coincides with that of *Fagus orientalis*.

Fig. 276. Single beech tree with widespreading crown and smooth bark. (Informationsdienst Holz, Germany)

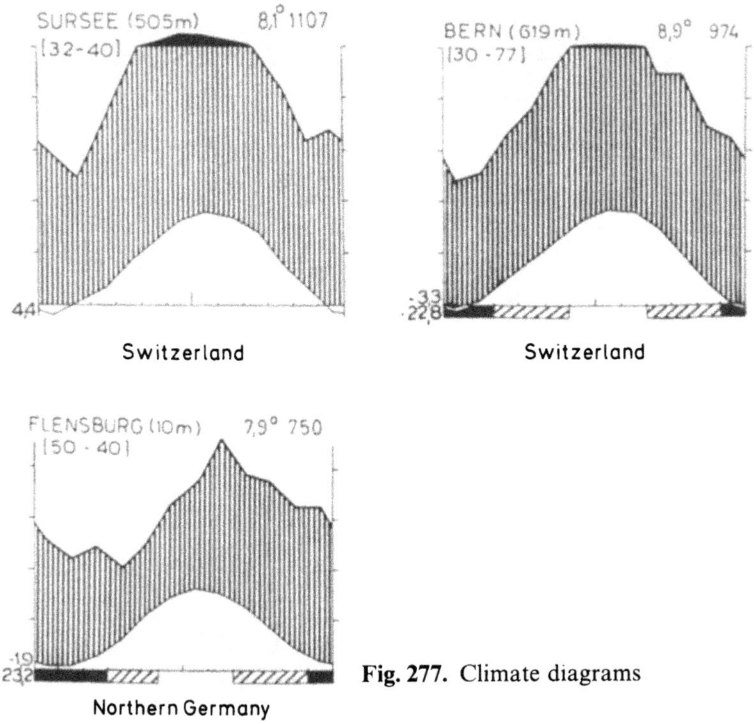

Fig. 277. Climate diagrams

Vertical: in the North and West, the species is mainly found in the plains; in the South and East more in the mountains and forest borders. The upper border lies in the Harz Mountains at an elevation of 650 m, in the Pyrenees and in the Alps at around 1200–1500 m, and in the Apennines at 1800 m.

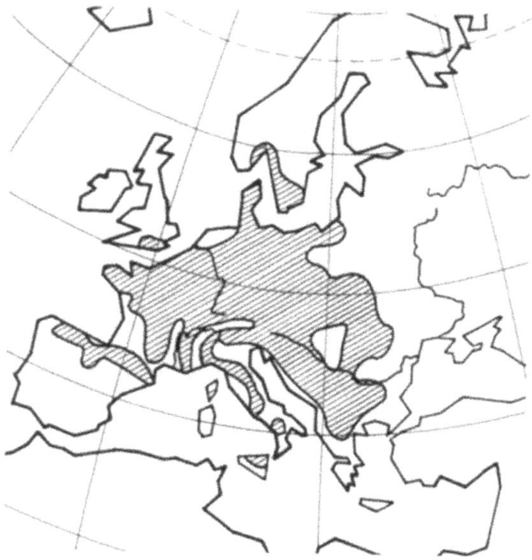

Fig. 278. Distribution of *Fagus sylvatica*

Wood

Generally without heartwood. Occasionally, a pathologic, red-brown, irregularly shaped heartwood. The wood is reddish to yellowish. The rays can be seen with the naked eye. The density is 490–880 kg/m^3.

Cross-section: diffuse-porous to semi-ring porous (Fig. 279a, b). Very numerous solitary and clustered vessels in the earlywood. Earlywood vessels often contain tyloses. Thick-walled fibers and fiber-tracheids. Tangential chains and oblique rows of parenchyma cells. The large rays are generally thickened at the growth ring boundaries.

Tangential section: uniseriate to multiseriate ray up to 0.5 mm (20 cells) wide.

Radial section: simple perforation plates, occasionally scalariform. Presence of transition forms from simple bordered pits and scalariform pittings. Homogeneous to slightly heterogeneous rays. The main tissue consists largely of libriform fibers.

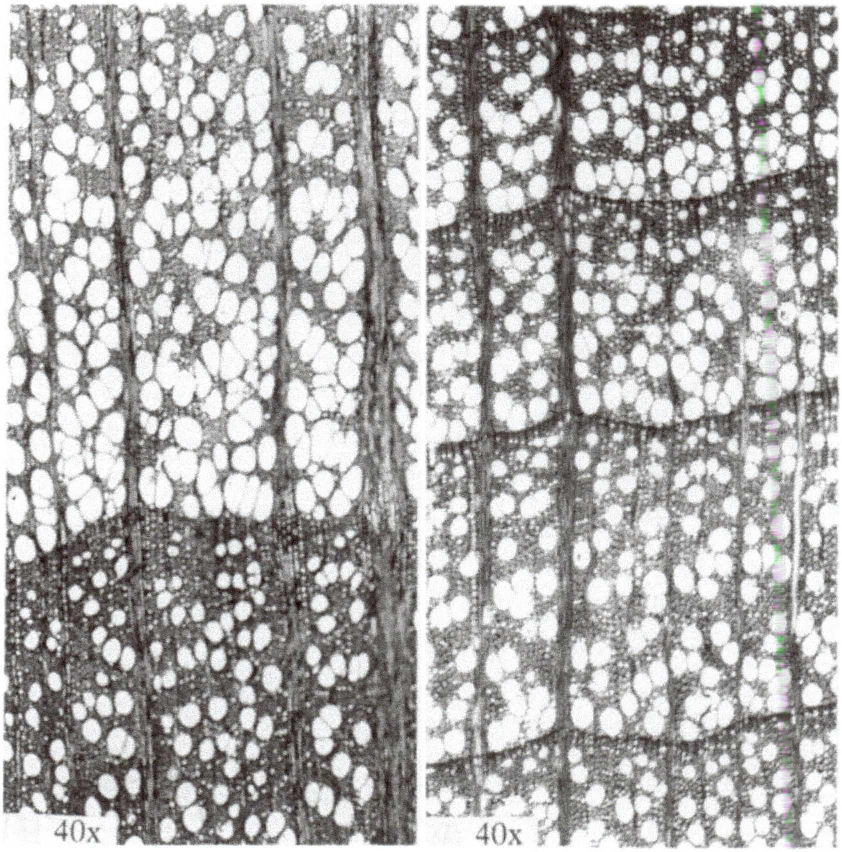

Fig. 279a,b. Cross-sections of *Fagus sylvatica*

Use in Dendrochronology

Suitability of the Species

Beech is suitable for all dendroecological studies. It has, at least in central Europe, a wide ecological spectrum, a relatively long lifespan, and annually reflects environmental conditions in distinct indicator years, abrupt growth reduction, and varying proportions of vessels. Beech is less suited for densitometric studies because its large rays render analysis of the wood difficult.

Previous Uses

Beech has seldom been used in dendrochronology.

Jazewitsch (1953) studied recent beech trees from the Spessart and constructed a chronology reaching back to 1654 A.D. Hollstein (1978) expanded the chronology back to 1320 A.D. with wood samples taken from old buildings in

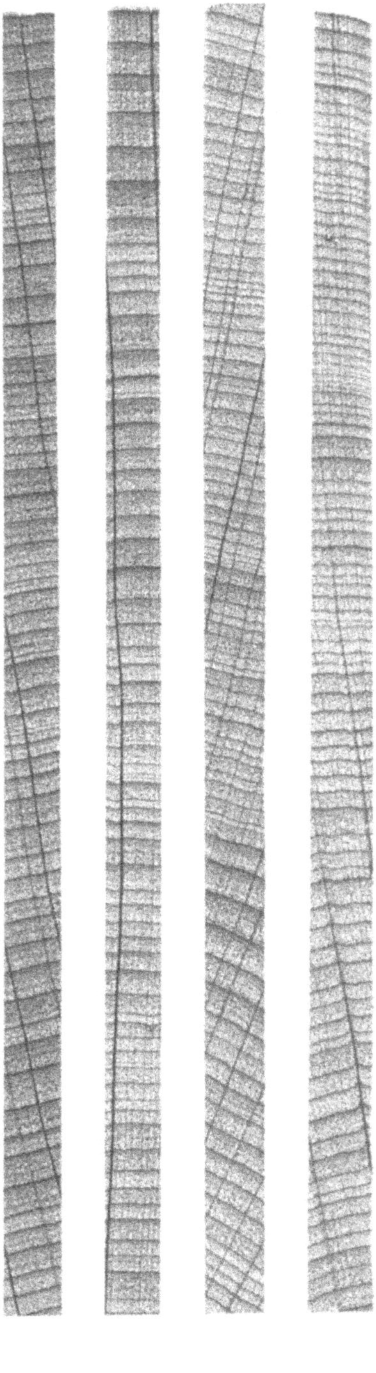

Fig. 280. Radiographs of *Fagus sylvatica*

central Germany. Klein (1983) has been able to date paintings with panels made out of beech wood.

According to Z'Graggen (1992) indicator years appear to be a good index of extreme weather conditions. Tree-ring width, latewood width, and proportion of vessels are all characteristics of pointer years.

Quercus pubescens Willd.

Pubescent oak, downy oak, white oak

Morphology

This deciduous, generally small (10–15 m) tree often has an irregularly shaped crown and stem, the latter with a diameter of barely over 50 cm.

Fig. 281. *Quercus pubescens.* (Hess et al. 1967)

Site, Site Conditions, Characteristics

Quercus pubescens is a tree or shrub of the Submediterranean and Mediterranean region. It is rather drought-resistant and is capable of withstanding Mediterranean summers as well as the cold winters of central Europe and Asia Minor. The species grows on dry, limestone to acid soils. In Mediterranean countries, it appears in mountain zones partly spotwise and partly in pure

Fig. 282. *Quercus pubescens* in a defoliated condition. (Rol 1962)

Fig. 283. A *Quercus pubescens* forest on a limestone-rich bedrock in the canton Schaffhausen, Switzerland. The species-rich vegetation is typical (Photo W. Keller, WSL)

stands (*Quercetum pubescentis*). In the submediterranean and temperate regions, for example north of the Alps, it grows on slopes with warm, southern exposure.

Quercus pubescens is, especially on limestone soils, associated with many small trees and shrubs, for example *Acer monspessulanum*, *Fraxinus ornus*, *Ostrya carpinifolia*, *Buxus sempervirens*, and *Cotinus coggygria*, along with others.

Distribution

Horizontal: southern and central Europe, from the Pyrenees to the Crimea and the Caucasus. North of the Alps only small areas are found.

Vertical: in the Mediterranean region, *Quercus pubescens* is predominately found at high altitudes above the evergreen angiosperms. North of the Alps it grows in shallow soil on dry southern slopes.

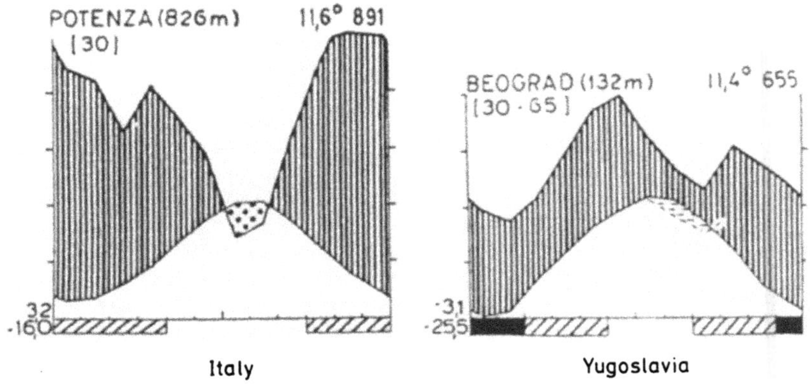

Fig. 284. Climate diagrams

Fig. 285. Distribution of *Quercus pubescens*

Wood

With heartwood. Similar to *Quercus robur* and *Q. petraea.* The tree rings are usually very small. The density averages 720–990 kg/m^3.

It is impossible to distinguish *Quercus pubescens* from *Q. robur* and *Q. petraea* on the basis of their anatomic characteristics.

Use in Dendrochronology

Suitability of the Species

Quercus pubescens is especially suited for dendroecological purposes. Because it is found in dry Mediterranean regions, it can be used to reconstruct past precipitation conditions. It is not very well suited for constructing long chronologies because it is only used locally as building material.

Previous Uses

Quercus pubescens, above all from southern France, is used by the dendrochronologic school in Marseille for dendroecological and statistical purposes (Tessier et al. 1982, Tessier 1986).

Quercus robur L. Syn. *Q. pedunculata* Ehrl.

European oak, English oak

Quercus petraea Liebl. Syn. *Q. sessiliflora* Salisb.

Durmast oak

The wood of these two species cannot be differentiated on the basis of their anatomic characteristics. Thus, they are treated together in dendrochronology. Hybridization is common.

Fig. 286. *Quercus petraea.* (Hess et al. 1967)

Fig. 287. *Quercus robur*, European oak

Quercus robur

English oak

Morphology

This deciduous tree grows to a height of 30 m and has a broad crown, thick, horizontally set branches and a strong, deep root system. Its trunk has a diameter of 3 m.

Fig. 288. Free-standing common oak in Ireland (Photo Baillie)

Fig. 289. Free-standing durmast oak in Emmental, Switzerland

Fig. 290. *Quercus robur* (*left*) and *Q. petraea* (*right*) as witnesses to a former intermediate forest. The accompanying beech trees are much younger and have often developed from sprout forests

Fig. 291. A free-standing durmast oak in a bush-hedge site. The lower branches of the crown are often pruned for feed. This pruning is indicated as an abrupt growth reduction in the tree rings. (Goers 1981)

Site, Site Conditions, Characteristics

English oak is a light tree able to tolerate many different climates. In western Europe, in oceanic climates, temperatures in July average around 14 °C and in January around −14 °C; annual precipitation measures 300 mm. The species requires more warmth than beech and *Quercus petraea*. English oak flourishes in deep, nutrient-rich, consistently fresh soils. In dry regions of the eastern area, demanding races exist.

Common oak forms forests in marine and cool summer sites in nutrient-rich, deep, fresh soil. On normal sites it is often accompanied by beech and on moist sites by *Fraxinus excelsior*, *Acer pseudoplatanus*, and *Carpinus betulus*. Common oak is also present in *Quercus pubescens* bush stands.

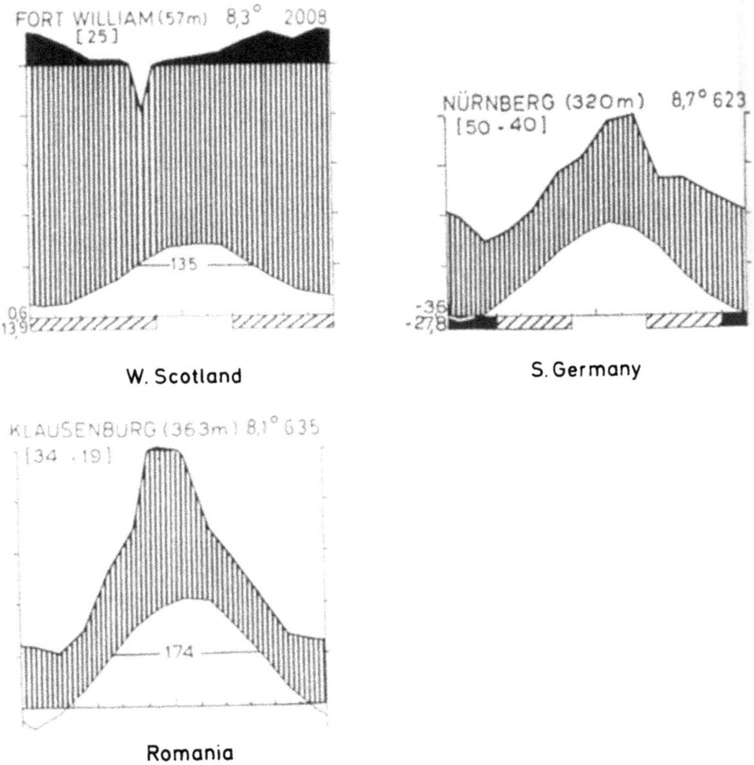

Fig. 292. Climate diagrams

Distribution

Horizontal: in Europe between the 40° and 60° northern latitudes.

Optimal forests occur in France, along the mid and lower Rhine and on the Danube plateau.

Vertical: *Quercus robur* forms forests at low and intermediate elevations. The upper border in the Harz Mountains lies at around 500 m, in the Vosges Mountains at 300 m, in the central Alps at 1000 m, and in the southern Alps at about 1300 m.

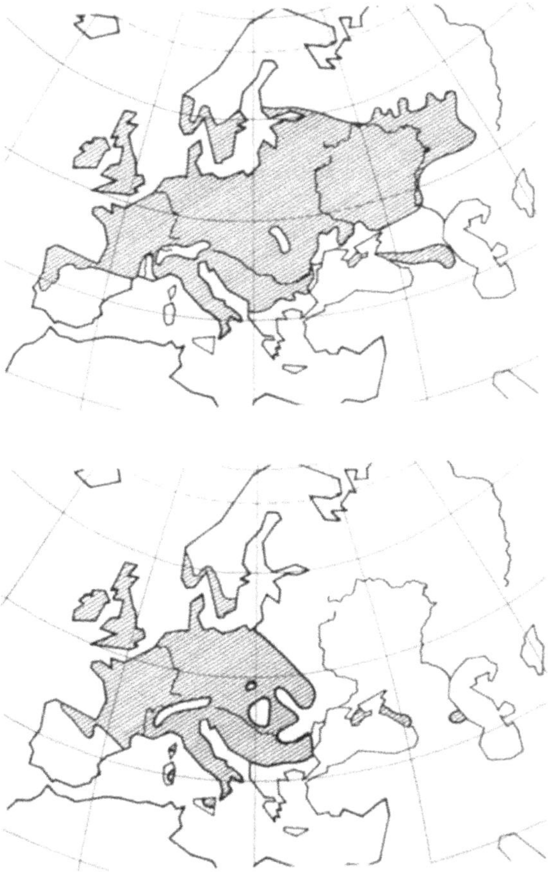

a

b

Fig. 293a,b. Distribution of **a** *Quercus robur*, **b** *Quercus petraea*

Wood

With heartwood. Yellow sapwood, yellow-green, brown heartwood. The tree rings are distinct and usually wide. The density is 430–930 kg/m^3. Subfossil wood is black.

Cross-section: ring-porous. Flame-like distribution of groups of vessels in latewood of the larger growth rings (Fig. 294a). Thin-walled tyloses in the vessels of earlywood. The libriform fibers are arranged in groups and have very thick walls in the latewood. Narrow parenchyma bands occur in the latewood.

Tangential section: uniseriate and multiseriate rays. Multiseriate rays are 0.5 to 1 mm wide (up to 30 cells) and 1–5 cm high.

Radial section: simple perforation. Homogeneous rays. Libriform fibers, tracheid fibers, and vascular tracheids are present.

Quercus petraea Liebl.

Morphology

Similar to *Quercus robur*. However, its crown is often narrower and more regular. Maximum stem diameter is 2 m. The tree has a thinner outline than English oak.

Site, Site Conditions, Characteristics

Similar to *Quercus robur* but with fewer climatic demands; also similar to beech. The eastern border of its area has mid January temperatures of −5 °C.

Quercus petraea demands warm summers and is sensitive to late frost. It requires less moisture than *Quercus robur* and prefers loose rather than hard soils.

The species forms forests on low sites with loose soil. Its main distribution area is in the oak-hornbeam forest. *Quercus petraea* is often associated with *Pinus sylvestris*. It avoids continental regions.

Oak, especially from dry sites, can grow to be ancient, over 1000 years. On sites with good water and nutrient supply, oaks with diameters of 1.5 m are barely over 200 years old.

Oak leaves are attacked by many insects. The two most common insect species are *Tortix viridiana* and *Melolontha melolontha/M. hippocastani*. *Tortix viridiana* infests oaks sporadically. *Melolontha* infestation occurs every 3–6 years while the leaves are budding (April–May). Each larger region exhibits its own cycle. Total defoliation is clearly reflected in the tree rings (Christensen 1987).

Oak stands (*Quercus robur* and *Q. petraea*) are mixed with various tree species, depending on the region and its geological foundation. For example, on acid, wet soils oak stands are mixed with *Pinus sylvestris* and in dry regions with *Pinus sylvestris*, *Pinus nigra*, and various oak species. In temperate regions, oak is accompanied by beech and in riparean areas by ash, maple, and elm.

Distribution

Horizontal: in western Europe between the 40° and 60° northern latitudes. Optimally on warm mountain sites in Germany, Transylvania, Hungary and Romania.

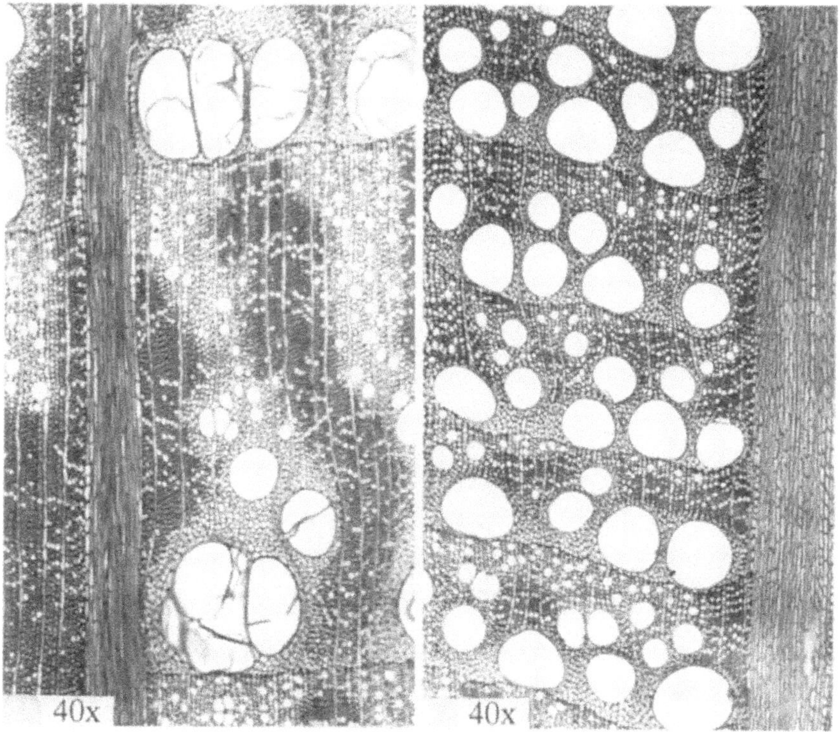

Fig. 294a,b. Cross-sections of *Quercus petraea*

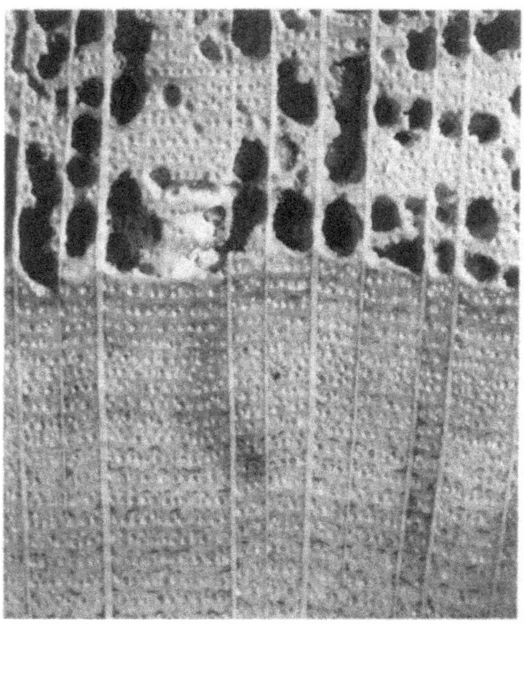

Fig. 295. Heartwood and sapwood. The sapwood rapidly decomposes in wood used as building material

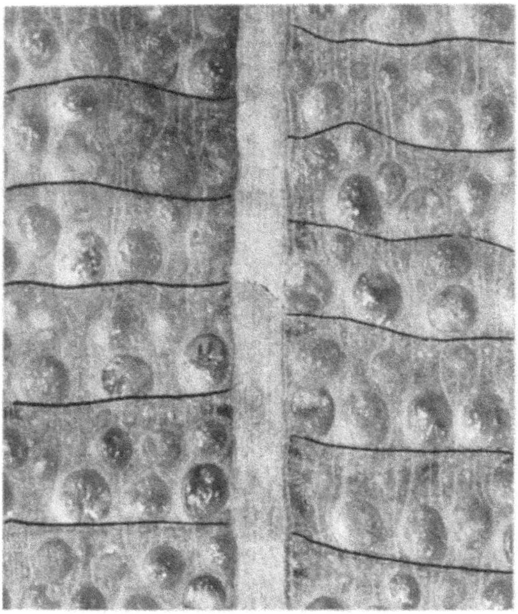

Fig. 296. Very narrow tree rings. The boundaries are difficult to recognize because the earlywood vessels of two rings nearly touch one another

Vertical: the species builds forests in intermediate mountain regions. In the Harz Mountains, the upper border is around 500 m, in the Black Forest at around 900 m and in the central Alps in Wallis at around 1200 m.

Wood

With heartwood. *Quercus petraea* tree rings are normally narrower than those of *Quercus robur*, but otherwise the wood is similar.

Use in Dendrochronology

Suitability of the Species

Oak wood is the classic wood of central European dendrochronology. The trees have long lifespans, cover wide areas, and have distinct tree rings. The wood is often used as a building material and preserves well.

Difficulties in measuring occur only in specimens with very narrow rings, where the earlywood vessels practically occur on top of one another. In such cases, it is often difficult to recognize the tree-ring borders. Furthermore, displaced tree rings often appear alongside large rays. Although oak has basically no uneven or missing rings, due to the displacement of the tree rings, several radii must be combined for synchronizing.

The ecological spectrum of both species is rather large at low and intermediate elevations, which provides many possibilities for studying site conditions. These possibilities have by no means yet been fully exhausted in dendrochronological research.

It is not clear what climatological information can be derived from oak. It is, however, anticipated that the method of image analysis will produce results indicating past climate conditions.

Previous Uses

Oak has, in an ecological respect, received only very limited attention.

However, there have been many dendrochronological studies with the species.

Historic time period: practically every European dendrochronology laboratory has constructed at least one regional 1000-year-oak chronology. The dating of these chronologies was mostly done with oak from the second century A.D. (Klein and Eckstein 1988). The range of application is large and extends from the dating of lakeside and river fortifications, ships, oak caskets, and construc-

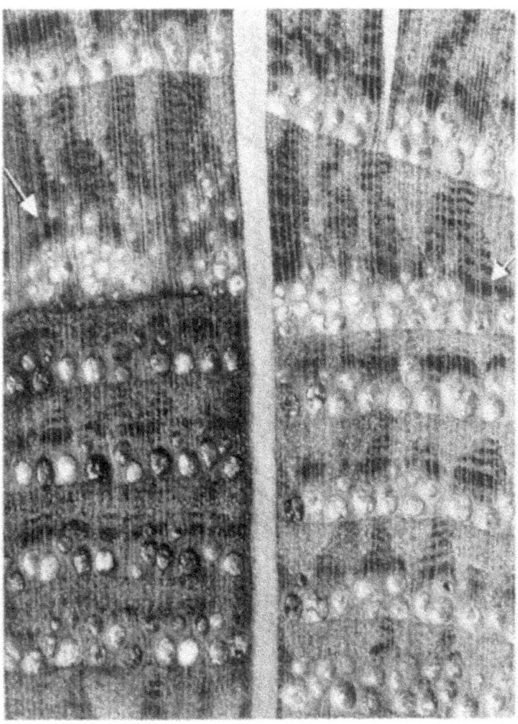

Fig. 297. An extended earlywood vessel group (*arrow*)

tion wood to art history objects of all kinds, such as furniture, statues, and paintings.

Archeological time period: oak wood has been the most popular lumber since as far back as the prehistoric period and has therefore become an important chronological mainstay in archeological research. Of special importance is the dating of lakeside, riverbank, and bog settlements in England and northern Germany (Eckstein et al. 1983) and in central Europe (Becker et al. 1985). The ceasing of settlements along central European lakes, for example, was determined through the datings of several phases of lake dwellings.

Geologic deposits: oak trunks found in the quarternary sedimentation of large central European rivers (Becker 1983) and bogs (Leuschner et al. 1987) as well as western European swamps and bogs (Baillie 1982) provide the foundations for the construction of chronologies spanning many centuries. The longest reaches back to the beginning of the oak migration into Europe in the 8th century B.C. Oak taken from bogs as well as rivers allows for the absolute dating of archeological wood in Europe and the standardization of the radiocarbon method (Baillie et al. 1983). Leuschner et al. (1987) were able to use datings of numerous oaks from northern German bogs to project the forest dynamics of the region in the post-glacial period.

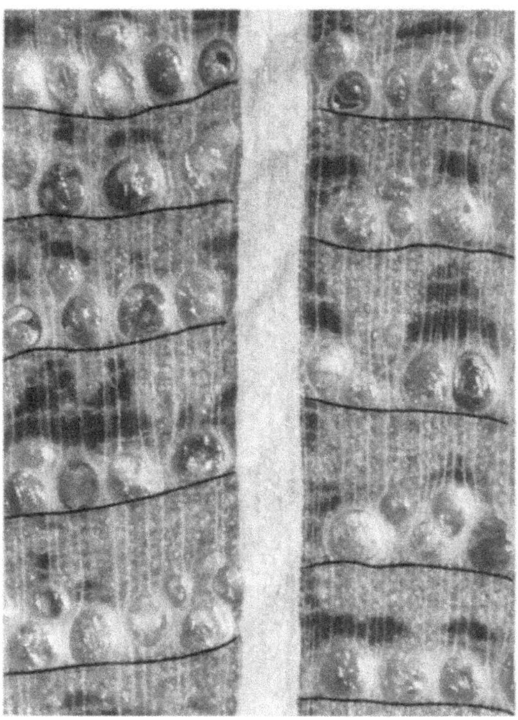

Fig. 298. Tree ring displacement. The tree rings along the rays are displaced

Quercus: Species from the Mediterranean Region

According to Flora Europaea there are an estimated 25 species of oak native to the Mediterranean region. They can be classified into three subgenera. The following are the most widely distributed species.

Subgenus

Sclerophyllodrys:	*Quercus ilex* (evergreen) evergreen oak
Cerris:	*Q. cerris* (deciduous) Turkey oak
Quercus:	*Q. faginea* (partly evergreen) Portuguese oak
	Q. infectoria (partly evergreen) Aleppo oak
	Q. suber (evergreen) cork Oak
	Q. trojana (partly evergreen) Macedonian oak
	Q. frainetto (deciduous) Hungarian oak
	Q. hartwissiana (deciduous)

Morphology

Corresponding to diverse site conditions and human influence, the trees have many shapes: widespreading in open stands, narrow, gnarled in forest stands. On dry, shallow soil sites, the trees reach maximal heights of 10 m, while on moist gorge sites they can grow up to 25 m.

Fig. 299. *Quercus ilex*

Site, Site Conditions, Characteristics

All the species have adapted to the Mediterranean climate with its drought period of several months and hot summer months. Most resistant to drought are the evergreen species *Quercus ilex* and *Quercus suber*. Temperatures in the coastal regions barely drop below 0 °C, while inland temperatures fall under −10 °C, for example in Anatolia and central Spain. Specially adapted races of evergreen species can survive in these regions. In the entire distribution area, the annual precipitation level lies under 1000 mm with maximum precipitation in the winter. Most of the species are relatively indifferent to soil acidity. *Quercus ilex* prefers limestone, *Quercus suber* limestone-poor foundations. The evergreen species are typically found in species-rich, shrub vegetation in altitudes under 1000 m.

Fig. 300. A free-standing, pruned cork oak

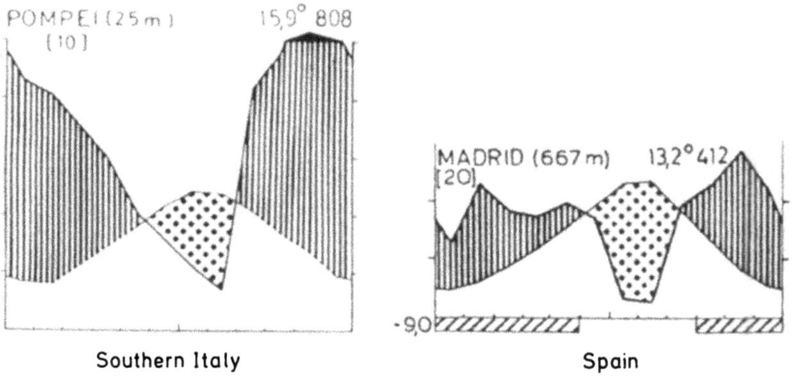

Fig. 301. Climate diagrams

Wood

The deciduous and partly evergreen species exhibit a typical ring-porous structure. See *Quercus robur* and *Q. petraea* for description.

It is not always possible to distinguish the tree-ring sequences of the evergreen species. A ring-porous structure is not always present.

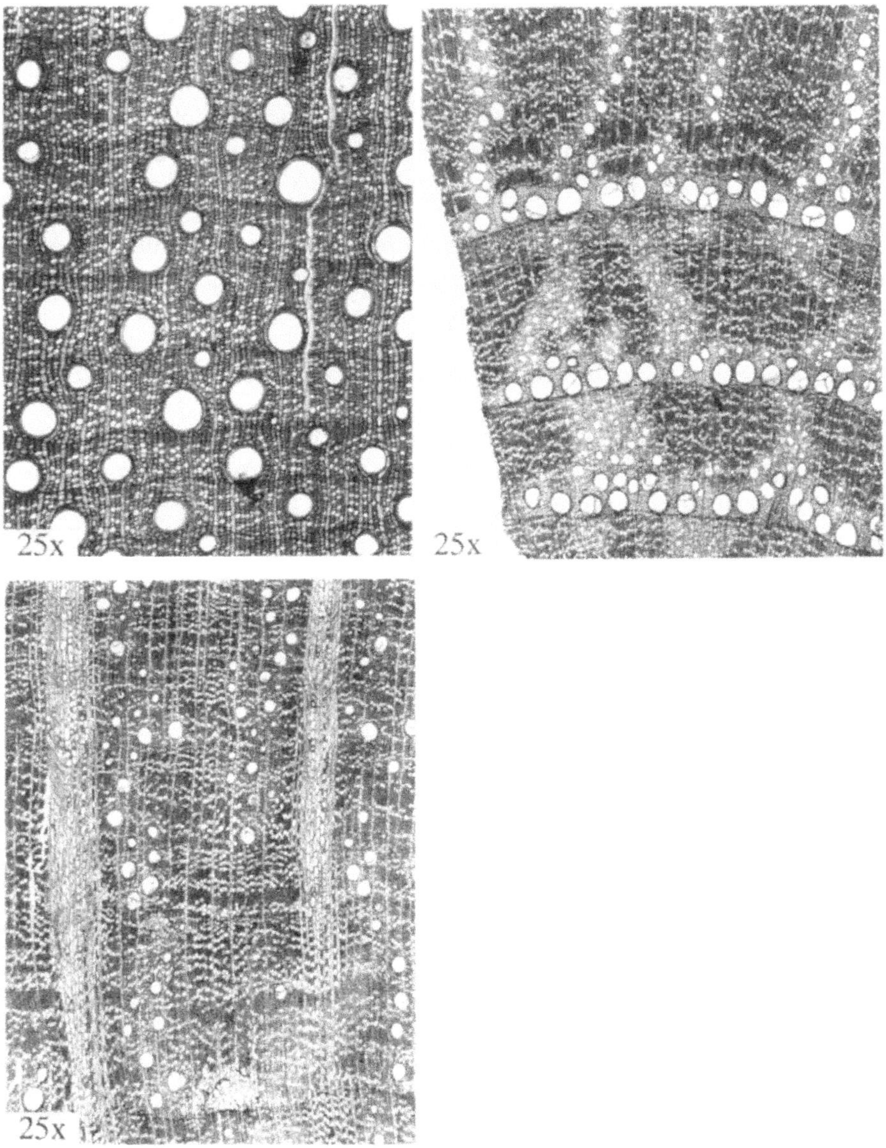

Fig. 302. **a** *Quercus ilex* 40 x. Single-ringed, with individual large earlywood vessels. The tree rings are usually distinct. **b** *Quercus ilex*. Broad ringed, with many small, distinct earlywood vessels, which are separated from the latewood. **c** *Quercus ilex*. The tree rings cannot be recognized because the earlywood vessels do not differ in size or quantity from those in the latewood

Use in Dendrochronology

Suitability of the Species

Although oaks preserve well, species from Mediterranean countries are not especially suited for dendrochronological purposes, because the generally denuded Mediterranean area has been stripped of its older trees. It is not always possible, especially in *Quercus suber* and *Q. ilex*, to distinguish the tree rings.

Radial growth of the partly evergreens appears to react to many ecological influences. The reactions are, however, not easily detected.

Most of the larger trees have been exploited by pruning, grazing, and logging.

The cambium of cork oak has often been damaged through improper treatment during cork harvest. Scars are therefore common.

Fires affect the growth of surviving trees, either through direct fire damage or through increased sunlight or ash fertilization in the years following the fire.

If ecological analysis is directed toward studying periodic disturbances, oak is a good source of information.

Previous Uses

According to Munaut and Serre-Bachet (1982), *Quercus faginea*, *Q. cerris*, and *Q. persica* have been used in dendrochronology. Kuniholm and Striker (1983) recorded a long chronology from the Aegean region. These authors proved that it is fundamentally possible to date oak (at least the deciduous species).

Ulmus glabra Huds.

Wych Elm, Scots Elm, Mountain Elm

Morphology

This deciduous species has an irregular crown, a deep root system, a trunk diameter of over 1 m, and reaches a height of 30 m.

Site, Site Conditions, Characteristics

Ulmus glabra is native to the temperate European climate. It has habits similar to beech, except that *Ulmus glabra* has a broader ecological amplitude and is, therefore, more widely distributed throughout continental Europe. Wych elm tolerates high summer and low winter temperatures. It grows, above all, on moist sites in fresh, deep, and loose soils. *Ulmus glabra* can be found together with *Acer pseudoplatanus* and *Fraxinus* sp. in gorges, riparian areas, and at the base of moist slopes. It rarely grows older than 100 years. Since the mid-20th

Fig. 303. *Ulmus glabra*

Fig. 304. A free-standing elm planted as a street border

century, European stands have been destroyed by the fungus *Ceratocystis ulmi*. *Ulmus glabra* is often planted as a street border.

Distribution

Horizontal: in Europe between the 40° and 60° northern latitudes (similar to *Quercus robur*).

Fig. 305. Distribution of *Ulmus glabra*

Vertical: grows best at low and intermediate sites. In the Erz Mountains, the upper border is located at around 700 m, in the central Alps at 1300 m.

Wood

With heartwood. The sapwood is yellow, the heartwood red-brown. The tree rings are distinct. Within the tree rings there are usually distinct, undulating, tangential bands. The density is 480–860 kg/m^3.

The three European species, *Ulmus campestris*, *U. glabra*, and *U. laevis* cannot be differentiated on the basis of their anatomic characteristics.

Cross-section: ring-porous. Groups of two to four vessels in more or less tangential bands in the latewood (Fig. 306a, b). Rarely tyloses in the earlywood. Groups of thick-walled libriform fibers in the earlywood. Abundant paratracheal parenchyma in earlywood and tangential bands in latewood.

Tangential section: multiseriate rays (four to five cells wide) less frequently one to three or six cells wide. Ray height: 30–60 cells.

Radial section: simple perforation plates. Homogeneous to heterogeneous (type 1) rays. Distinct spiral thickenings in vessels.

Use in Dendrochronology

Suitability of the Species

In most cases the tree rings are distinct and thus easy to measure. As Brett (1978) has proved, elm is well suited for dendroclimatological research, especially for reconstructing past precipitation. Because elm logs were earlier only seldom

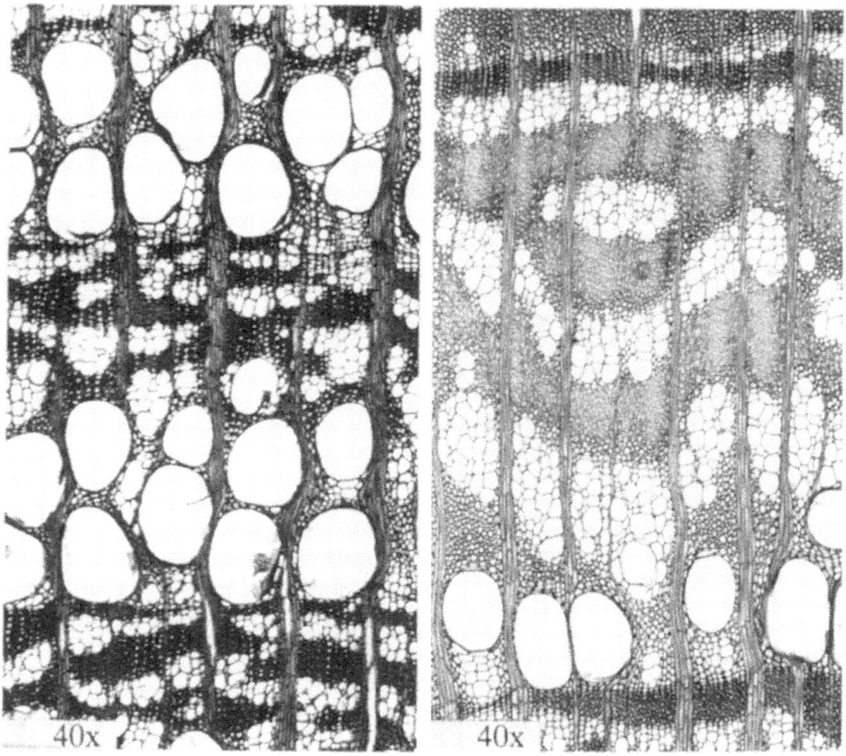

Fig. 306a, b. Cross-sections of *Ulmus glabra* (40 x)

used as building material, the possibilities for constructing long chronologies are minimal.

Previous Uses

Elm has practically never been used in dendrochronology. However, in the course of cutting down damaged elm stands, the species should acquire greater importance. Brenninkmeijer (1983) studied the concentration of the stable isotope deuterium $^{16}O/^{18}O$, $^{12}C/^{13}C$ in the tree rings of a Dutch elm.

Tree Species in North America

Abies balsamea (L.) Mill.

Balsam fir

Morphology

This conifer has a deep root system. Its crown varies considerably in shape and size. Trees reach average heights of 15 to 20 m and can, in extreme wind conditions, be dwarfed. From the bark exudes a clear, intensely odorous resin. The stems rarely have diameters larger than 60 cm. In the western part of its area, *Abies balsamea* hybridizes with *Abies lasiocarpa*.

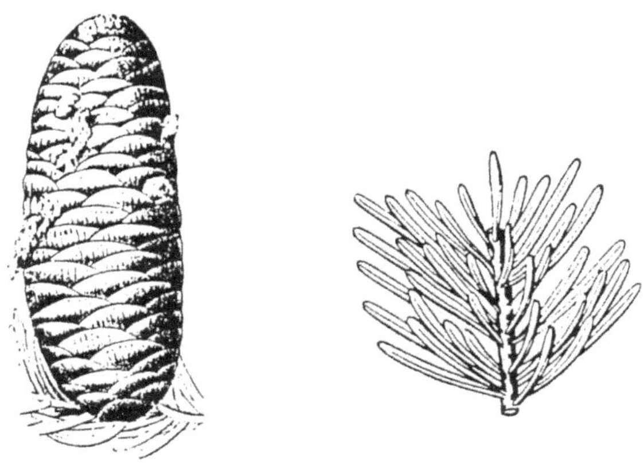

Fig. 307. Cone and needles of *Abies balsamea*

Site, Site Conditions, Characteristics

Abies balsamea is found in cool, moist climates. It flourishes in the northeastern United States where annual precipitation averages between 700 and 1000 mm.

Temperatures average around 20 °C in the summer and between −10 and −5 °C in the winter. *Abies balsamea* can withstand minimal temperatures of −40 °C. The frost-free period varies between 80–180 days. While the species flourishes in deep soil, it grows very slowly in shallow soil, sand, or moors.

Fig. 308. Up to 25-m-high balsam firs in a sugar maple-beech stand in southern Quebec, Canada

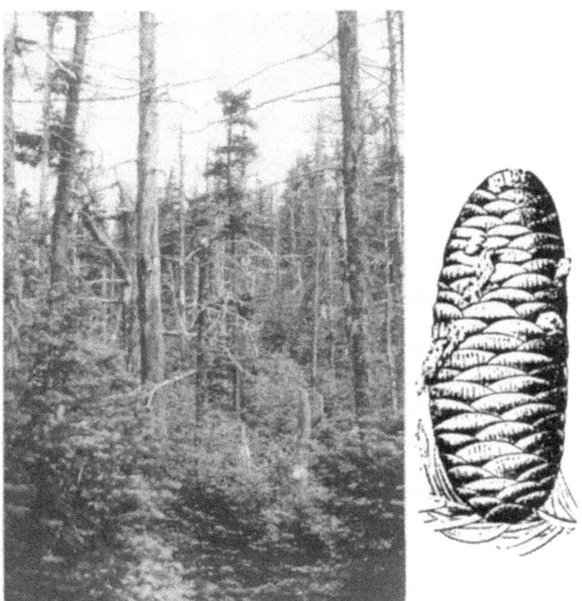

Fig. 309. Balsam firs at the highest elevation of the species on Mt. Washington, New Hampshire, USA. On wind-exposed slopes, the front line firs die off, resulting in a phenomenon referred to as fir wave mortality. A rear view of a wave mortality zone. In the *background* are dead 100-year-old firs; in the *foreground* new growth

Fig. 310. A balsam fir deformed by the wind on Mt. Washington, New Hampshire, USA. Under extreme wind conditions the trees are often dwarfed

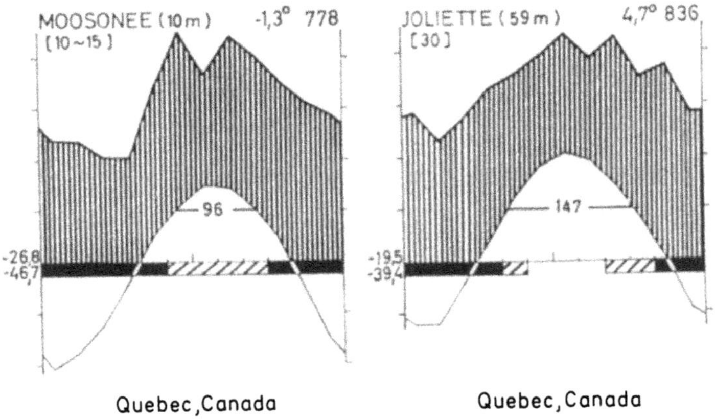

Fig. 311. Climate diagrams

Abies balsamea forms pure stands on mountain sites located in northeastern North America. In the north, *Abies balsamea* commonly associates with *Picea glauca*, *Picea mariana*, and in central regions with *Picea rubens*, *Pinus strobus*, *Betula alleghenensis*, *Betula papyrifera*, and various maple species. In managed stands, the trees seldom live longer than 60 years. On extreme, unmanaged sites, the trees may grow to be over 250 years old. *Abies balsamea* is especially subject to spruce budworm (*Choristoneura fumiferana*) infestation, which has sporadically destroyed thousands of hectares.

Distribution

Horizontal: *Abies balsamea* is found in the southern part of the boreal zone, from Labrador to Alberta, and in the deciduous belt from New York to Minnesota. In the south, distribution is scattered throughout the mountain ranges.

Fig. 312. A cross-section of a 220-year-old, extremely thick-barked balsam fir from the northernmost distribution border in Quebec, Canada. The average tree-ring width is 0.2 mm

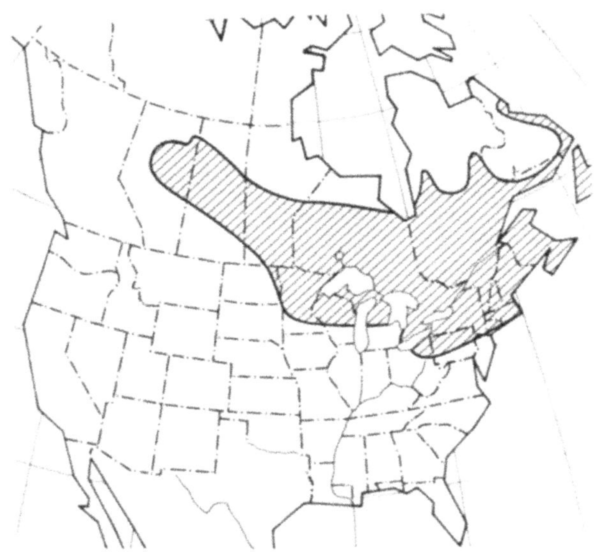

Fig. 313. Distribution of *Abies balsamea*

Vertical: in Labrador, the area extends from sea level to 500 m. On Mt. Washington in New Hampshire *Abies balsamea* grows between 800 to 1500 m. In the continental climate of the west, it is mainly found at low altitudes.

Wood

See *Abies alba* for description. The earlywood-latewood transition is not as distinct as in the European species. There are no crystals in the rays. The wood has a density of 400–500 kg/m^3.

Use in Dendrochronology

Suitability of the Species

These trees, which normally do not grow very old, are best suited for forest dynamic studies. Extreme climatic fluctuations as well as alterations in soil and canopy conditions are recorded in the tree-ring sequences as abrupt growth changes. The species may also be studied to reconstruct past spruce budworm infestations. Only by analyzing tree rings is it possible to chronicle the development of fir wave mortality, a phenomenon that occurs only in extremely windy regions of eastern North America.

Previous Uses

Begin et al. (1989) used shore line fluctuations to date rises in sea level or, more likely, downward tectonic movements in New Brunswick, Canada. Sprugel (1976), Marchand et al. (1986), and Stoeckli (1990) (unpubl.) studied fir wave mortality in New England, USA.

Abies lasiocarpa Nutt.

Subalpine fir

Morphology

This evergreen conifer normally has a narrow, conical crown. On windy sites, or in regions with heavy snow, the trees can develop into prostrate shrubs. The trees grow to heights of 20–30 m and have stem diameters of 75 cm. *Abies lasiocarpa* hybridizes with *A. grandis*.

Fig. 314. *Abies lasiocarpa.* (*Drawing by Hirzel*)

Site, Site Conditions, Characteristics

Like Engelmann's spruce, subalpine fir is native to cold-moist climates. It grows within a broad average temperature range: January −5 to +4 °C; July +5 to +15 °C. It can, however, tolerate extreme temperatures from −40 °C to +30 °C. *Abies lasiocarpa* is a very undemanding species. It settles especially in subalpine sites where it forms the forest border together with either *Tsuga mertensiana* or *Picea engelmannii*, depending on the region. It prefers acid soils. In stands that are relatively untouched by industry, trees can live for longer than 250 years.

Fig. 315. Old fir trees with narrow crowns on the snowy timberline in Washington State. (Krüssmann 1972)

The species is very sensitive to fire. It suffers from spruce budworm (*Choristoneura fumiferana*) infestation. Wood-decomposing fungi, such as *Melampsorella cerastii*, frequently attack older stems.

Fig. 316. A fir tree blown lopsided (flag-shaped) by the wind. Front Range, Colorado

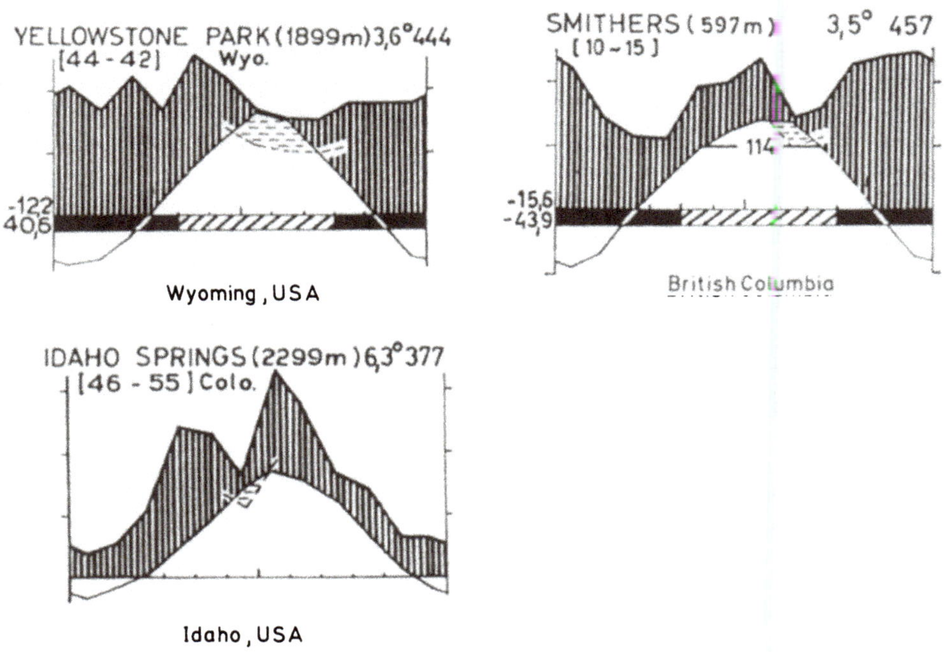

Fig. 317. Climate diagrams

Distribution

Horizontal: in western North America from southern Alaska to Arizona.

Vertical: in Alaska from sea level up to an elevation of 1000 m, in the northern Rocky Mountains from 1500–2700 m and in the central and southern Rocky Mountains between 2700 and 3500 m.

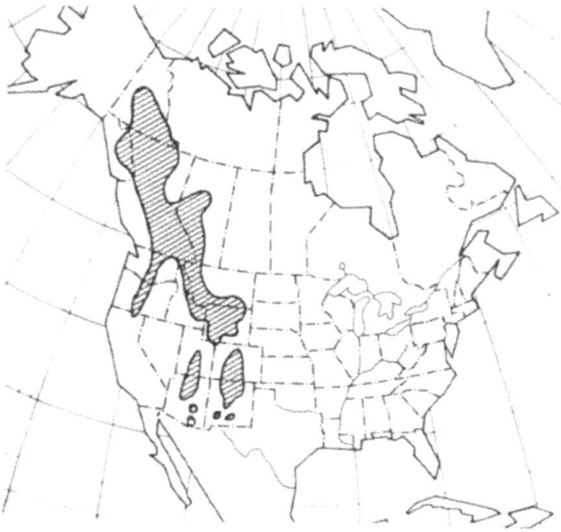

Fig. 318. Distribution of *Abies lasiocarpa*

Wood (see *Abies alba*)

Abies species share very similar wood structures.

Two groups can be distinguished, characterized by the existence or absence of crystals in the marginal ray walls.

With crystals: *Abies procera*, *A. grandis*, *A. magnifica*, and *A. concolor*

Without crystals: *Abies amabilis*, *A. lasiocarpa*, and *A. balsamea*

Use in Dendrochronology

Suitability of the Species

As it grows in cold-humid climates, the species is especially suited for densitometric research. Because the trees generally do not grow to be very old and the core often breaks apart during increment boring, its use in dendrochronology is rather limited. Since subalpine fir is on many sites the secondary tree

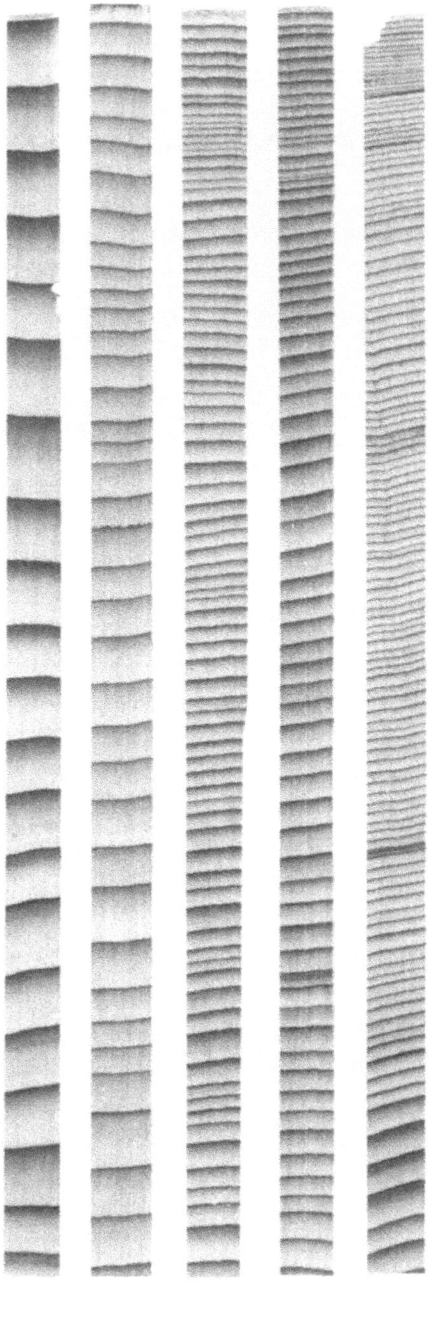

Fig. 319. Radiographs of *Abies lasiocarpa*

species, it is possible to use the age and abrupt growth changes of the trees to determine special events in the life of a forest, such as fires and destructive storms.

Previous Use

The species has received little dendrochronological attention. In the western American sample grid from Fritts (1965), subalpine fir is represented in a few recent chronologies. The potential use of American *Abies* species in dendroecological studies has yet to be determined.

Juniperus deppeana Steud.
Juniperus monosperma (Engelm.) Sarg.
Juniperus osteosperma (Torr) L.
Juniperus scopulorum Sarg.

Alligator juniper
Cherrystone juniper
Utah juniper
Rocky mountain juniper

Only the most important species found in southwestern America will be covered.

Fig. 320. *Juniperus deppeana.* (Drawing by Hirzel)

Morphology

Most juniper species have compact, bushy, often spindle-formed crowns. Older trees can reach heights of 10 m, with stem diameters up to 1 m and can, in extreme cases, grow to be 3000 years old.

Site, Site Conditions, Characteristics

Juniperus species have ecological demands similar to those of the pinyon species, such as *Pinus monophylla*. At low, arid forest borders these two species are commonly associated with one another.

For further ecological characteristics see *Pinus monophylla*.

Fig. 321. A typical *Juniperus osteosperma*-pinyon landscape in southwestern USA. The stands are scattered. (Fowells 1965)

Fig. 322. A mature *Juniperus scopulorum* on a shallow limestone soil, located at an elevation of 2500 m in the Sierra Nevada

Fig. 323. A young *Juniperus scopulorum* on a former pasture in semiarid southwestern USA. (Fowells 1965)

Fig. 324. *Juniperus deppeana* at a dry, lower forest border in Arizona

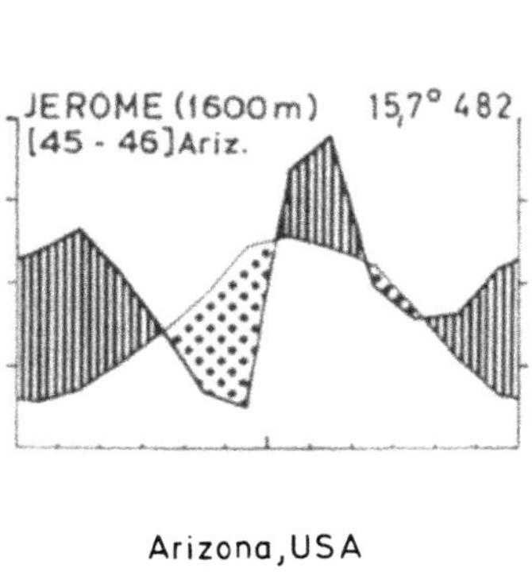

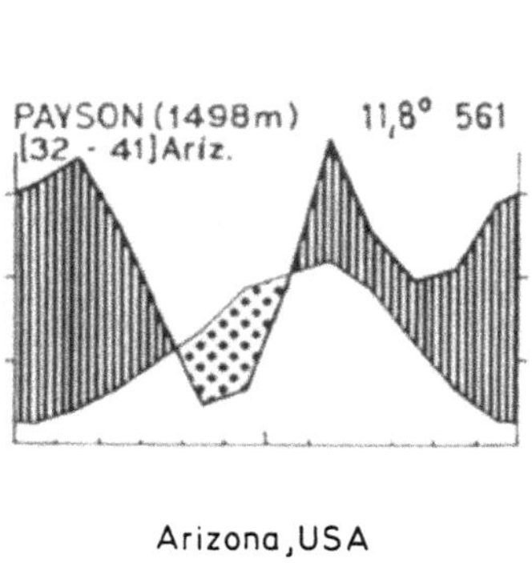

Fig. 325. Climate diagrams

Distribution

Horizontal: the three species are all mainly found growing in the states of Arizona, New Mexico, Nevada, Utah, and Colorado. The growing range of *Juniperus deppeana* and *Juniperus monosperma* reaches into northern Mexico, while *Juniperus scopulorum* extends into southern British Columbia.

Vertical: every species grows at the low, arid forest border which, in the south, lies at an elevation of 1400–2000 m.

Fig. 326. Distribution of the species *Juniperus monosperma, J. osteosperma* and *J. scopulorum. J. deppeana* is distributed further south into Mexico

Wood

The heartwood is red to wine-red to brown and the sapwood is yellow-white. The wood has an intense odor, has spotted coloring on its cross section and a density of 450–550 kg/m^3.

Cross-section: resin canals are missing. The extractives in the parenchyma cells in the heartwood are brown. The latewood is usually quite narrow and the transition from earlywood to latewood is continuous. (Fig. 327a–c)

Tangential section: the ray height averages two to seven cells.

Radial section: there are no treacheids in the rays. Pits in the crossfields are usually taxodioid, i.e., with wide apertures. The horizontal ray cell walls are thick and rather smooth, while the tangential walls have nodules. The various *Juniperus* species are microscopically indistinguishable.

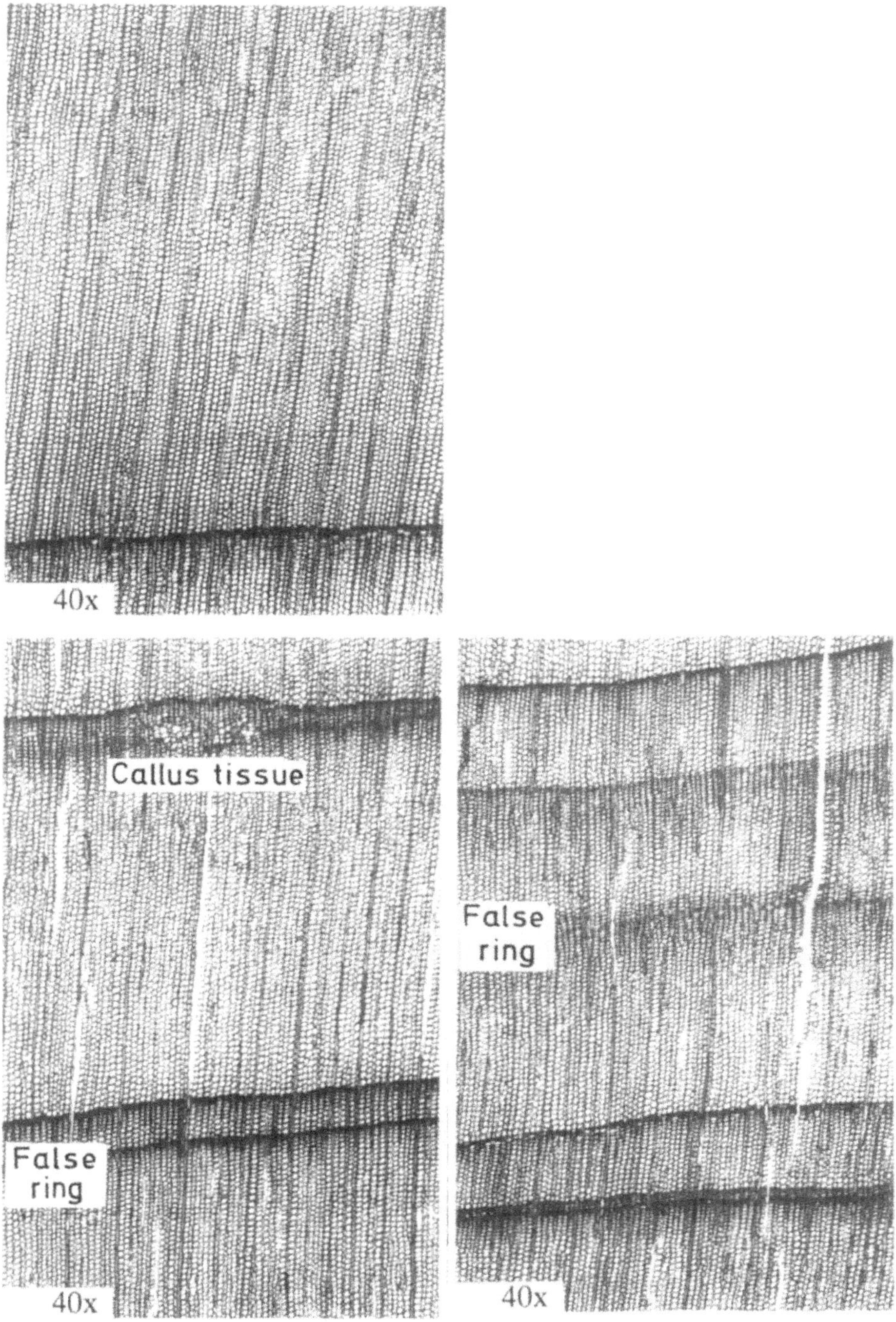

Fig. 327a–c. Cross-sections of *Juniperus deppeana*. **a** Callus tissue. **b** False ring. The narrow tree ring is most likely a "false ring". **c** False ring. A distinct density fluctuation in the middle tree-ring almost resulted in the forming of a "false ring"

Fig. 328. An exposed trunk with adventitious roots. (Photo Clark)

Fig. 329. A buried *Juniperus deppeana* forest in northern Arizona. The trees in the plains continue to live although only their tree tops project above the ground. (Photo Dean)

Use in Dendrochronology

Suitability of the Species

Juniperus species often grow to be 2000 years old. *Juniperus scopulorum*, growing in the mid and northern regions of the Rocky Mountains appears to be the most easily synchronized. In contrast, all species found in the southern region, at the arid forest border, form many false rings, making synchronization practically impossible. For this reason, it appears that *Juniperus osteosperma*,

J. deppeana, and *J. monosperma* cannot be dated. Dendroclimatological information in the presentday sense cannot be expected. The species are suitable for density fluctuation analysis. When possible, western American dendrochronologists try to avoid using *Juniperus* species.

Previous Uses

Juniperus woods are, along with other species, common to Native American settlements. Dean (1969 pers. comm.) dated floods from tree species in northern Arizona and in southern Utah. During heavy storms, complete forests were flooded, covered with sediment, and eventually, often centuries later, exposed through erosion. Using trees with exposed roots, Alexander and McCord (1987) calculated the mean rates of erosion.

Larix laricina (Du Roi) K. Koch

Tamarack, eastern larch

Morphology

This deciduous conifer has a shallow root system and normally a narrow crown. Older trees are often cedar-like in form. On optimal sites, in the southern part of its area, the species grows rapidly and can reach heights of 25 m. On extreme sites, along the northern border, trees may live for hundreds of years, but reach heights of only 8 to 10 m.

Fig. 330. *Larix laricina* (Krüssmann 1972)

Site, Site Conditions, Characteristics

Larix laricina is a species of open stands in the North American boreal zone. It has a broad ecological spectrum. Within its natural area, it usually grows in

Fig. 331. a A solitary larch on the edge of a bog at the northern border of its area in Quebec. The bush-shaped crown indicates prior insect infestation. **b** Larch-spruce stands on the northern forest border in Quebec

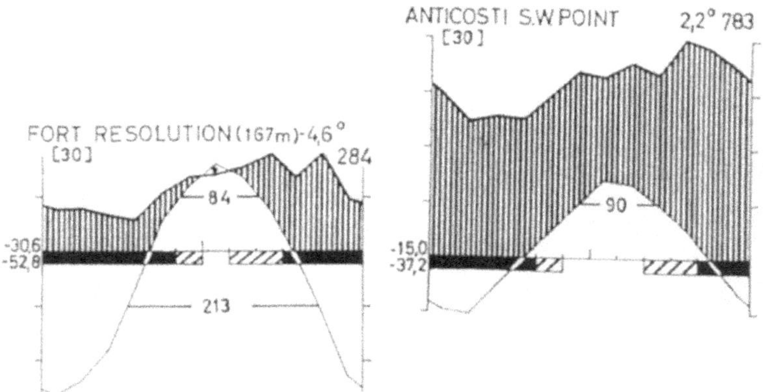

NWT, Kanada

Quebec, Kanada

Fig. 332. Climate diagrams

bogs, but does not necessarily avoid drier sites. It flourishes on sites with deep soil and plenty of sunlight. *Larix laricina* grows in continental regions with precipitation less than 250 mm and average January temperatures of −25 °C. It also grows in marine cool climates with annual precipitation of more than 1000 mm and January temperatures of −2 °C. In its northern area, *Larix laricina* is often accompanied by *Picea mariana*. In the south it associates with *Populus tremuloides*, *Betula papyrifera*, and *Abies balsamea*. Since approximately the beginning of the 20th century, *Larix laricina* has been attacked by the introduced larch sawfly (*Pristiphora erichsonii*). Trees as old as 300 years are common at the northern forest border.

Distribution

Horizontal: north of the 48 parallel, the species distribution is practically transcontinental. The western border is at the Mackenzie.

Vertical: in northern regions the species is found predominantly at low elevations, (sea level to 500 m). In the south, the area extends to 1200 m.

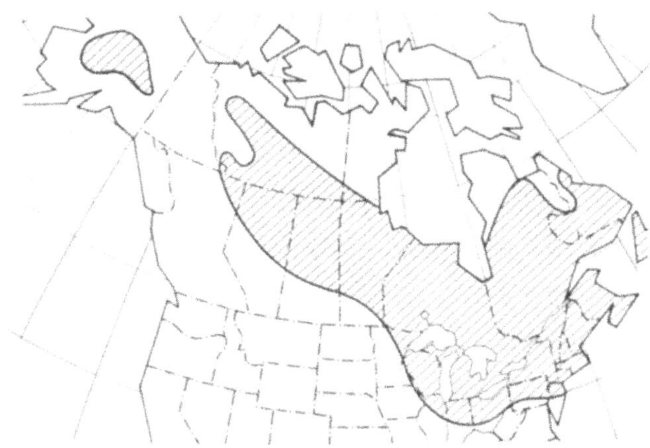

Fig. 333. Distribution of *Larix laricina*

Wood

With heartwood. Yellow sapwood, green-brown heartwood. Distinct tree rings. The wood density varies considerably between sites. Wood from bog sites has a density of 500–600 kg/m^3.

Anatomically, *Larix laricina* cannot be distinguished from *Larix decidua*.

Fig. 334. Tree-ring sequences of a *Larix laricina* with noticeably narrow tree rings and narrow latewood. Both are very likely the results of a sawfly attack. (× 4)

Use in Dendrochronology

Suitability of the Species

Due to its ring width sensitivity, this species is suitable for climatic reconstruction in the boreal zone. Sawfly-effects inflicted in the 20th century may interfere with climatological signals. Therefore, any potential studies must incorporate a comparison with spruce. The common occurrence of fire scars at the base of the trunk makes it possible to date forest fires.

Previous Uses

Studies have been undertaken only in northern Quebec. Morin and Payette (1984) used the age structure of stands to date the expansion of the northern forest border over the last 50 years. Payette (1980) used scars to determine the ice level in a river in Nouveau, Quebec.

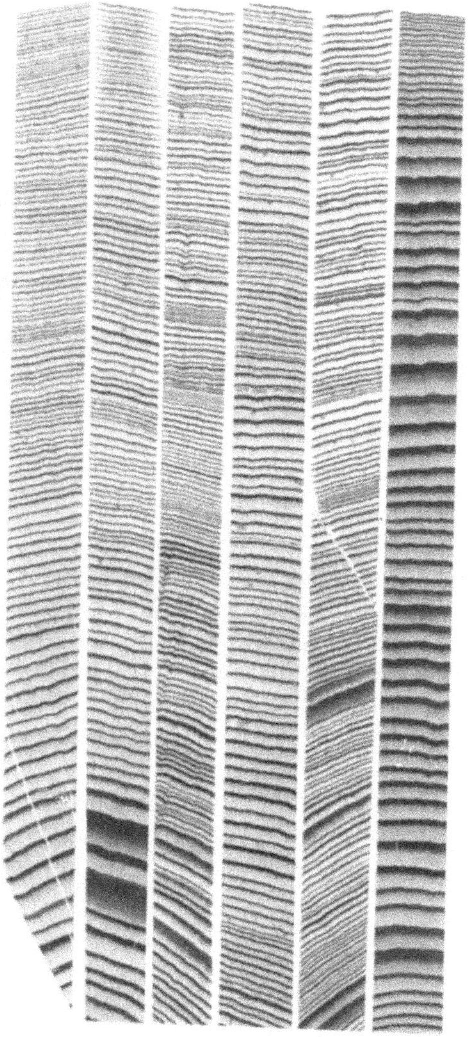

Fig. 335. Radiographs of *Larix laricina*, Manitoba, Canada

Picea engelmannii Parry

Engelmann Spruce

Morphology

This conifer has a shallow root system and a dense, narrow crown. It grows to a height of 30 m with a stem diameter of 1–1.5 m. It has no geographical races and only rarely crosses with other spruce species, such as *Picea glauca*.

Fig. 336. *Picea engelmannii.* (Krüssmann 1972)

Site, Site Condition, and Characteristics

Picea engelmannii is a cold-humid climate tree. It grows at the highest altitudes with low winter temperatures, the average temperature in January being between −6 and −10 °C and with heavy snow. Summers are cool, the average temperature in July being around 6 to 11 °C. In all regions, the annual level of precipitation exceeds 700 mm. At extreme sites, the frost-free period lasts only 1–2 months. The species grows in moderately well-drained, optimally basic soil. Engelmann spruce forms the subalpine forest border, either in pure stands or together with *Abies lasiocarpa*. Depending on the region, Douglas-fir, lodgepole pine, and blue spruce may, at lower sites, accompany Engelmann spruce. At higher sites, it may be associated with whitebark pine or bristlecone pine. At the

Fig. 337. Dwarfed Engelmann spruce at the windy subalpine forest border of the Colorado Front Range

Figs. 338, 339. A typical Engelmann spruce forest at the subalpine forest border. The trees are often over 300 years old. Both sites have an elevation of 3000 m. (**Fig. 338** Cedar Breaks, Utah, **Fig. 339** Silverton, Colorado)

Fig. 340. A *Picea engelmannii* site located on a windy ridge near the timberline in New Mexico, USA (Sandia Crest)

forest border, the trees often grow to be 300 or, in rare cases, even 600 years old. Except for stem-decomposing fungi, the species suffers attack from only very few pests.

Distribution

Horizontal: widely dispersed in western North America. In the north, in the region of British Columbia, the species borders on and mixes with *Picea glauca.* In the south, the distribution has a mosaic-like pattern.

Vertical: in the coastal range of British Columbia (from 700–1200 m) and in the central and southern Rocky Mountains, Engelmann spruce often forms the highest subalpine border (between 2500 and 3300 m).

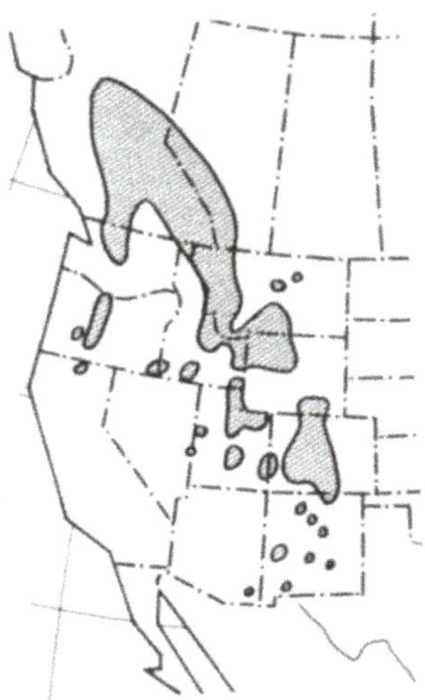

Fig. 341. Distribution of *Picea engelmannii*

Wood

Without heartwood. The wood has a yellow-white to light brown color with a smooth longitudinal section. The density is 350–400 kg/m^3.

Cross-section: the resin canals are lined by 8–12 rather thick-walled secretory cells. The latewood is usually narrow with a continuous earlywood to latewood transition. The latewood is not as pronounced as in *Picea abies* (Fig. 343a, b).

Tangential section: the average ray height is between 10–25 cells. Resin canals are lined by 7–12 epithelial cells.

Radial section: there are ray tracheids in the rays. Pinoid pits in the earlywood result in slit-formed openings. The difference between minimum and

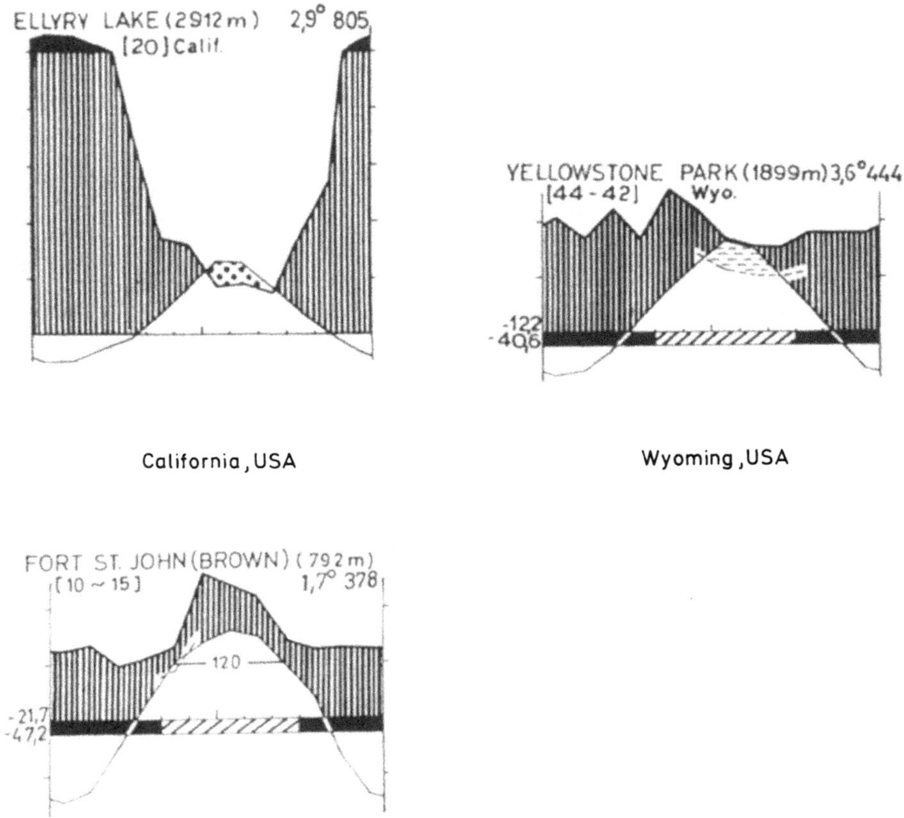

Fig. 342. Climate diagrams

maximum density within a ring is less than in *Picea abies*. The adaptability to climatologic integration ability is, nevertheless, very good.

The various species of spruce cannot be distinguished from each other on the basis of their anatomic characteristics.

Use in Dendrochronology

Suitability of the Species

This borderline-forming, cold-humid climate species is extremely suitable for dendroclimatological and dendroecological research. Due to the frequency and regularity of its tree ring sequence, the species lends itself well to the radiodensitometric techniques of temperature-relationship reconstruction. Because high altitude forests in the mountains of western America have often remained old growth forests, many sites have spruce trees over 300 years old.

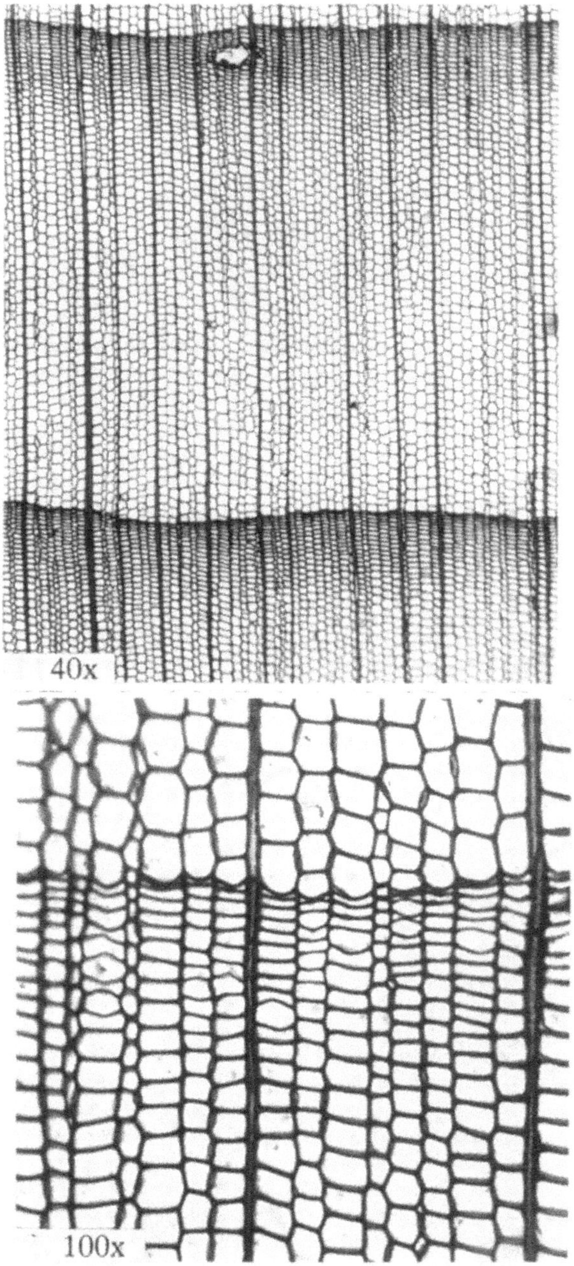

Fig. 343. a Cross-section of *Picea engelmannii.* **b** Latewood with low maximum density (cold summer). The walls of the latewood cells are not thicker than those of the earlywood

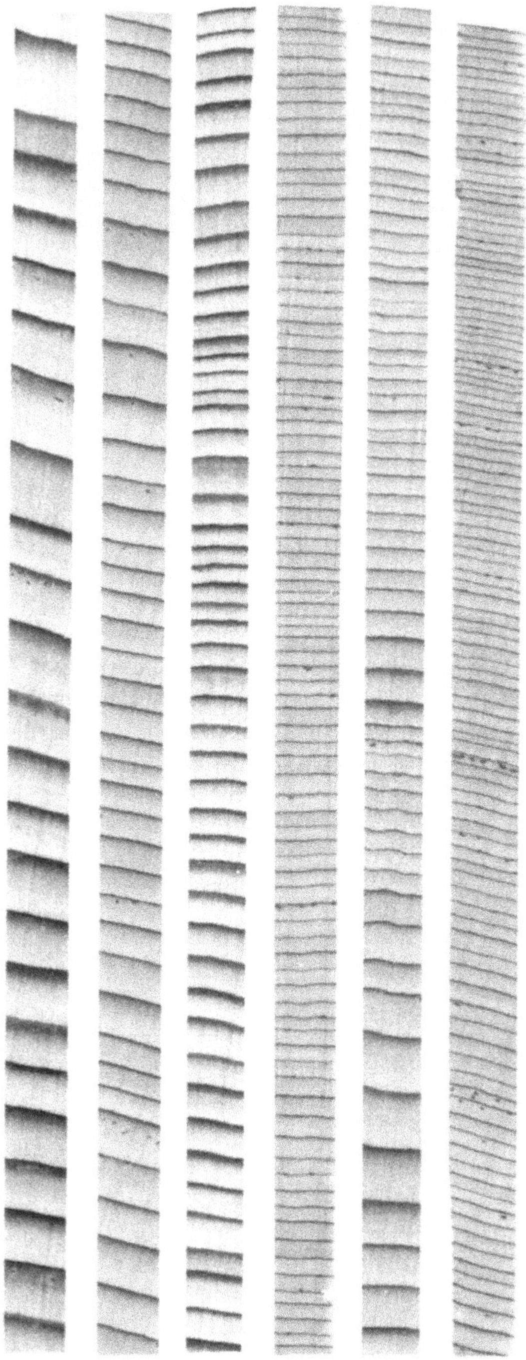

Fig. 344. Radiographs of *Picea engelmannii*, Arizona

The minimal amount of research done on this species is by no means representative of its potential for use in dendrochronology.

Previous Uses

Parker and Henoch (1971) determined that a significant amount of information on summer temperatures could be derived from the latewood densities of Engelmann spruce growing in Jasper National Park. This study has been the initial point of all recent dendroclimatological work done at high altitudes and northern latitudes. The Western American sample grid by Schweingruber (1987) was taken mostly from *Picea engelmannii*; Johnson (1987) gathered data using tree-damaging avalanche slides, and Brubaker (1987) used Engelmann spruce for studies on forest dynamics.

Picea glauca **(Moench) Voss**

White spruce

Morphology

This shallow-rooted species normally has a regular, narrow, conical crown. It reaches an average height of 25–30 m, sometimes up to 35 m. The stem usually has a diameter of around 40–80 cm, sometimes up to 1.2 m. The species has various races, each exhibiting considerable morphological and physiological differences.

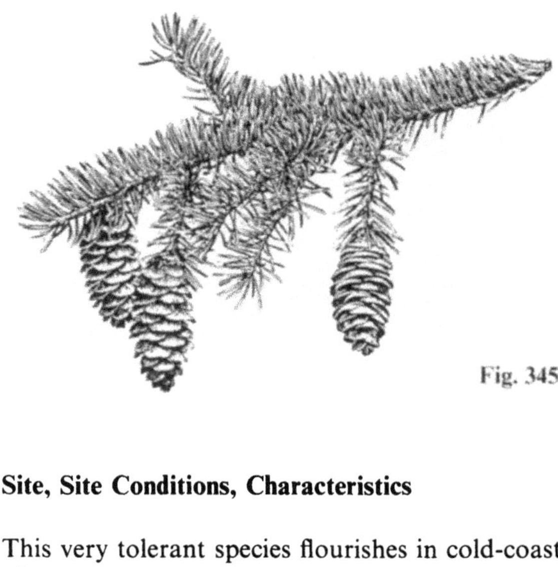

Fig. 345. *Picea glauca.* Krüssmann (1972)

Site, Site Conditions, Characteristics

This very tolerant species flourishes in cold-coastal as well as in cold-semiarid climates. It can withstand temperatures from −50 °C in January to 10–20 °C in

Fig. 346. *Picea glauca* on the Tanana River in Alaska. Trees growing on the banks of large rivers usually reach a maximum age of 300 years. The banks are frequently washed away by erosion, leaving the trees to fall into the river

Fig. 347. *Picea glauca* at the northern forest border

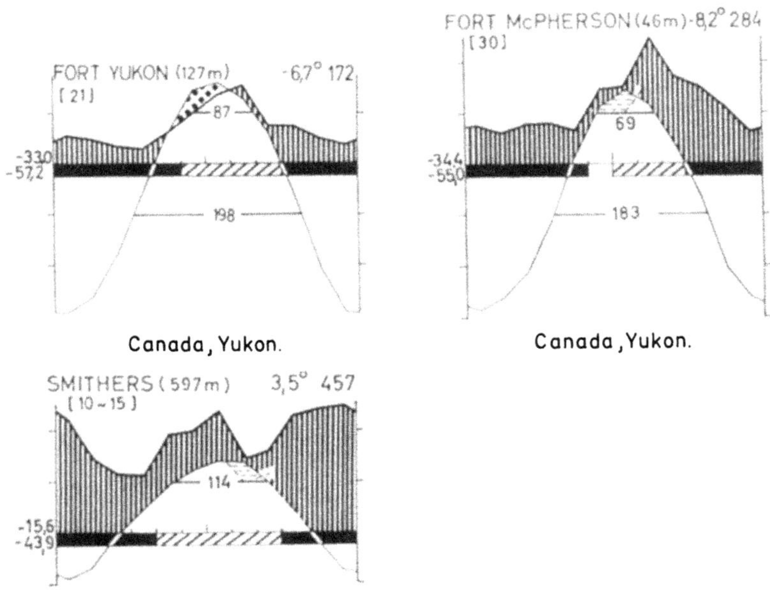

Fig. 348. Climate diagrams

July. It grows in regions with annual precipitation levels of 300 mm just as well as in regions with annual levels of 2500 mm. White spruce tolerates both silicate and limestone soils and develops best in podzol. In the east, the species is often associated with black spruce and balsam fir, in the west with birch and quaking aspen. At the northernmost sites, not used for forestry, trees over 100 years old are common. Stands mixed with balsam fir are often decimated by the spruce bud worm (*Choristoneura fumiferana*).

Distribution

Horizontal: transcontinental distribution throughout northern America.

Vertical: in Labrador from sea level up to an elevation of 600 m and in the Rocky Mountains up to 1500 m.

Fig. 349. Distribution of *Picea glauca*

Wood

The wood of *Picea glauca* is similar to that of the *Picea engelmannii*. The density is 400–500 kg/m^3.

Use in Dendrochronology

Suitability of the Species

Picea glauca is widely distributed throughout the boreal zone of North America, and reaches a high age, which makes it very suitable for dendrochronological

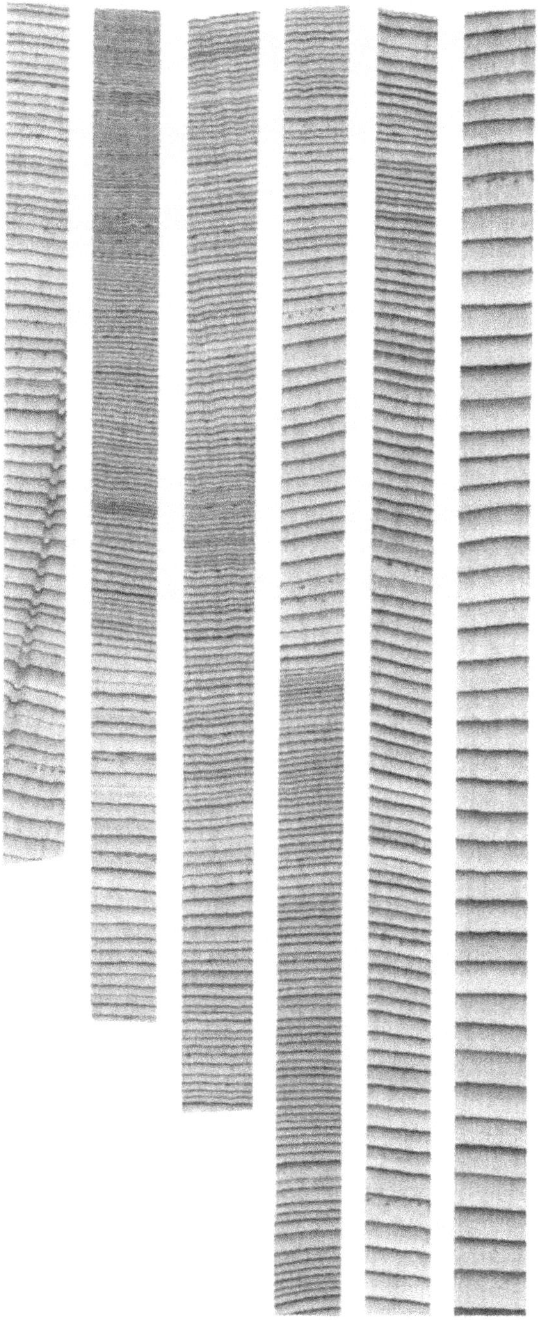

Fig. 350. Radiographs of *Picea glauca* from Yukon

studies. Because the species normally grows in temperate areas, it also lends itself well for dendroclimatological purposes, especially for constructing a network. *Picea abies* is, however, even more suitable. Damage caused by red rot and forest fires significantly limits the age of the trees. Only rarely can a tree older than 300 years be found (maximum 500 years). Tree rings are frequently quite narrow. High sensitivity is found in the maximum density of the latewood and very low in the tree-ring width. Dating of the wood is, however, still possible, even with skeleton plots (Giddings 1938).

Previous Uses

The stems of this species have been used for many different purposes. Giddings (1938) dated abandoned Eskimo settlements in Alaska within the last 5 centuries. Delwaide and Filion (1987), using abrupt growth rate recoveries in trees located near abandoned Native American settlements, studied migration behavior over the past 30 years. Jacoby and Cook (1981) and Schweingruber (1987) reconstructed summer temperatures from latewood densities using radiodensitometrical methods. The oldest chronologies reach back to circa 1500 A.D. Using trunk injuries, Parker and Jozca (1971) were able to reconstruct ice accumulation in the McKenzie River for the past 100 years.

Picea mariana (Mill.) B.S.P.

Black spruce

Morphology

This slow-growing conifer has a shallow root system and a sparse, narrow crown. On good sites the trees can grow up to heights of 25 m but in poor soils only up to 12 m. The stems are rarely thicker than 60 cm. Black spruce hybridizes with red and white spruce. At the northern and subalpine forest border, the species often grows as a prostrate shrub.

Site, Site Conditions, Characteristics

The undemanding black spruce has a very broad ecological spectrum within cold, subhumid climates. The average temperature in January is in the north around −30 °C and in the south around −6 °C. The average temperature in July averages in the north around 10 °C and in the south around 20 °C. The frost-free period lasts for 2–4 months. The eastern distribution area has an annual rainfall level of 1400 mm (in northwest Canada only 250 mm). The species can be found growing in mineral and organic, dry and wet, acidic and neutral soils which are partly located above permafrost. *Picea mariana* forms

Fig. 351. *Picea mariana.* (Drawing by Sebek)

Fig. 352. *Picea mariana* on the timberline in southern Alaska at an elevation of 600 m on a high swamp expanse. In the background is the Alaska pipeline

either a pure stand or is accompanied by white spruce, jack pine, balsam fir or tamarack. In Canada the species forms the northern forest border. The oldest trees grow to be 500 years. Black spruce is sensitive to floods and fire. It propagates itself, mainly vegetatively, along the northern forest border. Besides trunk-decomposing fungi, black spruce is especially prone to damage from dwarf mistletoe (*Arceutobium pusillum*).

Fig. 353. *Picea mariana* site at a side river of the McKenzie in northern Canada, NWT

Fig. 354. Deformed by the wind, *Picea mariana* shrubs are shown here with flag-formed crowns on Mt. Washington in New Hampshire, USA

Distribution

Horizontal: transcontinentally distributed throughout North America. (Distribution see map under *Picea glauca*)

Vertical: the area of distribution is located at an elevation of 150–750 m. In eastern Canada it extends down to sea level. In the Rocky Mountains it is found at elevations of up to 1500 m.

Wood

Similar to that of *Picea engelmannii* (see there for description). The wood has a density of 400–500 kg/m^3.

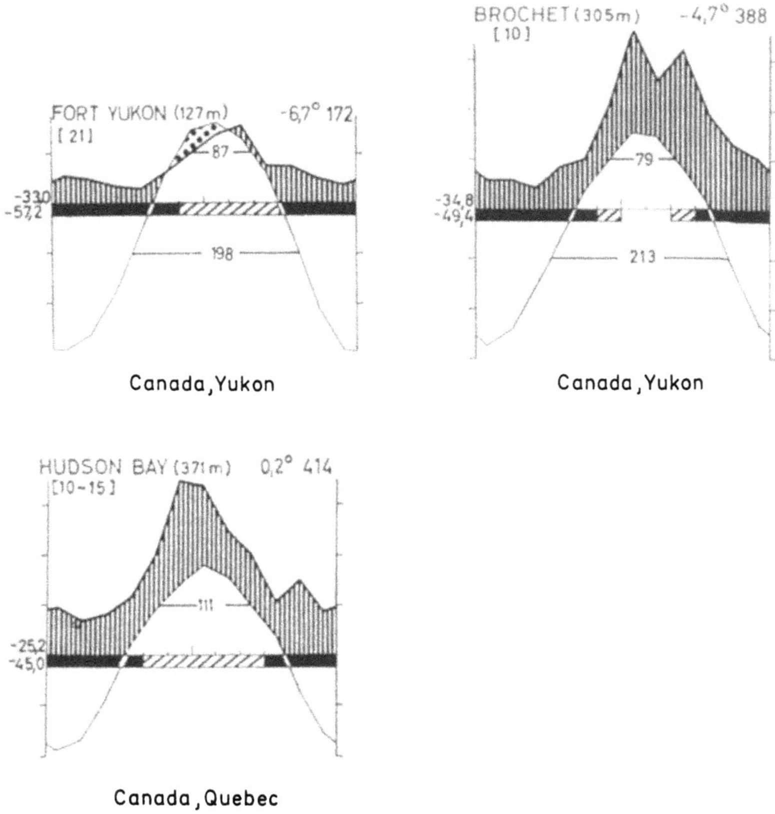

Fig. 355. Climate diagrams

Use in Dendrochronology

Suitability of the Species

The species is widely distributed and fills a diverse ecological spectrum. It is suitable for studies on geomorphological processes, i.e., the building of sedimentation caused by dunes (Filion and Morin 1988) and landslides (Begin and Filion 1985), along with the reconstruction of ecological conditions, for example the time period during which the site is covered with snow (Morin and Payette 1986).

Previous Uses

The oldest chronologies from Payette and Filion (1985) reach back to 1398. The "light rings" determined by Filion et al. (1986) correspond to low summer temperatures.

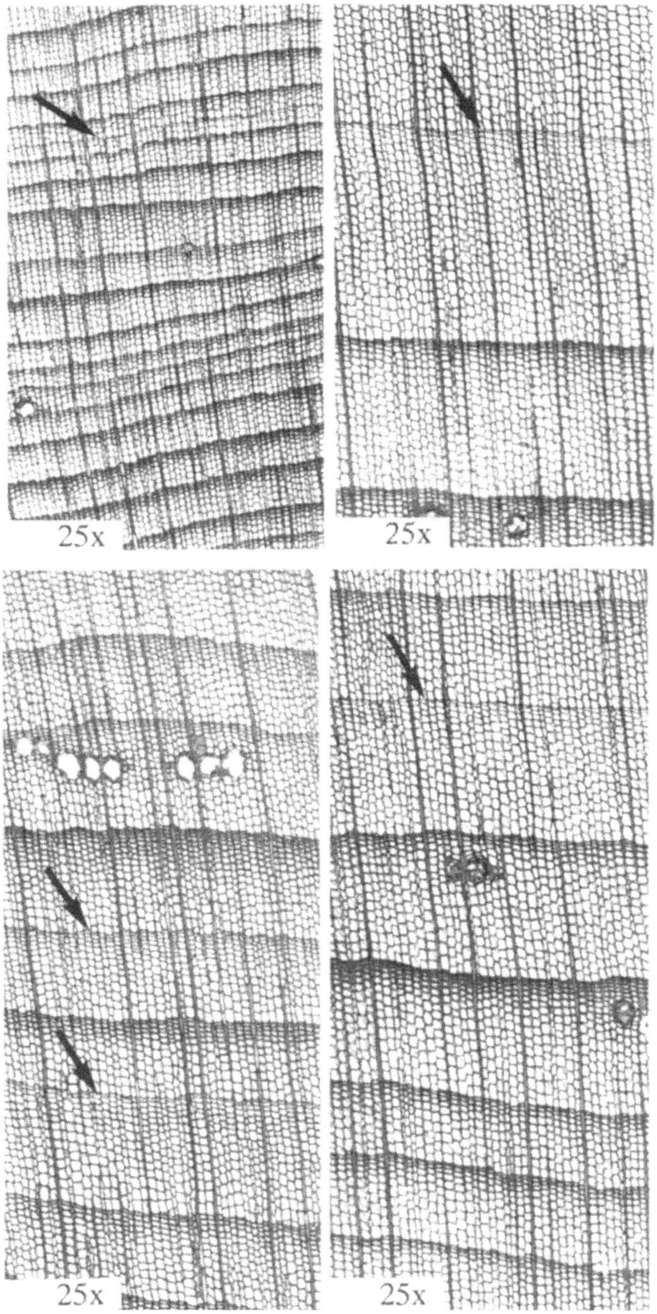

Fig. 356. a–d. "Light rings". Tree rings with very thin-walled latewood cells. In photomicrographs **a** and **b** the latewood cells are made up of one to two rows. In **c** and **d** they are made up of multiple rows. *Picea mariana* from a dwarfwood zone in northern Quebec, Canada

Picea rubens Sarg.

Red spruce

Morphology

This shallow-rooted conifer has a tall, generally conical crown. The trees normally grow to heights of 20–30 m, exceptionally as high as 45 m, and their stems generally have diameters between 40–60 cm, sometimes as large as 140 cm.

Fig. 357. *Picea rubens* (Elias 1980)

Site, Site Conditions, Characteristics

This species flourishes in the moist-cold climate of the subalpine zone of the Appalachian Mountains. The average temperature in January is between 0 and

Fig. 358. *Picea rubens* – *Pinus strobus* stand in the Laurentides in Quebec. Along the shore grows *Thuya occidentalis*

−10 °C. Temperatures in July are between 15 and 20 °C. The frost-free period lasts for at least 2 to 3 months. Annual rainfall is measured at 1000–2000 m. Red spruce requires well-aired, acid soil with a thick cover of raw humus. The species forms pure stands at the supalpine forest border. At lower sites it is accompanied by balsam fir, sugar maple, and yellow birch. Trees older than 300 years exist only on a few sites. The species is very sensitive to fire and is epidemically threatened by the spruce budworm. At high altitudes, many sites are currently suffering from "spruce decline".

Fig. 359. Dying *Picea rubens* in New Hampshire (red spruce decline)

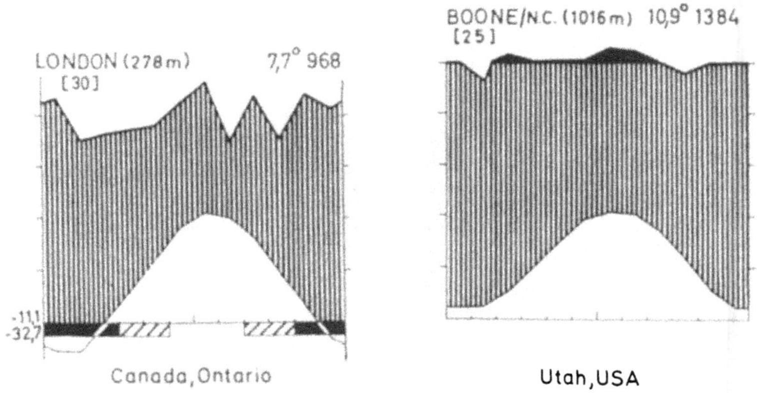

Fig. 360. Climate diagrams

Distribution

Horizontal: the main area of distribution is in the northeastern United States.

Vertical: in northern regions, the species extends from sea level up to an elevation of 1400 m, in southern regions at elevations between 1100 to 1400 m.

Fig. 361. Distribution of *Picea rubens*

Wood

See *Picea abies*

Use in Dendrochronology

Suitability of the Species

This species has a considerable lifespan. It grows at high altitudes in the Appalachian Mountains from Quebec to Tennessee. It has regular tree ring sequences. Species from mountain sites are suitable for both radiodensitometric-climatological and ecological studies.

Previous Uses

Picea rubens reflects average summer temperatures in its maximum latewood density. In this respect, it behaves like *Picea abies* from high altitudes and northern latitudes in Europe (Conkey 1979, 1986) and *Picea engelmannii* in western USA.

Many studies on the decline of *Picea rubens* in northeastern USA show a pattern of growth reduction beginning in the 1960s. Hornbeck et al. (1987) and Federer and Hornbeck (1987) see a normal aging process as a possible cause. Other trees, however, including species with short lifespans, i.e., *Acer saccharum*,

Fig. 362. Radiographs of *Picea rubens* in Maine, USA

do not follow this trend. LeBlanc et al. (1987) attribute the decline to climatic factors: an extreme drought in the 1960s and an abnormally warm winter in the late 1970s. Air pollution (ozone) and acid rain have also been mentioned as contributing factors in the red spruce decline in northeastern USA.

Harrington (1986) analyzed tree-ring sequences in trees located on wind-stressed sites in New Hampshire, USA.

Picea sitchensis (Bong.) Carr.

Sitka spruce

Morphology

This fast-growing conifer has a shallow root system and a broad, open crown of horizontally set branches. The species normally grows to a height of 35–45 m, exceptionally 65 m, and has a stem diameter between 90–150 cm, sometimes as large as 250 cm. Hybrids with *Picea glauca* can be found in Alaska.

Fig. 363. *Picea sitchensis.* (Drawing by Hirzel)

Site, Site Conditions, Characteristics

Sitka spruce flourishes in very moist, overcast marine climates where the temperatures in January average around 0 °C, and in July around 10 °C. Annual rainfall is high (1500–3500 mm). The species grows best in moderately drained, acidic soils with a thick cover of raw humus. It forms stands together with *Tsuga heterophylla*. In the north, the stands may be mixed with *Picea engelmannii*, on the coast with *Pinus contorta*, and in the south with *Pseudotsuga menziesii*. In northern natural stands some trees can grow to be 800 years old with stem diameters of 2.5 m. Strong winds frequently destroy the stands. Red rot (*Fomes piniocola*) reduces the stability of the stands. Many insect species epidemically threaten Sitka spruce.

Distribution

Horizontal: distributed along the northwest coast of North America.

Vertical: in northwestern Alaska from sea level up to 300 m in southern Alaska and in Washington State up to an elevation of 750 m.

Fig. 364. Distribution of *Picea sitchensis*

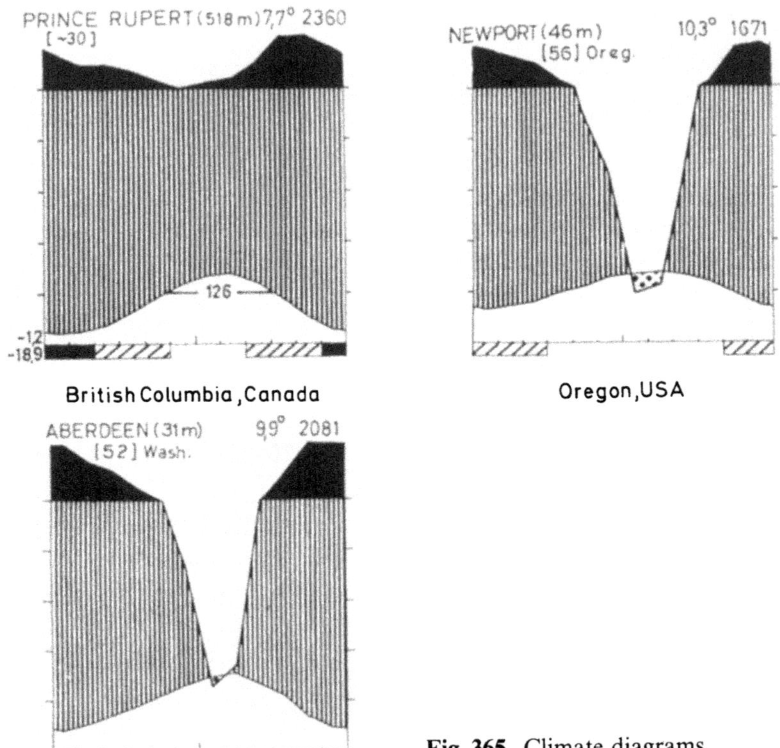

Fig. 365. Climate diagrams

Wood

Heartwood is absent. The outer zones are light yellow and those inside yellow-red. The growth rings are quite distinct, and the density is about 400 kg/m^3.

Fig. 366. A wood sculpture from a huge Sitka spruce trunk in Valdez, Alaska

Fig. 367. A Sitka spruce, old growth forest with a large quantity of dead wood and new growth

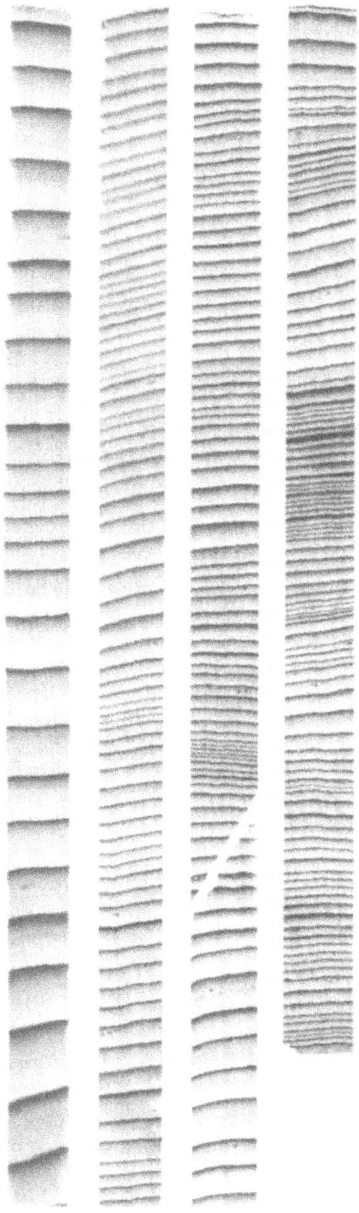

Fig. 368. Radiographs of *Picea sitchensis*, British Columbia

Only through the red coloring can a distinction be made between the wood of *Picea sitchensis* and that of *P. glauca*. It is impossible to distinguish between the two species on the basis of their wood anatomy. For further description, see *Picea engelmannii*.

Use in Dendrochronology

Suitability of the Species

Like all conifers in the boreal zone and in marine climates, Sitka spruce is best suited for radiodensitometric studies. Latewood densities are good indicators of late summer temperatures. In the highly tectonic coastal regions of Oregon and Alaska, the spruce lends itself well to the reconstruction of the ecological history of the site, (ground swelling, ground sinkage, relative changes in sea level, tsunamis, earth quakes, volcanic activity) Kaiser and Kaiser-Bernhard (1987).

Previous Uses

Until recently, only very few studies have included *Picea sitchensis*. Radiodensitometrical analysis of trees from British Columbia show that summer temperatures are reflected by the maximum latewood densities. From pointer years in the tree-ring sequences of *Picea sitchensis* in Kodiak, Alaska, Kaiser and Kaiser-Bernhard (1987) recognized damage caused by the volcanic eruption of Katmai in 1912 and tsunamis in 1964. He attributes the continuous growth reduction to tectonic sinkage, which raises the ground water level.

Pinus aristata Engelm.
Pinus longaeva D.K. Bailey
Pinus balfouriana Grev. et Balf.

Bristlecone pine
Bristlecone pine
Foxtail pine

Morphology

These three species are taxonomically very similar. They were earlier combined within the collective species *Pinus aristata*. They have short, twisted or crooked trunks and irregular crowns with spreading, partially dead branches. The stems have diameters of 2 m. *Pinus longaeva* has a maximum age of 4600 years. *Pinus aristata* can grow to be 2400 years old. It is possible for *Pinus balfouriana* to grow as old as *Pinus longaeva*.

Fig. 369. *Pinus longaeva.* (Drawing by Hirzel)

Site, Site Conditions, Characteristics

The three species grow in regions with extremely severe climatic conditions. At the timberline, the minimum temperatures can drop to as low as −45 °C. Precipitation in *Pinus aristata* stands in the Front Range is measured at 500–800 mm, while the two other species in the rain-protected areas of the

Fig. 370. *Pinus aristata* on a southern slope in the Front Range, Colorado at an elevation of around 3400 m. The tree is about 700 years old

Sierra Nevada receive annually only 30–250 mm of precipitation most of it as snow. Snow can remain on the southern slopes of the Front Range for over 6 months. The stands of *Pinus balfouriana* and *Pinus longaeva* in the Sierra Nevada are very dry in the summer. This dryness results in sparse ground vegetation. Daily temperature variations are extreme. The low night temperatures may limit the vegetation period to 2–3 months.

The three species grow especially on southern, exposed slopes and ridges at the timberline, at an elevation of 3500–3800 m. They generally form very scattered pure stands. *Pinus aristata* usually associates with *Picea engelmannii*

Fig. 371. A wind-eroded *Pinus longaeva*, dead for centuries. Located in the White Mountains, California

Fig. 372. *Pinus longaeva* site in the White Mountains, California. Some of the trees on this site are over 4000 years old

Fig. 373. A sign at the border of the Bristlecone Pine Forest in the White Mountains, California. (Photo Häsler, WSL)

Fig. 374. A *Pinus longaeva* twig from the White Mountains in California. Characteristic are the varying needle lengths of the individual annual shoots

and *Pinus flexilis*, while *Pinus longaeva* and *Pinus balfouriana* grow together with *Pinus albicaulis*, *Pinus monophylla*, and *Pinus contorta*.

Distribution

Horizontal: the three species grow in different areas: *Pinus aristata* occurs in Front Range, Colorado, *Pinus longaeva* in the mountains east of the Sierra Nevada in California and Nevada, and *Pinus balfouriana* in the Sierra Nevada.

Vertical: all of the three species are found growing at the timberline at an elevation of 3200–3800 m.

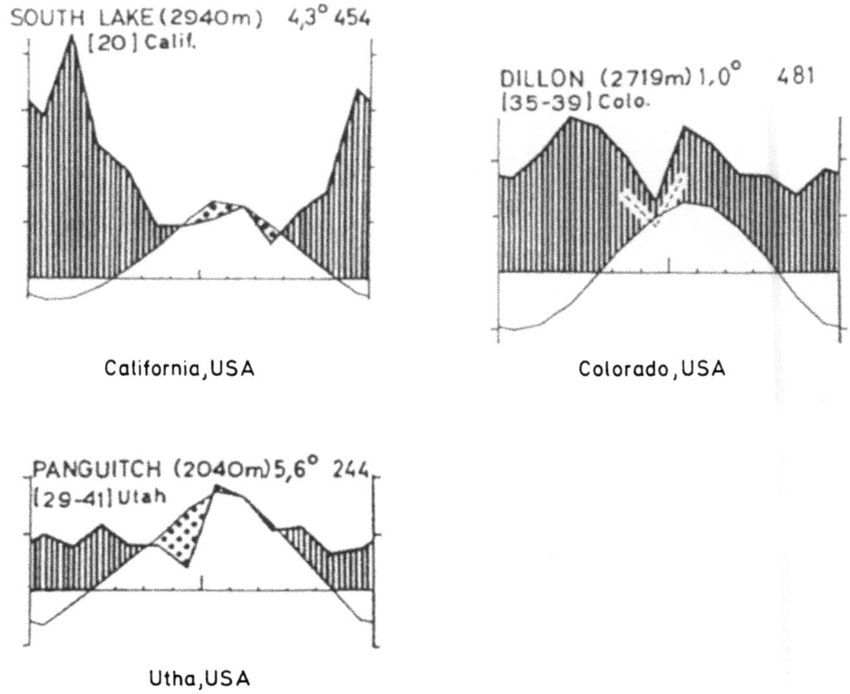

Fig. 375. Climate diagrams

Fig. 376. Distribution of *Pinus balfouriana*, *P. longaeva*, and *P. aristata*

Wood

The latewood is yellow. The heartwood is red to red-brown. The density is around 350 kg/m^3. The wood has a strong resin odor.

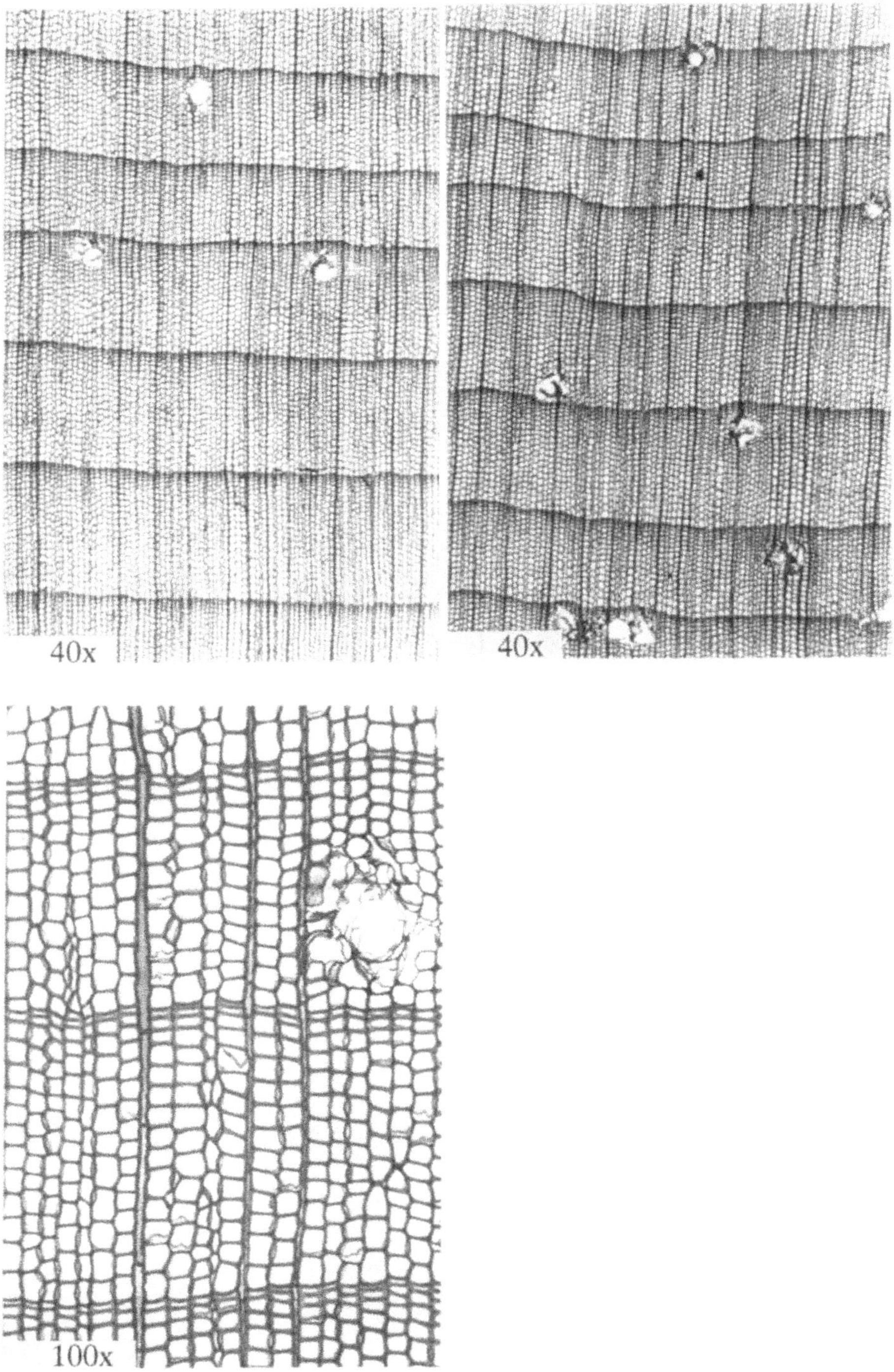

Fig. 377. a–c. Cross-sections of *Pinus aristata*

Fig. 378. *Pinus longaeva* at the timberline. The Salt Lakes are recognizable in the background

Fig. 379. *Pinus longaeva* sharing a site with *Pinus edulis*, in the White Mountains, California

Cross-section: the resin canals are bordered by thin-walled secretory cells. In general the tree rings are narrow with exceptionally narrow latewood. The transition from earlywood to latewood is continuous (Fig. 377a–c).

Tangential section: the average ray height is 5–15 cells.

Radial section: there are smooth-walled, transverse ray tracheids in the rays and one to two large pinoid pits per ray parenchyma in the cross-field.

It is impossible to differentiate between the *Pinus aristata* group and the pinyon group on the basis of their anatomic characteristics.

Use in Dendrochronology

Suitability of the Species

The forest border species are generally suitable for climatological analysis. Because summer dryness can in part stunt the growth, trees growing on southern slopes contain relatively little information useful in dendrochronological studies (LaMarche and Stockton 1974). The information base of the *Pinus aristata* area in western Colorado is especially low. Unfortunately, all five-needled pine (Haploxylon type) is equally poorly suited for radiodensitometrical research. The latewood is naturally invariable and the narrow tree rings are often unable to withstand the rigors of technical procedures. According to Alexander and McCord, (1987) *Pinus longaeva* is suitable for determining patterns of soil erosion over the centuries.

Previous Uses

Edmund Schulman started building bristlecone pine chronologies in 1954 (Schulman 1956). Ferguson and Graybill (1983) made a chronology available which today stretches over 8300 years. The species was mostly used for the standardization of the ^{14}C dating method. Recently, Graybill (pers. comm. produced a 1000-year chronology from the entire area of distribution of the three species. *Pinus balfouriana* has been climatologically studied by Scuderi (1984).

Pinus contorta Dougl. ex Loud.

Lodgepole pine

Morphology

Pinus contorta has a diverse habitus. On the coast the trees are small, contorted, and bushy while inland they grow tall and pole-like (lodgepole) with narrow, conical crowns. In the Sierra Nevada the trees have short and thick stems and large crowns. In avalanche zones or in windy regions the trees are dwarfed.

The species has the following varieties:

Pinus contorta var. *contorta*	shore lodgepole pine
P. contorta var. *latifolia*	inland lodgepole pine
P. contorta var. *murrayana*	Sierra lodgepole pine

Fig. 380. *Pinus contorta*. (Drawing by Hirzel)

Site, Site Conditions, Characteristics

The ecological amplitude of *Pinus contorta* corresponds to its wide distribution and genetic variability. The widely distributed inland form can be found in the

Fig. 381. Dwarfed coastal form in Oregon. (*Pinus contorta* var. *contorta*)

following climates. Best development occurs in regions with an annual precipitation level exceeding 600 mm. Pure stands can, however, be found in regions with only 500 mm precipitation. Winter precipitation occurs mainly as snow, and winter temperatures are very low, dropping to $-50\,^{\circ}C$ in deep winter. The species can withstand summer temperatures up to $40\,^{\circ}C$. The vegetation period is short, lasting only 60–100 days. *Pinus contorta* normally

Fig. 382. Up to 500-year-old trees growing at the subalpine forest border at an elevation of 2800 m in Colorado. Shown here is a *Pinus contorta* ssp. *murrayana-P. flexilis-Picea engelmannii* stand

Fig. 383. A dense new-growth stand on a former fire area. This type of stand is common. (*Pinus contorta* var. *latifolia*). (Fowells 1965)

grows in well-drained soils over practically all geological foundations. It can also be found in wet soils. The ecological spectrum is comparable to that of the *Pinus sylvestris* in Eurasia. After fires and clearings, the inland form forms extended pure stands. *Pinus contorta* associates with the following species:

At high altitudes: *Abies lasiocarpa, Pinus flexilis.*

In the north to the Yukon: *Picea glauca.*

On central sites: *Larix occidentalis, Pseudotsuga menziesii. Pinus ponderosa, Abies magnifica*, along with others.

On dry sites: *Pinus ponderosa, Pinus jeffreyi, Juniperus* species, along with others.

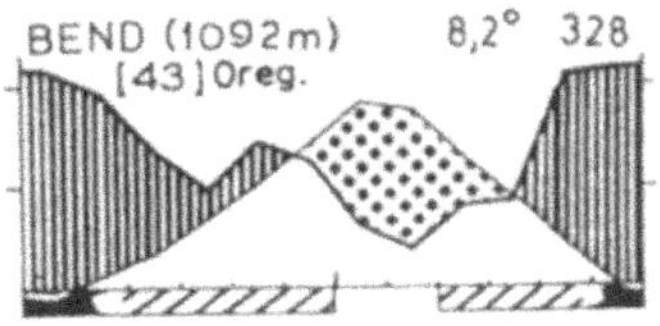

Oregon, USA

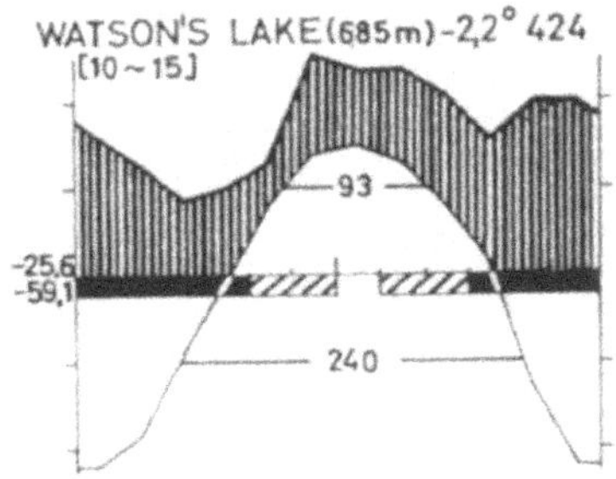

Yukon, Kanada

P. contorta ssp. *latifolia*

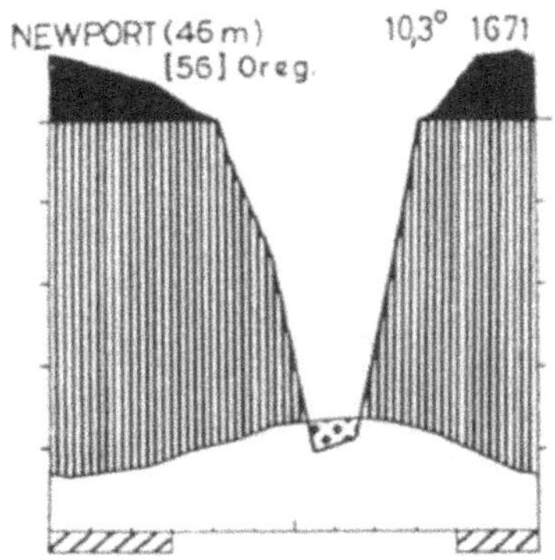

Oregon, USA

ssp. contorta

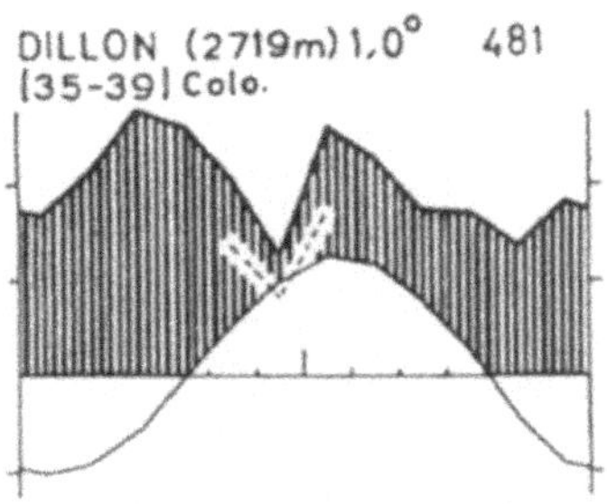

Colorado, USA

ssp. *murrayana*

Fig. 384. Climate diagrams

The oldest trees can be 600 years. Trees in pole stands rarely live longer than 100 years. Dense secondary stands are often decimated by *Dendroctonus* species (pine bar beetle) and dwarf *misteletoe* (*Arceutobium americanum*).

Distribution

Horizontal: in the Rocky Mountains from the 35° northern latitude to close to the Arctic Circle in the Yukon.

Vertical: in the South at high altitudes to the forest border at an elevation of 3200 m. In the west, the coastal form is found on foggy, low sites. The area of distribution is limited to the coastal regions of Oregon and Washington. In the north the inland form, sometimes forms the forest border together with *Picea glauca*.

Fig. 385. Distribution of *Pinus contorta*

Wood

The heartwood is light yellow to light brown. The sapwood is light yellow. There is often no difference between the heartwood and sapwood. The density is 400–450 kg/m^3. The wood has a distinct resin odor.

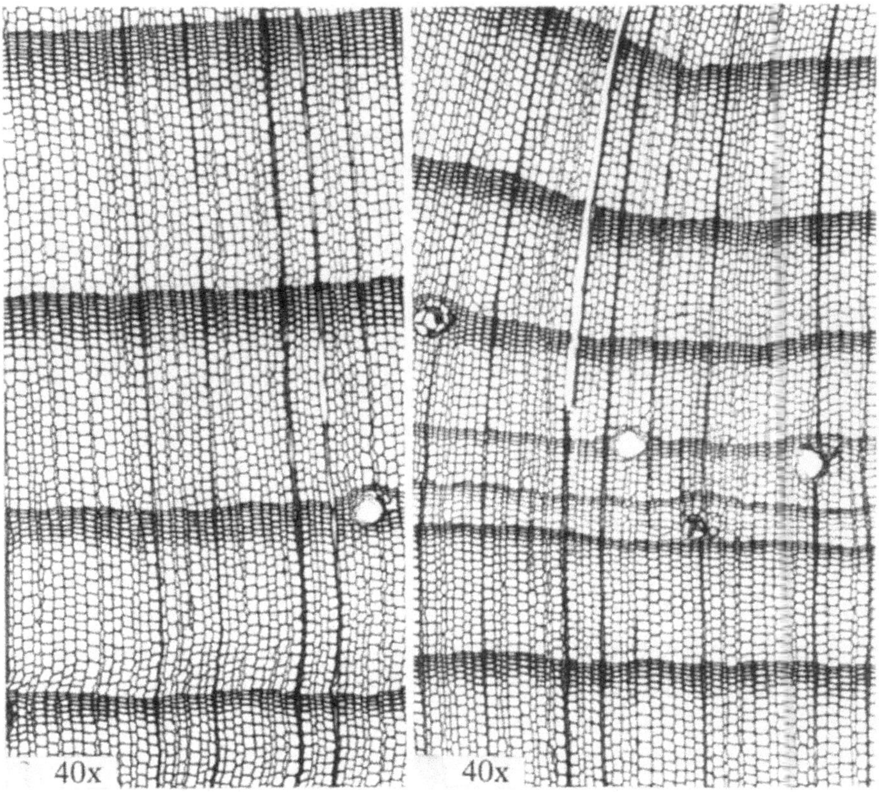

Fig. 386. a,b. Cross-sections of *Pinus contorta*

Fig. 387. *Pinus contorta* var. *latifolia* with a flag-like crown, shaped by the wind in the Shinouk region of the Colorado, Front Range

Fig. 388. *Pinus contorta* var. *murrayana* in Yosemite National Park. Note the fire scars on the base of the trunk

Pinus contorta cannot be separated from *P. ponderosa* and *P. jeffreyi* on the basis of its anatomic characteristics. As a rule, *Pinus contorta* has small resin canals (Fig. 386a,b).

The different species can be distinguished from one another on the basis of their extractives.

For an anatomic description, see *Pinus ponderosa.*

Use in Dendrochronology

Suitability of the Species

Pinus contorta is suited for dendroecological and climatological research. The long-living variety *murrayana*, from the Sierra Nevada, is especially suitable for radiodensitometrical analysis. The width of its latewood is variable. The relatively young stands in regions with large areas of forest damage from the 19th century lend themselves well to forestry studies. It is strange that the dendroecological potential of this variety has until now remained neglected.

Previous Uses

By radiodensitometric means, past summer temperatures can be reconstructed from the maximum density of trees growing at high altitudes. Fire histories have frequently been reconstructed using *Pinus contorta*, for example by Pruden et al. (1987) and Sheppard and Lassoie (1987). Using trunk injuries, Johnson (1987) determined the frequency of avalanche slides in the Canadian Rockies.

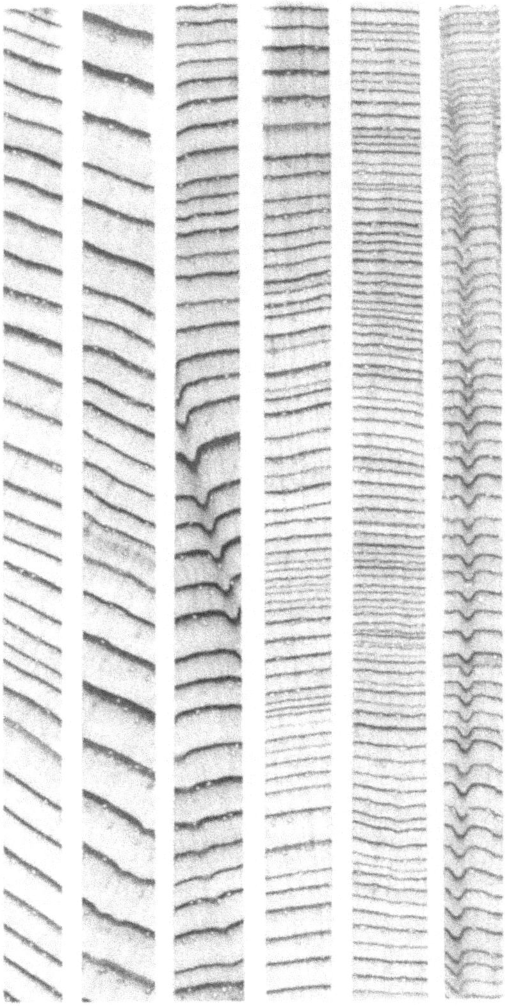

Fig. 389. Radiographs from *Pinus contorta* var. *murrayana*, Yosemite National Park

Pinus flexilis James

Limber Pine

Morphology

This conifer is small, growing up to only 15 m in height. It generally has an open crown and a short, often twisted stem and many branches. The oldest trees live for 2000 years.

Fig. 390. *Pinus flexilis.* (Drawing by Hirzel)

Site, Site Conditions, Characteristics

Pinus flexilis is a timberline tree from regions with limited cloud coverage. Like *Pinus aristata*, it can withstand extreme conditions. Within its range, the level of annual precipitation, mostly snow, is 400–800 mm. Limber pine can tolerate minimal temperatures in the winter of −45 °C and maximal summer temperatures of 35 °C. Like *Pinus aristata*, *P. flexilis* grows predominately on slopes

Fig. 391. A dwarfed tree at the windy forest border in the Colorado Front Range

and ridges over mostly granite foundations. Dwarf forms develop in windy regions. The stands are normally scattered and accompanied by *Pinus albicaulis*, *P. aristata*, *P. contorta*, *Picea engelmannii*, and *Tsuga mertensiana*.

Fig. 392. A *Pinus flexilis* stand at the subalpine level of Colorado at an elevation of 2800 m; mixed with *Pinus contorta* and *Picea engelmannii*

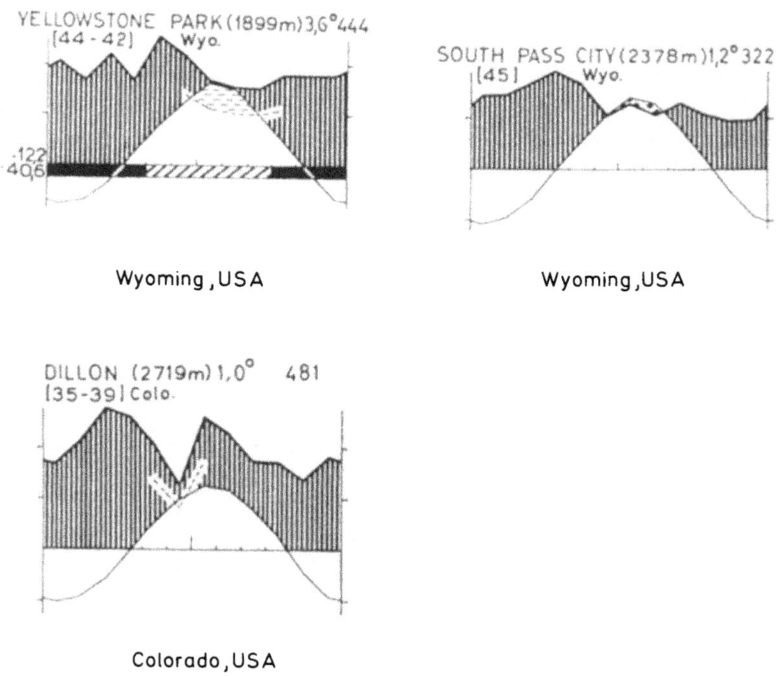

Fig. 393. Climate diagrams

Distribution

Horizontal: in many small pockets of distribution located in the central and eastern parts of the Rocky Mountains from 35° to 53 °N.

Vertical: at the timber line in the south up to 3600 m, in the north up to 2400 m. In Canada the distribution area is in lower regions and especially in dry valleys.

Fig. 394. Distribution of *Pinus flexilis*

Wood

Without heartwood. The density averages 350–400 kg/m^3. The wood has a strong, pleasant resin odor.

Cross-section: resin canals with large, thin-walled secretory cells. The transition from earlywood to latewood is continuous. The latewood cells are normally quite small (Fig. 395). Tree rings from trees growing on mountain sites are as a rule very narrow.

Tangential section: the average ray height is 5–15 cells.

Radial section: smooth-walled, transverse ray tracheids. One to two large pinoid pits per ray parenchyma cross-field.

Pinus flexilis cannot be distinguished anatomically from *P. albicaulis*, *P. monticola*, *P. strobus*, *P. peuce* and *P. cembra*.

For illustration see *Pinus cembra*.

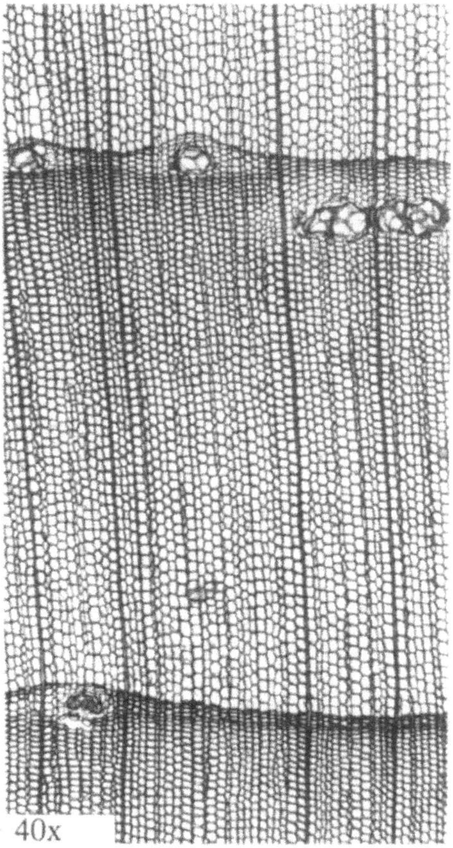

Fig. 395. Cross-section of *Pinus flexilis*

Use in Dendrochronology

Suitability of the Species

This species can live for 2000 years and its tree-ring series can be synchronized. The tree-ring sequences from trees growing on southern slopes at high altitude forest borders are generally uniform. In areas below the forest border, the sequences are very variable, yet often so narrow and irregular that dating is impossible. *Pinus flexilis* is not well suited for densitometric analyses because its maximum latewood thickness displays limited variability.

Previous Uses

A few limber pine chronologies can be found in the western American sample grid from Fritts (1976). Graybill (1987) developed a network of old *Pinus* species from timberline stands in southwestern USA, which included *Pinus flexilis.*

Pinus monophylla Torr et Frém.
Single leaf pinyon

Pinus edulis Engelm.
Pinyon (two needles per shoot)

Pinus cembroides Zucc.
Mexican pinyon

Pinus quadrifolia Parl. ex Sudw.
Parry pinyon (four needles per shoot)

Pinyon pine (the four species are combined here to the pinyon group).

Morphology

Pinyon is a small tree with a maximum height of 15 m. Its crown is made up of many branches and its root system either buries itself deep into the ground or spreads itself out over the surface. The trees have, as a rule, stems with diameters of less than 50 cm. The established maximum age is 973 years. Normally the trees reach ages of 200–300 years.

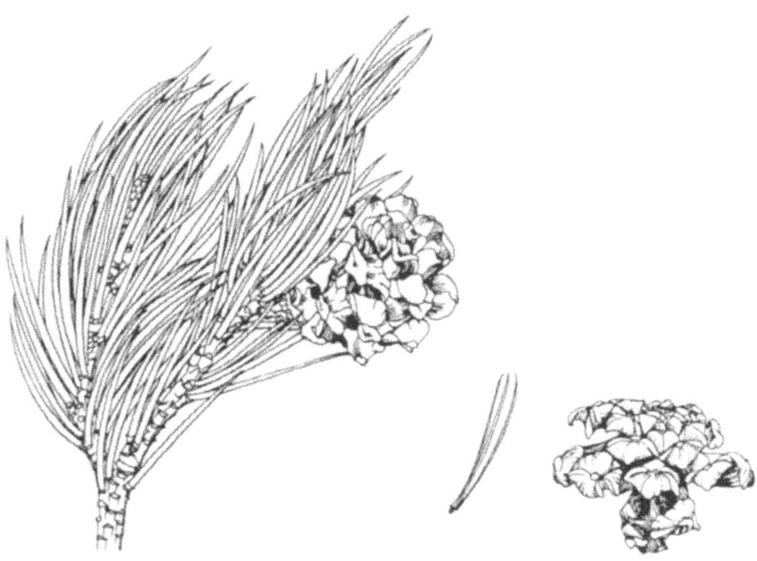

Fig. 396. *Pinus monophylla* and *Pinus cembroides.* (Drawing by Hirzel)

Site, Site Conditions, Characteristics

Pinyon is a species group from the semiarid climates of southwestern USA. Together with *Juniperus* species and especially with the shrub *Artemisia tridentata*, the trees in the pinyon group form the lower forest border. The stands

Fig. 397. *Pinus cembroides* on volcanic rock in Arizona

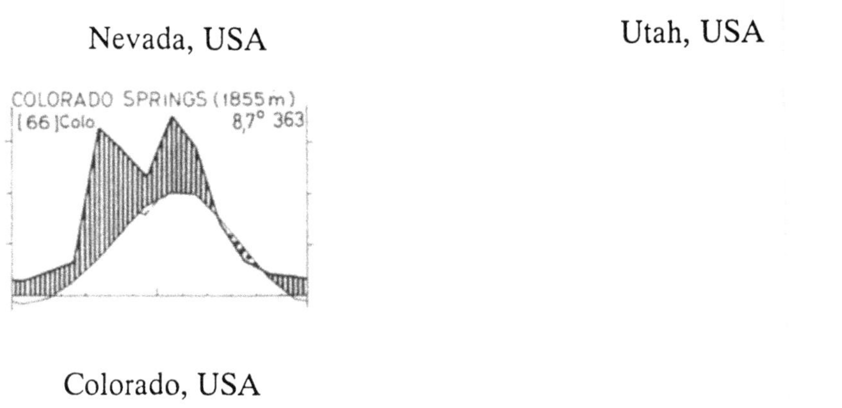

Fig. 398. Climate diagrams

are always scattered. Annual precipitation in the area is only 300–500 mm, with the majority falling as snow. The slight summer precipitation has little effect because it either fails to reach the soil, remains only on the soil surface, or evaporates. Temperatures in the winter fall to −15 °C and in the summer often climb higher than 40 °C. The daily temperature fluctuations are extreme: from 5 °C in the night, climbing up to 45 °C at midday. The vegetation period lasts from April to October. Tree growth occurs mainly during April and May. Pinyon is a very undemanding species, flourishing on all geological foundations, even on lava streams.

Distribution

Horizontal: mostly distributed through southwestern USA.

Vertical: dry forest borders, usually at an elevation of 1400–2000 m.

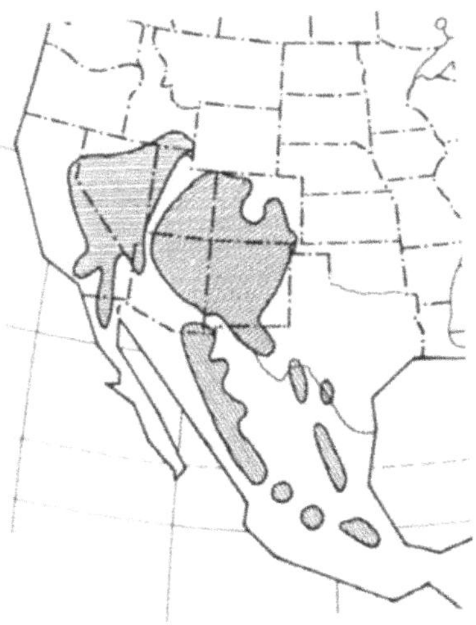

Fig. 399. Distribution of the pinyon group

Wood

This species group cannot be differentiated from the *Pinus aristata* group on the basis of its anatomic characteristics. See *Pinus aristata* for description.

Fig. 400a,b. Cross-sections from *Pinus monophylla*

Use in Dendrochronology

Suitability of the Species

Each of the four species grows at the arid forest border. The tree-ring sequences are generally very sensitive. Information on precipitation can be gathered from the ring widths. Due to the longevity of the four species, it is possible to use it in reconstructing past geomorphological processes. Dating is also feasible but at times rather problematic due to missing rings.

Previous Uses

About a fifth of all dendroclimatologically interpreted chronologies from southwestern USA are derived from these species. Fritts (1976) and Stockton (pers. comm.) have been able to reconstruct precipitation and temperature relationships for the last 400 years. Because many Native American settlements were situated on forest border sites, pinyon pine was often used as building material.

Dean (1969) was therefore able to date many settlements and settlement foundations. Using trees with exposed root systems, Alexander and McCord (1987) determined mean annual erosion levels.

Pinus ponderosa Dougl. ex Laws. *P. jeffreyi* Grev. et Balf.

Ponderosa pine
Jeffrey pine

Morphology

Pinus ponderosa and *Pinus jeffreyi* are morphologically very similar. The latter is distinguished from the ponderosa pine above all by its larger cones and thicker needles. Both species have tight, thin crowns when they are young and more open, crowns when they are older. On optimal sites, the trees can grow to be 70 m tall with stem diameters reaching up to 3 m. They can live for over 1000

Fig. 401. *Pinus ponderosa.* (Sudworth 1967)

years. The bark on older trees is thick. *Pinus ponderosa* is divided into different races, namely in a southern Arizona form and a coastal form. It hybridizes easily with *Pinus jeffreyi*.

Site, Site Conditions, Characteristics

Ponderosa pine and Jeffrey pine are species of the moderate, dry summer climates of southwestern America. The ecological amplitude correlates with the extensive area. The annual precipitation level varies from 300 to 2000 mm. In the summer there is practically no precipitation. The species can withstand extreme winter temperatures, down to $-40\,^{\circ}C$ and long periods of summer maxima of over $40\,^{\circ}C$. The species grow in soils with normal drainage on all geological foundations, even on volcanic lava. They form pure stands or are associated with a variety of species, such as:

In the north with *Larix occidentalis*, *Pseudotsuga menziesii*, and *Abies concolor*.

In dry areas with *Pinus lambertiana*, *Abies concolor*, *Calocedrus decurrens*, and *Pinus contorta*.

On the coast with *Pseudotsuga menziesii*, *Sequoia gigantea*, and various oaks.

At the lower, arid forest border with *Juniperus* and pinyon species.

With their thick bark, the older trees are protected against fire. A large number of insects damage the species, especially the *Dendroctonus* species (pine beetle). *Fomes annosus* (red rot) also attacks the trees. The most damaging pest is dwarf mistletoe, notably *Arceutobium vaginatum*.

Little (1979) recognizes only *Libocedrus decurrens* and rejects the new genus *Calocedrus* for incense cedar. However, there are indications in recent literature that *Calocedrus* might be justified.

Fig. 402. *Pinus jeffreyi* (left) and *P. ponderosa* stands (right) in northern California. The distribution pattern of the old and young trees is generally a result of fire influences. Usually, only the old, larger trees survive fires, while the younger trees are eliminated. (Fowells 1965)

Fig. 403. A so-called catface fire scar on an old ponderosa pine. Catface scars build the foundation for fire history reconstruction in southwestern, USA

Fig. 404. A site with two age groups

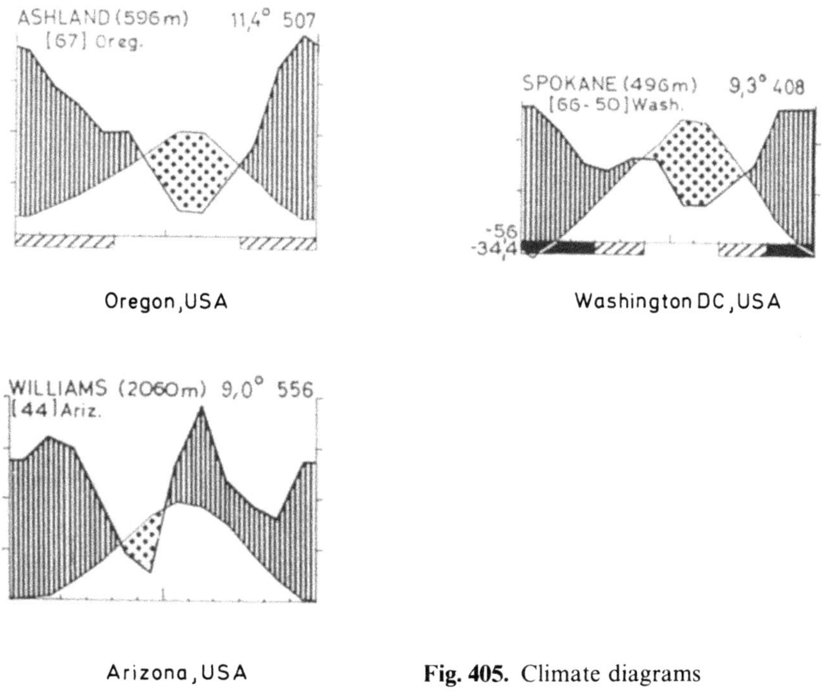

Fig. 405. Climate diagrams

Distribution

Horizontal: western North America, from Mexico at 24 °N to southern British Columbia at 55 °N. The species is not found in desert regions.

Vertical: from sea level in Washington up to 2700 m in California, Colorado, and Arizona.

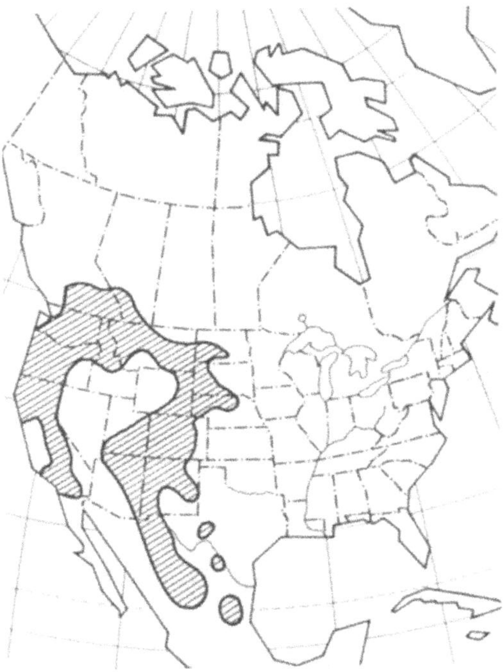

Fig. 406. Distribution of *Pinus ponderosa* and *Pinus jeffreyi*

Wood

The heartwood is red to red-brown, sapwood is yellow. The wood has a distinct resin odor and a density of 450–600 kg/m^3.

Cross-section: the resin canals are bordered by thin-walled epithelial cells. There is a distinct latewood zone, and the transition from earlywood to latewood is abrupt (Fig. 407a,b). Ring width and latewood widths are extremely variable. Density fluctuations are frequent.

Tangential section: the average ray height is 6–12 cells.

Radial section: the rays have ray tracheids. The transverse tracheid walls are tooth-shaped. Ray parenchyma cells have large, open pits. There are three to five pinoid pits per cross-field.

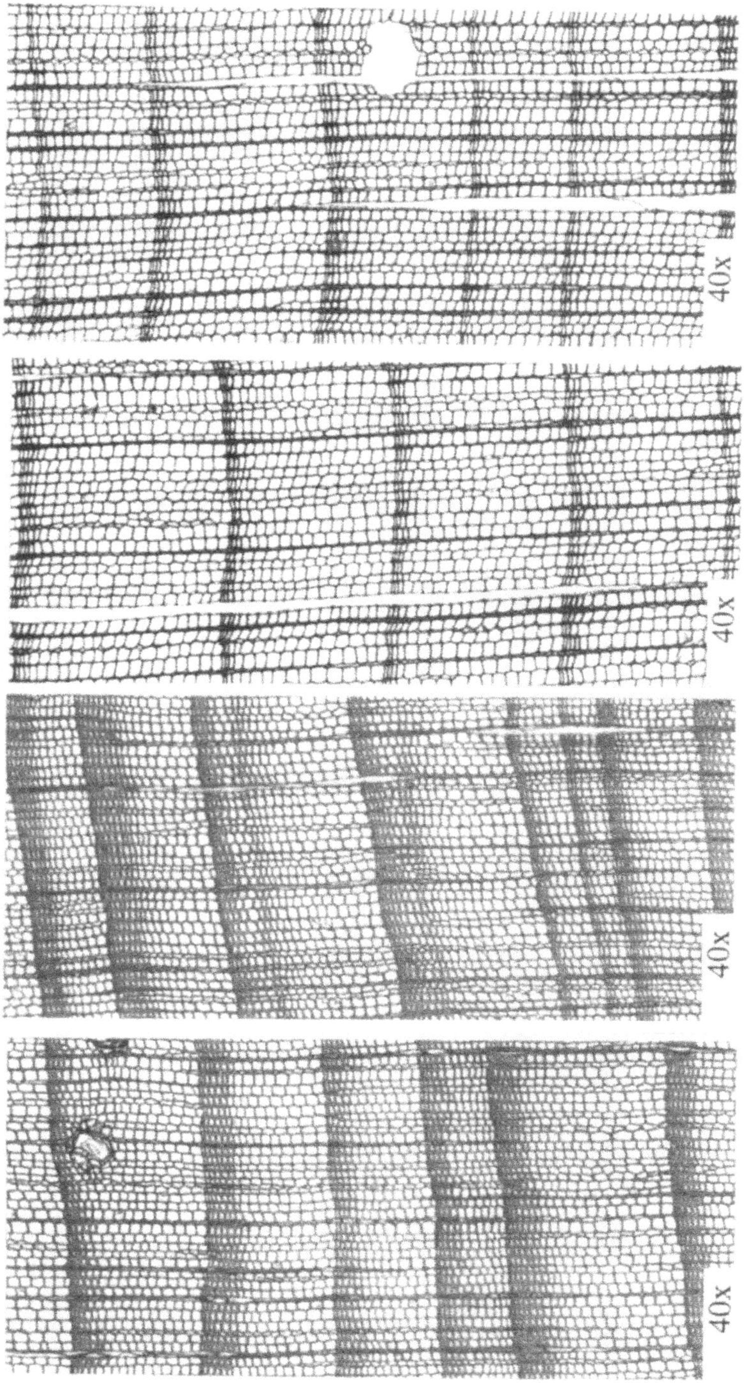

Fig. 407. Cross-sections of *Pinus ponderosa*

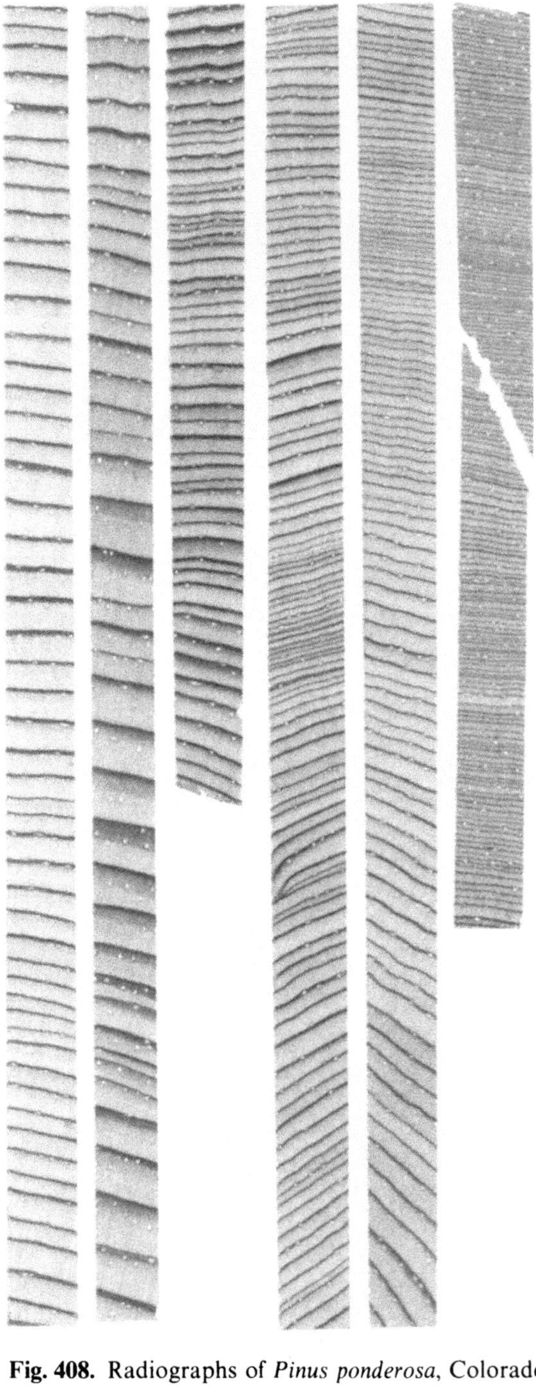

Fig. 408. Radiographs of *Pinus ponderosa*, Colorado

It is impossible to distinguish between *Pinus jeffreyi* and *P. ponderosa* on the basis of their anatomic characteristics. There is, however, a difference in the chemical composition of their oleoresin. Ponderosa pine contains monoterpenes, while Jeffrey pine contains heptane.

Use in Dendrochronology

Suitability of the Species

The two species are extremely well suited for dendrochronological research. They are widely distributed, settle in many different sites, develop well-formed stems, live for 700 years, and are frequently used as lumber. The species are mainly an indicator of past precipitation conditions. The tree-ring sequences of trees from dryer sites are very sensitive both with respect to ring width and latewood. Missing tree rings are, however, common.

Previous Uses

The first experiment with dendrochronology was made by A.E. Douglass with the *Pinus ponderosa* from Flagstaff in northern Arizona. A fifth of all dendroclimatologically used chronologies in the USA were obtained with *Pinus ponderosa*. Stockton (1976) succeeded in reconstructing the river water flow of the Colorado River over 300 years. Fritts (1965) used *Pinus ponderosa* for spot reconstructions of both temperature and precipitation in the USA.

In the majority of Native American settlements in western USA, it is possible to find dateable carbonized and uncarbonized wood. Most archeological datings at the Laboratory of Tree-Ring Research in Tucson, Arizona, involve *Pinus ponderosa*.

The forest fire history of western North America has been largely elucidated by dating of fire scars (Stokes and Dietrich 1980). Sheppard and Jacoby (1987) dated earthquakes along the St. Andreas Fault in California. Leavitt and Long (1987) demonstrated the possibility of reconstructing climate relationships by $^{12}C/^{13}C$ measurements. A relationship between tree ring widths and ozone concentration in city areas could not be clearly proven (Peterson et al. 1987).

Pinus strobus L.

Eastern white pine

Morphology

This is a large tree, up to 40 m high, with an irregularly shaped crown and a deep, widespreading root system. *Pinus strobus* usually grows much higher than its accompanying deciduous trees, such as *Acer saccharum* and *Betula*

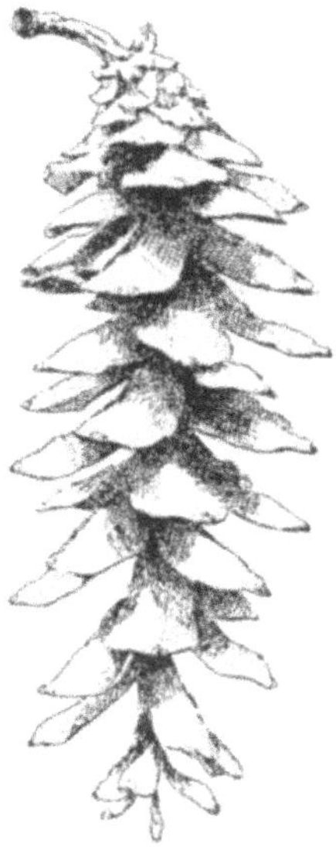

Fig. 409. *Pinus strobus*. (Hosie 1973)

Fig. 410. A large, solitary *Pinus strobus* on a deep soil site. The irregular sometimes flag-shaped arrangement of the branches is typical. (Hosie 1973)

alleghenensis. The stems commonly are without branches for the first 20 m. Old trunks can have diameters larger than 1.5 m.

Site, Site Conditions, Characteristics

Pinus strobus is a typical tree of the cool, moist climate of northeastern North America. The species grows on a great variety of sites. It can be found in shallow, sandy soil as well as in deep, nutrient-rich soil. The growing requirements for *Pinus strobus* are for the most part similar to those of *Acer saccharum*: 600–1500 mm annual precipitation, average January temperatures between −5 and −10 °C. The species seldom forms pure stands. It is usually associated with deciduous species such as *Acer saccharum*, *Betula alleghenensis*, and *Quercus rubra*. *Pinus strobus* has a maximal life expectancy of 300 years.

Fig. 411. Old *Pinus strobus* on a granite site in New York State.

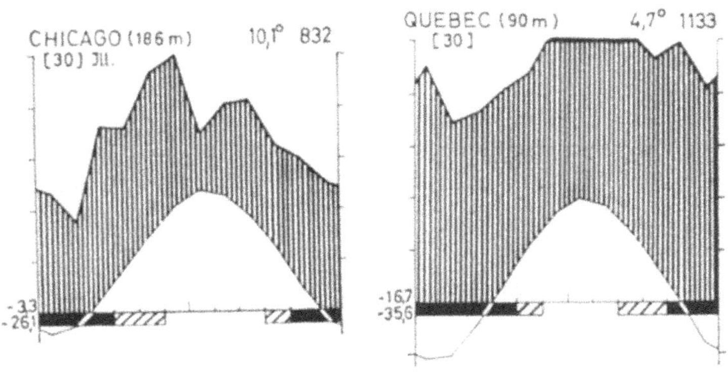

Fig. 412. Climate diagrams

Distribution

Horizontal: northeastern North America. The area is restricted between 50 °N and the South Carolina State border. In the west, the border is set approximately at the Mississippi. The distribution is strongly influenced by human use.

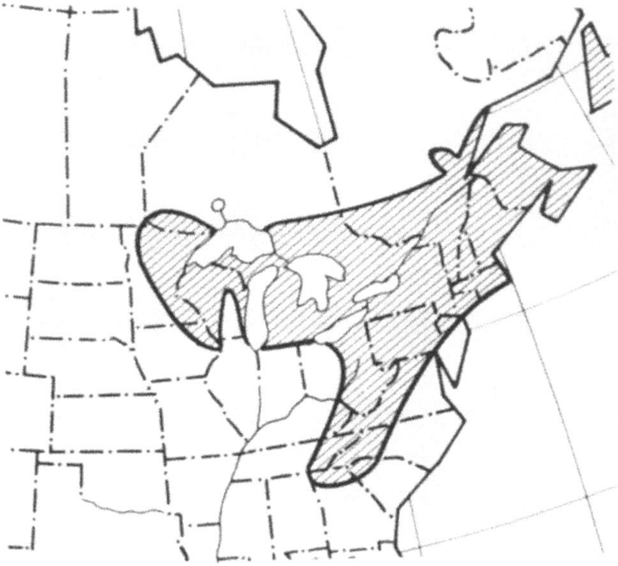

Fig. 413. Distribution of *Pinus strobus*

Vertical: in the north, distribution extends from sea level to 500 m. In the southern Appalachians, however, *Pinus strobus* can be found at elevations as high as 1200 m.

Wood

The heartwood is yellow-red to light brown. The sapwood is yellow-white. The density is 350–450 kg/m^3. The wood differs from that of *Pinus cembra* only in the width of its tree rings. Microscopically, *Pinus strobus* cannot be distinguished from *Pinus cembra*. See *Pinus cembra* for description.

Use in Dendrochronology

Suitability of the Species

Pinus strobus is suitable for climate reconstruction. The event years provide important information. Using fire scars it is possible to reconstruct the frequency of forest fires. The wood was used for building material by European settlers and thus can be used to date historic structures. During the Holocene, forests in the region of the St. Lawrence River were flooded, and the trunks were buried in sediments. It should be possible to compose floating chronologies. Because the latewood widths and densities remain relatively constant, the wood has only minimal use for densitometric purposes.

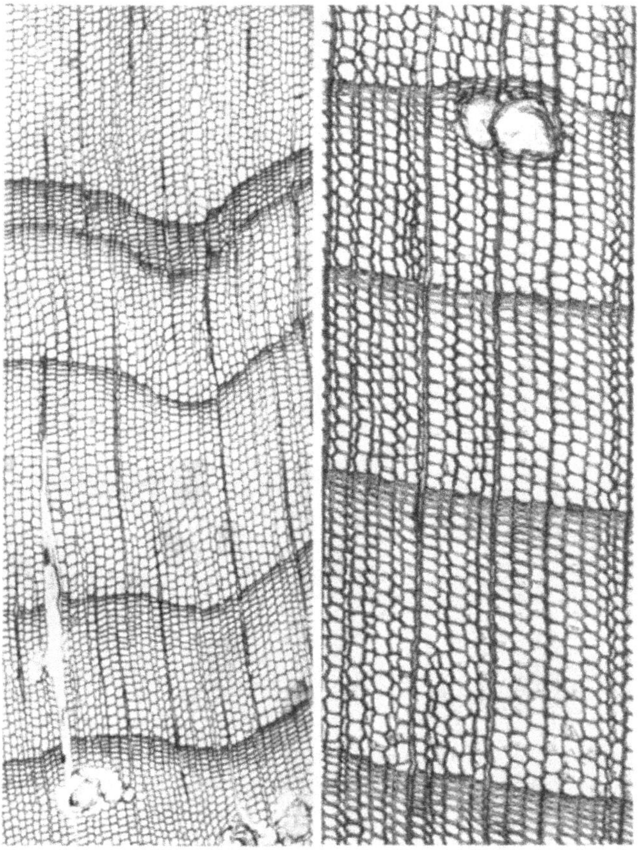

Fig. 414. Cross-sections of *Pinus strobus*. (Courtesy of P. Baas, Leiden)

Previous Uses

Lyon (1941) determined pointer years in several trees and their relation to lack of precipitation.

Pseudotsuga menziesii (Mirb.) Franco

Douglas-fir

Morphology

This large conifer grows rapidly on moist sites but slowly on dry sites. It has a thick bark and a deep root system. Old stands are dense, with trees with a

Fig. 415. *Pseudotsuga menziesii.* (Hess et al. 1967)

maximum stem thickness of 4 m and a height of 90 m. The trees remain branchless up to 35 m. In average old stands on the west coast, Douglas-fir often grows to be 60 m high and 1.5 m in diameter. On the coast with humid climates, the needles are normally green, in the dry climates of the Rocky Mountains they are different shades of blue.

Site, Site Conditions, Characteristics

Douglas-fir flourishes in temperate, moist climates and in generally deep, lightly acidic soils with normal drainage. The average temperatures in January are 1–4 °C; and in July are 15–18 °C. In extreme cases the trees can withstand minimal temperatures of −29 °C and maximum temperatures of over 40 °C. In the Rocky Mountains, Douglas-fir requires 82–187 frost-free days and on the coast 197–234 days. Annual precipitation is relatively high with 900 mm falling in the Rockies, and as much as 2500 mm on the coast. On southern sites, Douglas-fir associates with *Pinus ponderosa*, *P. lambertiana*, *Calocedrus decurrens*, and various oaks. In the north, the species is accompanied by *Tsuga heterophylla*, *Thuja plicata*, and *Picea sitchensis*. In primary stands it is not unusual for trees to live for 500 years, maximum age is over 1000 years. Western spruce budworm and tussock moth decimate the stands. The thick bark protects the older trees from fire. The pith of many trees, especially in the South, is attacked by *Fomes pinii*, making the trees susceptible to storm and wind damage.

Fig. 416. Douglas-fir in an area of high precipitation on the Oregon coast. (Fowells 1965)

Fig. 417. Douglas-fir on a semiarid site in northern Arizona. (Betatakin valley)

Fig. 418. A Colorado Douglas-fir with needles eaten by the Western spruce budworm

Fig. 419. An over 50-m-high Douglas-fi in its optimal growing region. (Oregon)

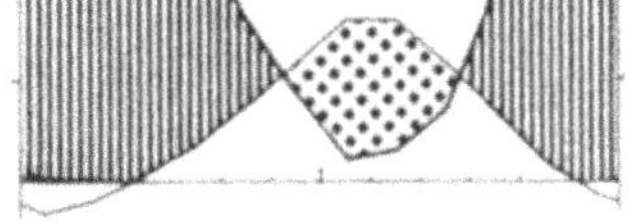

California, USA

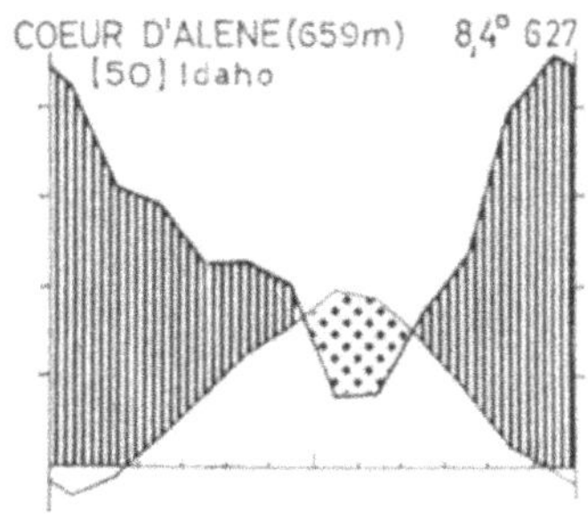

Idaho, USA

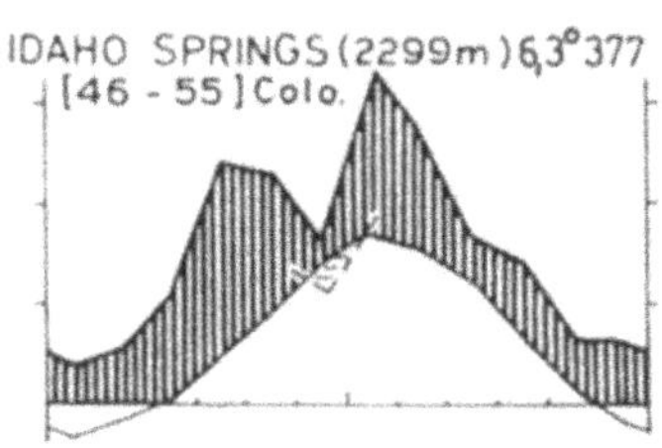

Colorado, USA

Fig. 420. Climate diagrams

Distribution

Horizontal: on the coast from 35° north, in the East from 23° to 60° north.

Vertical: in the South the species is restricted to high altitudes. In Arizona, the upper border occurs at an elevation of 2700 m, at 50° at an elevation of 1500 m and at the northern border at 800 m.

Fig. 421. Distribution of *Pseudotsuga menziesii*

Wood

The heartwood is red-yellow to orange red to redbrown. The sapwood is yellow. The wood has a fresh resin/grapefruit odor and a density of 450–550 kg/m^3.

Cross-section: resin canals, which are relatively sparse, are bordered by 7–12 thick-walled epithelial cells. The transition from earlywood to latewood is mostly sharp (Fig. 422c). Parenchyma is limited and its contents are colorless.

Tangential section: the average ray height is 18–15 cells. The rays are generally round.

Radial section: the rays have transverse ray tracheids. There are piceoid and taxodioid pits in the cross-fields of the earlywood. The transverse walls of the ray cells are thick and smooth; tangential walls are often denticulated. Thin spiral thickenings form a 10–15° angle to the vertical axis.

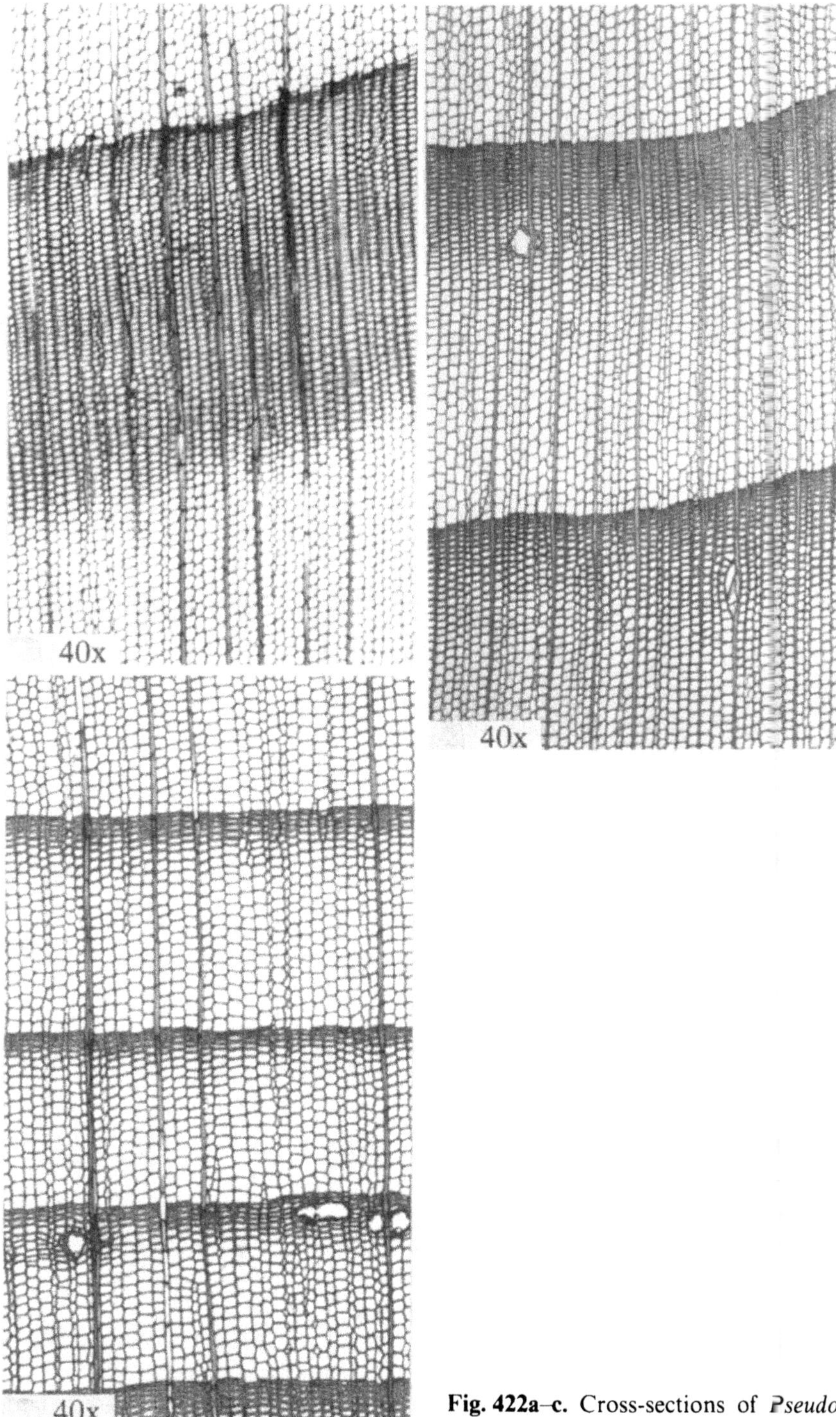

Fig. 422a–c. Cross-sections of *Pseudotsuga menziesii*

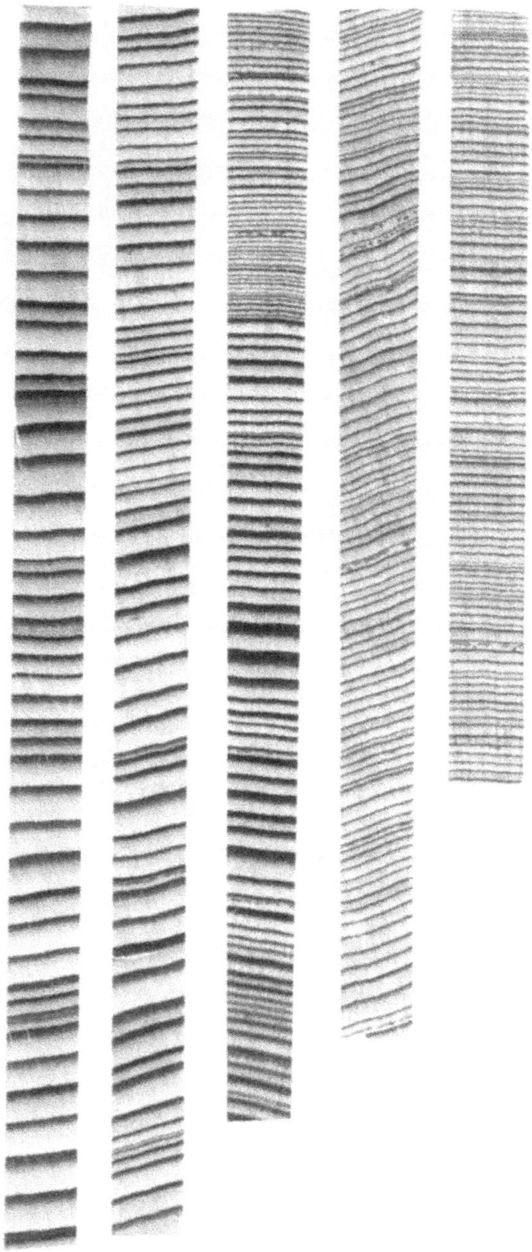

Fig. 423. Radiographs of *Pseudotsuga menziesii*, Arizona

Use in Dendrochronology

Suitability of the Species

This species lends itself exceptionally well to dendrochronological research. It has distinct tree-ring borders, a wide geographical and ecological range, grows quite old and has always been frequently used as lumber. Because Douglas-fir grows not only on moist sites but also on normal or dry sites, regional or zonal chronologies are necessary for dating and dendroclimatological analyses. The tree-ring sequences from trees growing on dry sites are very sensitive, yet tree rings, especially in narrow-ringed samples, are prone to be missing. A wide use in ecological studies is also possible.

Previous Uses

Around a third of all chronologies used in dendroclimatic research in western America involve Douglas-fir, and this is true for archeological research as well. Exact datings of many Native American settlements in arid southwestern America have been made with Douglas-fir and *Pinus ponderosa* wood found on the site. Repeated periods of spruce budworm infestation are shown in the tree-ring sequences of extensive chronologies from Swetnam (1987). The seemingly extreme situation in the eastern Rocky Mountains has been re-evaluated by Swetnam (1987). Wickman (1978) and McLean (1981) came to the conclusion that the presentday infestation is, in the long run, only a phase in the natural development process of the forests. Yamaguchi (1987) was able to show abrupt growth declines in old Douglas-fir in the region of Mt. St. Helens after an eruption in 1480. Through the use of radiodensitometric methods, Douglas-fir provides for the reconstruction of summer temperatures in the highest areas of its distribution. Like *Pinus ponderosa*, the fire-resistant Douglas-fir is an important source in fire history research. Brubaker (1987) incorporated Douglas-fir in forest dynamic studies.

***Sequoiadendron giganteum* (Lindl.) Buchh.**

Giant sequoia

Morphology

This tree grows up to 100 m high and with a diameter of 12 m. It has a long and narrow crown and a very deep root system. *Sequoiadendron giganteum* is recognized as the largest tree species on earth and the largest of all living things.

Fig. 424. *Sequoiadendron giganteum.* (Drawing by Hirzel)

Site, Site Conditions, Characteristics

Sequoiadendron giganteum grows in the mild regions of western America. Precipitation averages 100–1500 mm annually, with the majority falling as snow in the winter. Summers are quite dry. Temperatures in winter normally fall to around 5 to −10 °C and in summer rarely climb above 30 °C. The growing period lasts from June through September. The species normally flourishes in acid soils over a granite foundation. Early growth is mainly found in clearings. The species forms forests together with *Pinus lambertiana*, *P. ponderosa*, *Abies*

Fig. 425. The base of a giant sequoia trunk in Yosemite Park, California

concolor, *Calocedrus decurrens*, and various oaks. The giant sequoia, however, towers above the other species. The trees grow to be very old; the oldest giant sequoia has an established age of 3200 years. The majority of the ancient trees are today found in national parks. Thanks to their thick, soft bark, the old trees are protected from fire.

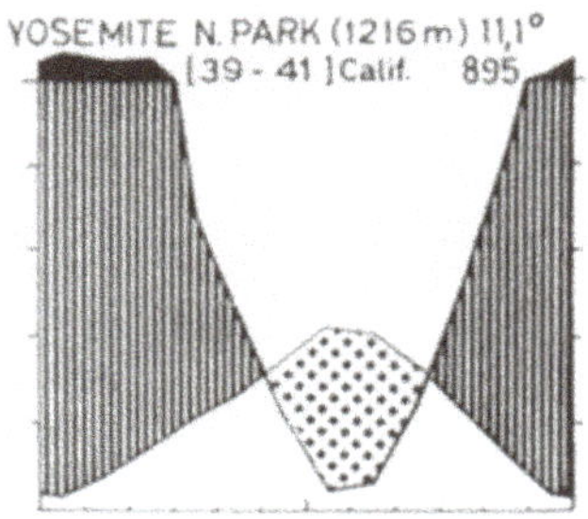

Fig. 426. Climate diagram, California, USA

Distribution

Horizontal: individual small areas in central California, Sierra Nevada.

Vertical: generally between 1500–2000 m. Sometimes *Sequoiadendron giganteum* can be found at 1000 m or at 2700 m.

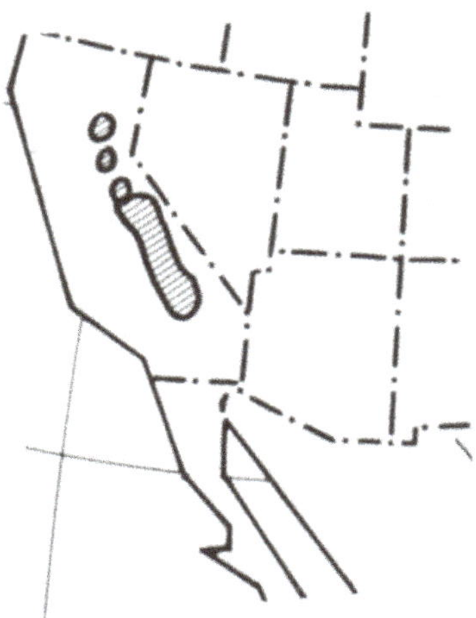

Fig. 427. Distribution of *Sequoiadendron giganteum*

Wood

The wood of *Sequoiadendron giganteum* is similar to that of *Sequoia sempervirens* although that of giant sequoia is softer and more brittle and of little commercial value. The heartwood of the giant sequoia is also darker, and the latewood zones are narrower. See *Sequoia sempervirens* for description.

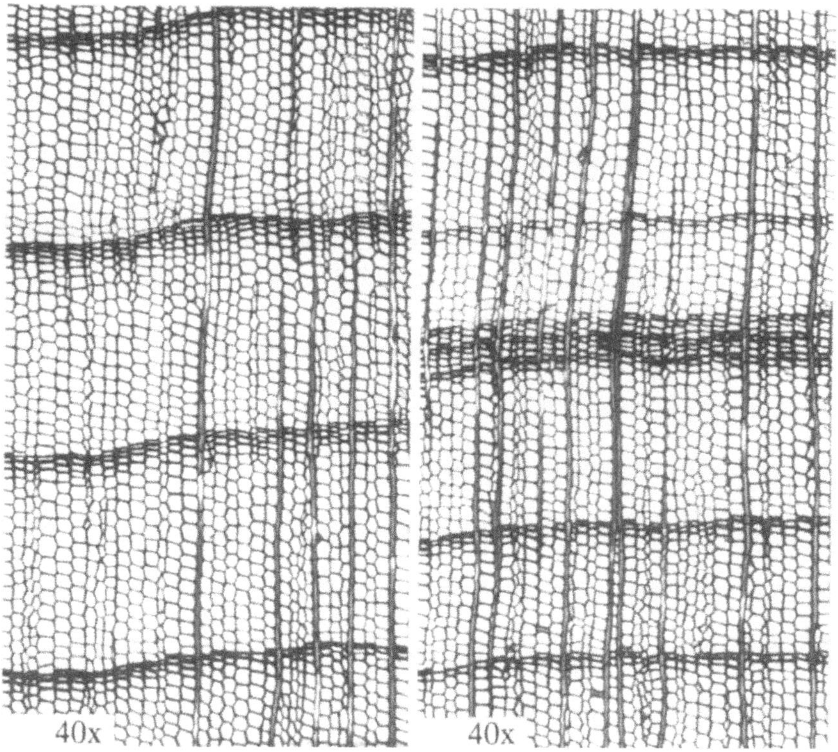

Fig. 428a,b. Cross-sections of *Sequoiadendron giganteum*

Use in Dendrochronology

Suitability of the Species

This ancient tree is limited to the relatively wet climate of the Sierra Nevada. The tree-ring sequences are generally not variable. Synchronizations are, however, possible through the use of pointer years. It remains unclear whether radiodensitometric analysis would be more beneficial for dendroclimatological research than measuring the tree-ring widths.

Previous Uses

The species was examined in the early years of the 20th century, by Douglass, but was never again used for dendrochronological purposes. LaMarche and Wallace (1972) used varying tree widths to date earthquakes in the St. Andreas Fault region. The Tree-Ring Laboratory in Tucson, Arizona, uses tree-ring sequences from old stumps to reconstruct extreme weather events and leveling forest fires (Hughes 1989).

Sequoia sempervirens (D. Don.) Endl.

Redwood

Morphology

This massive tree grows over 100 m high and has a stem diameter of 6 m. It is the tallest of all trees. The root system spreads over the surface.

Fig. 429. *Sequoia sempervirens.* (Drawing by Hirzel)

Site, Site Conditions, Characteristics

Sequoia sempervirens is a tree from extremely humid, Pacific regions. Average regional temperatures are between 10 and 15 °C. Winter temperatures rarely drop below −5 °C and temperatures in the summer rarely rise above 30 °C. The frost-free period lasts for 6–11 months. Precipitation is high, varying between 700 and 3100 mm, most of it in the winter. In the summer, the sites are often covered with clouds. The species grows best where it can be reached by fog drifting inland from the Pacific. Redwood grows generally in acid soils over the mesozoic marine sandstone. On prime sites, *Sequoia sempervirens* forms its own forest, apart from other species. On other sites it is accompanied by *Pseudotsuga menziesii.* Redwood trees have a long life span, up to 2000 years. Frequent forest fires cause damage to the basal part of the stem.

Fig. 430. An old redwood site on the Pacific coast. (Fowells 1965)

Fig. 431. Redwood trees on the foggy Oregon coast

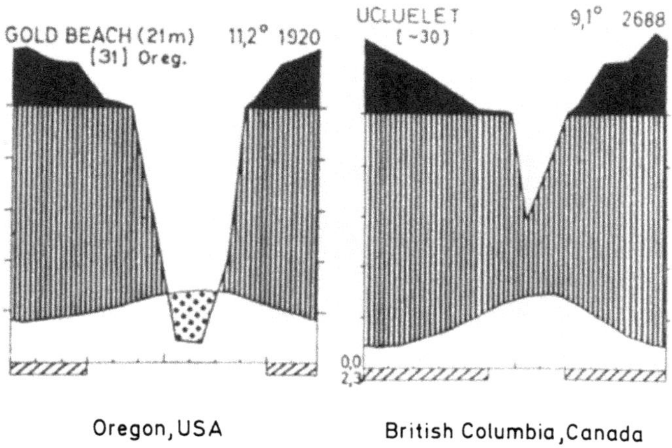

Fig. 432. Climate diagrams

Distribution

Horizontal: on the Pacific coast between 35° and 42 °N.

Vertical: From sea level to a maximum elevation of 900 m, usually on western slopes.

Fig. 433. Distribution of *Sequoia sempervirens*

Wood

The heartwood is red to red-dark brown. The sapwood is white. The wood has a density of around 400 kg/m^3.

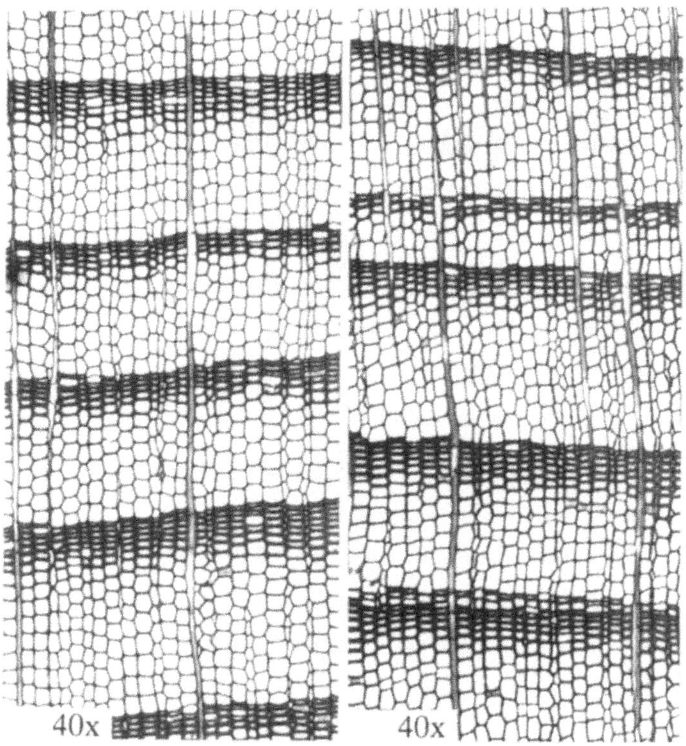

Fig. 434a,b. Cross-sections of *Sequoia sempervirens*

Cross-section: there are no resin canals. The transition from earlywood to latewood is abrupt (Fig. 434a,b). The amount of latewood is generally quite small. The parenchyma cells contain dark extractives.

Tangential section: the average ray height is 10–15 cells, sometimes up to 40 cells. The horizontal walls of the axial parenchyma are smooth.

Radial section: the rays are without tracheids. Large, open pits in the cross-field (taxodioid). Rather thick horizontal ray walls. The tangential walls are smooth and thin. Bordered pits on the tracheid walls are often in double rows.

Use in Dendrochronology

Suitability of the Species

These long-lived (up to 2200 years) trees have irregularly formed trunks, resulting in irregular tree-ring sequences. In addition, the series are quite regular. Synchronization, therefore, is practically impossible. Climatological information from trees growing in the extremely humid Pacific coast regions is, at the most, minimal. Radiodensitometric research could, however, be of some value.

Previous Uses

The species has practically never been used.

Taxodium distichum (L.) Rich.

Baldycypress, swamp cypress

Morphology

This species has sparsely set, widespreading branches and an extremely conical trunk, especially trees with raised roots. The trees have a maximum age of 1600 years (as far as it is possible to determine from the irregular growth rings), a diameter of 2 m, and a height of 40–50 m.

Site, Site Conditions, Characteristics

Swamp cypress flourishes in the warm, humid climate of southeastern USA. For a short period of time, it can withstand winter temperatures on the northern border of its natural distribution area down to −15 °C. *Taxodium distichum* is limited to level, very wet to overflooded, lightly acidic, and lightly alkaline soils. On these sites the species forms pure stands. On dry sites it is accompanied by *Chamaecyparis thyoides*, *Pinus elliottii*, *Nyssa aquatica*, *Ulmus americana*, various types of oak, and *Diospyros virginiana*. *Taxodium* grows to be over 1600 years old. *Taxodium distichum* is mainly attacked by trunk-core-decomposing fungi (Peckiness).

Fig. 435. *Taxodium distichum*. (Drawing by Sebek)

Fig. 436. *Taxodium* in an overflooded Florida bog. (Clark 1982)

Fig. 437. *Taxodium distichum* with large, board-like roots. (Krüssmann 1972)

Fig. 438. A stem base with board-like roots

Fig. 439. An extremely old *Taxodium distichum* tree at Ebeneedzer Creek, Georgia. Despite the obviously damaged crown, the tree-ring sequences are suitable for dendrochronological studies. (Photo Cleavland)

Distribution

Horizontal: in southern and southeastern USA.

Vertical: Practically only at elevations from 0 to 200 m.

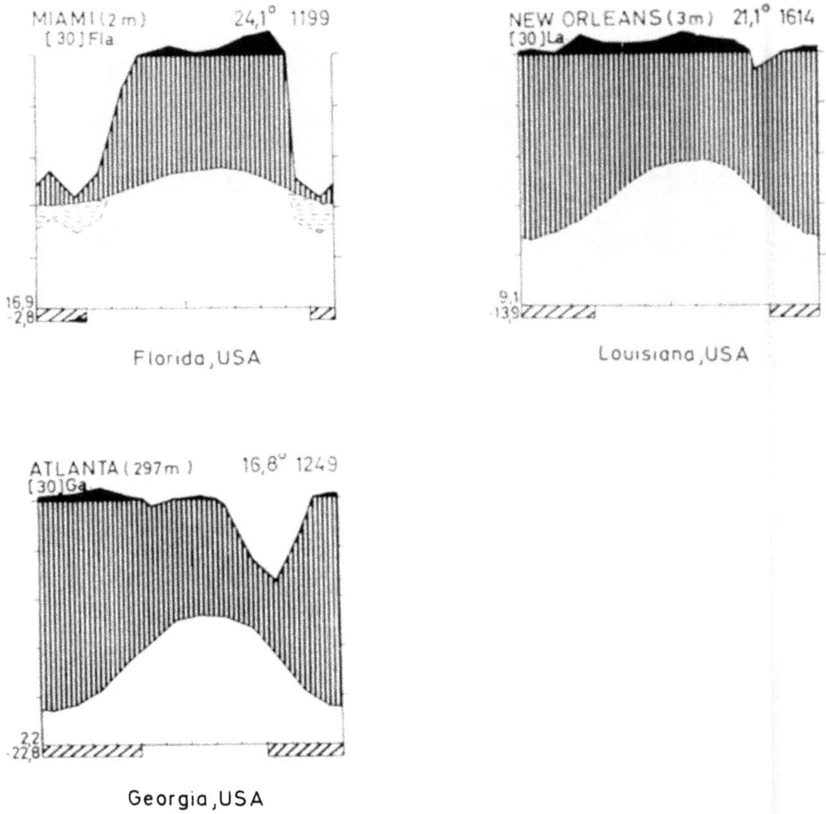

Fig. 440. Climate diagrams

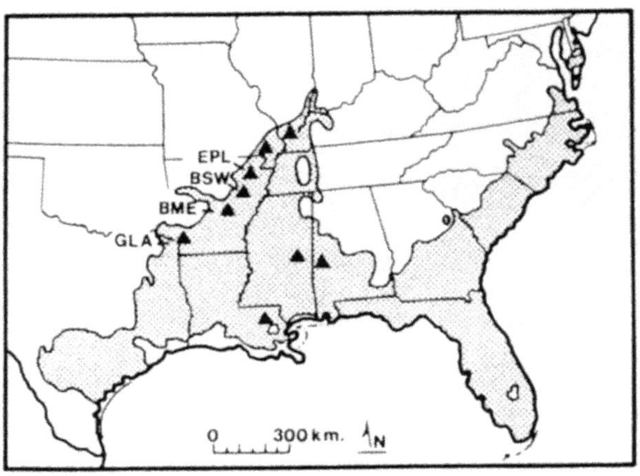

Fig. 441. The distribution area of *Taxodium distichum* in southeastern America and specific sites from which extensive chronologies have been made. (Stahle et al. 1985)

Wood

The heartwood is extremely variable, yellowish to redbrown and often spotted; the sapwood is bright yellow. The wood has a density of 450–500 kg/m^3.

Cross-section: heartwood is absent. The transition from earlywood to latewood is abrupt (Fig. 442b–d). The tree-ring boundaries are frequently unclear. There are generally many density fluctuations and false tree rings (Fig. 442c). The parenchyma cells contain colored substances.

Tangential section: average ray height is 15–30 cells, occasionally up to 60 cells. The cross-walls of the axial parenchyma cells have nodules.

Radial section: the rays are without ray tracheids. Pits in the cross-field have large openings (cupressoid to taxodioid). Both walls are smooth. Opposing bordered pits are frequent in the longitudinal tracheids. The horizontal ray walls are relatively thick; tangential walls are thin.

Use in Dendrochronology

Suitability of the Species

This species can be used for dendrochronological purposes. It is the only long-living species in the subtropical regions of southeastern North America from which yearly climatological information can be obtained. According to Stahle et al. (1985), it is possible to find well-preserved trunks and stumps from the last glacial period to the first Christian century (10 570 B.C. to 725 A.D.). It is possible to synchronize the individual stems with recent chronologies. Hopefully, in time, this material will be used to create chronologies spanning thousands of years. The stems from the early Quaternary described by Attolini et al. (1988) in Apulia, Italy, provide climatological information on an annual basis. Because the species from the early Quaternary has become extinct in Europe, longer chronologies cannot be obtained.

Dating is, in many respects, problematic. The trees from subtropical areas frequently have growth zones which do not conform to annual periods; false rings develop, especially in very old, slow-growing trees. Irregular tree-ring sequences in an irregular, board-like stem base also make dating difficult. According to Stahle et al. (1985), however, dating is still possible through the use of the skeleton-plot method.

Previous Use

According to Stahle et al. (1985), the oldest trees of this species are over 1600 years. By using live trees, the authors have made available chronologies from Arkansas, Mississippi, Louisiana, and North Carolina that span over 500 years. The longest of these chronologies reach back to 365 A.D., and from 670 A.D. four trees have already been integrated. According to Blasing et al. (1988), lack of precipitation in the months of April and May restrict radial growth.

Duever and McCollon (1987) have suggested a number of dendroecological possibilities.

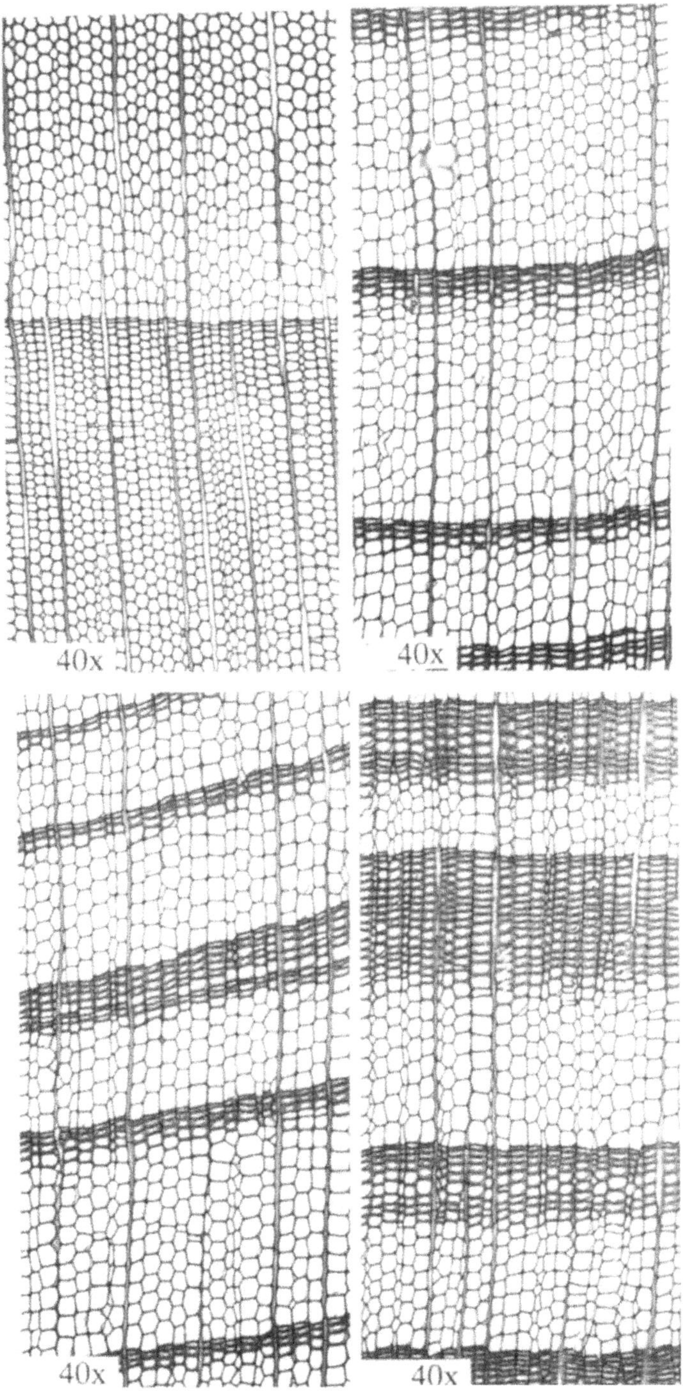

Fig. 442a–d. Cross-sections of *Taxodium distichum*

Fig. 443. A section from a tree-ring sequence with irregular growth. Of particular interest here are the wedging rings. Example from Florida

The age and height of a tree are related to water content levels and peat thickness.

The age and height of a tree are correlated with site conditions. *Taxodium* grows poorly in clay soils containing limestone.

When salt water enters fresh water zones, the ring widths in *Taxodium* increase.

During dry periods, i.e., periods of low water levels, radial growth is minimal.

Fluctuations in water level due to flooding appear in the tree-ring sequence.

Thuja occidentalis L.

Northern white cedar

Morphology

This shallow-rooted tree has a height of 10 m to a maximum of 20 m. The trunks rarely have diameters larger than 50 cm. The crowns are pyramid-shaped. The basal part of very old trees is often prostrate. Older stems are commonly malformed.

Fig. 444. *Thuja occidentalis.* (Krüssmann 1972)

Site, Site Conditions, Characteristics

Thuja occidentalis is a tree of cool, moist climate, growing mainly on moist and wet sites. Only in regions with high humidity can it grow in shallow soil. It

Fig. 445. *Thuja* on a lake shore in Quebec. *Thuja occidentalis* grows especially well along rivers and lakes

commonly forms the forest edge along rivers, lakes, and fjords. In its main area, precipitation averages 600–1200 mm annually. It can withstand January temperatures averaging −5 °C. The northern white cedar very rarely forms pure stands on dry site, instead it usually associates with *Picea mariana* and *Larix laricina* or with *Tsuga canadensis*, *Betula alleghenensis*, *Acer saccharum*, *Pinus strobus*, and other species. Normally the trees do not grow to be very old. On sites untouched by humans, the trees can live for 800 years.

Fig. 446. Thuja with a tilted stem on a river bank. The tops of trees bent over by erosion or storms eventually re-erect themselves. Prostrate trees continue to live, maintaining the same position sometimes for hundreds of years

Fig. 447. A typical *Thuja occidentalis* trunk. The bark loosens in long stripes and the base of the trunk is usually bent. (Hosie 1973)

Distribution

Horizontal: in northeastern North America and in the Appalachian Mountains. In the north, the species reaches to the southernmost point of the Hudson Bay

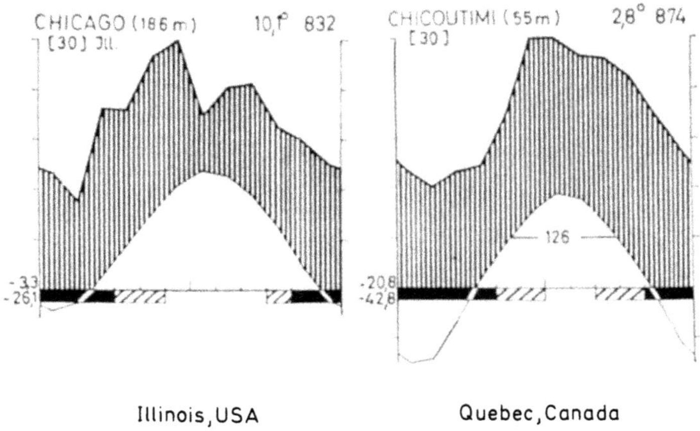

Fig. 448. Climate diagrams

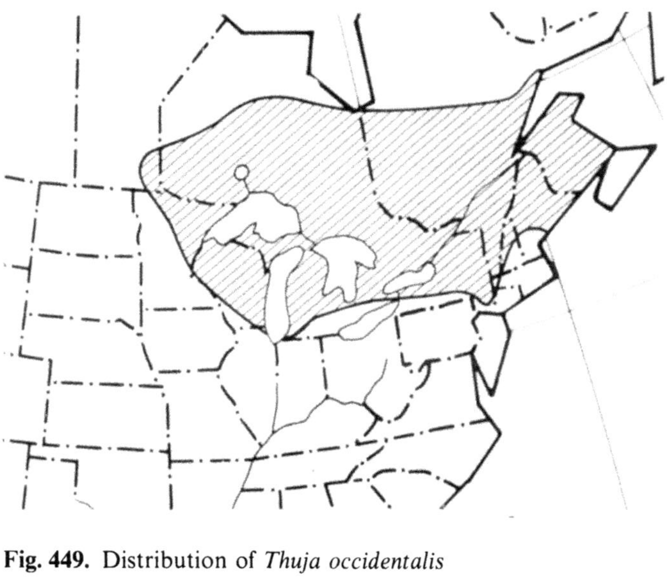

Fig. 449. Distribution of *Thuja occidentalis*

Fig. 450a, b. Cross-sections of *Thuja occidentalis*

and rarely extends further west than western border of Ontario. The southernmost trees grow in Pennsylvania.

Vertical: in the north, *Thuja occidentalis* is found exclusively at low altitudes. In the south, the area extends to the mountain zone.

Wood: the heartwood is red-brown, pink to purple, often with light stripes. The sapwood is yellow. The density is 350–400 kg/m^3. Microscopically *Thuja occidentalis* cannot be distinguished from *Thuja plicata*. See *Thuja plicata* for description.

Use in Dendrochronology

Suitability of the Species

As in all other Cupressaceae species, the cambium reacts strongly to environmental changes. Alterations in the stand or in the topography show up clearly in the tree-ring sequences. On stable sites, the tree-ring sequences are very even. Cross dating is only possible using the few event years. Only these years can ultimately be interpreted climatically. Because the latewood widths and densities vary only minimally, this species has limited suitability for densitometric purposes.

Previous Uses

Archambault (pers. comm.) found 800-year-old trees. The radial growth may have been restricted by low precipitation.

Thuja plicata Donn ex D. Don

Western redcedar

Morphology

This species normally grows to be quite large; up to 60 m high and 4.5 m in diameter. The trees can live for 1000 years. On good sites, the crowns are pyramid-shaped, beginning 20–30 m above ground. In contrast, on wet sites, the trees have narrow crowns with partially dead branches. *Thuja plicata* has a shallow root system.

Fig. 451. *Thuja plicata.* (Drawing by Hirze)

Fig. 452. A 500-year-old *Thuja plicata* in the rain forest of Washington. Typical of the species are the massive roots

Fig. 453. *Thuja plicata* in the superhumid climate of Vancouver Island. New seedlings grow on sites where older trunks have fallen

Fig. 454. Totem poles from the coastal region extending from Washington to Alaska are almost exclusively made from the stems of redcedar

Fig. 455. Practically all old stems in the eastern part of the distribution area have rotten cores

Site, Site Conditions, Characteristics

Thuja plicata grows in climatic regions similar to those of *Picea sitchensis*. It flourishes in mild, wet marine climates along the Pacific coast. Pure stands are often found in the foggy coastal region from Oregon to southern Alaska. For a more detailed climate description, see *Picea sitchensis*. *Thuja* grows best in deep, moderately moist, fairly acidic soils. On the coast it grows mainly on level terraces and in valleys along the banks of rivers. Only on extremely humid sites with the spray of Pacific salt water, does the species form pure stands. Along the coast, *Thuja plicata* generally associates with *Picea sitchensis*, *Pseudotsuga menziesii*, *Tsuga heterophylla*, *Abies grandis*, and poplars. Inland, in Idaho and Montana, *Tsuga* is more likely to be accompanied by *Larix occidentalis*, *Pinus contorta*, *Picea engelmannii*, and *Abies lasiocarpa*.

The life expectancy of the tree is limited by core rot.

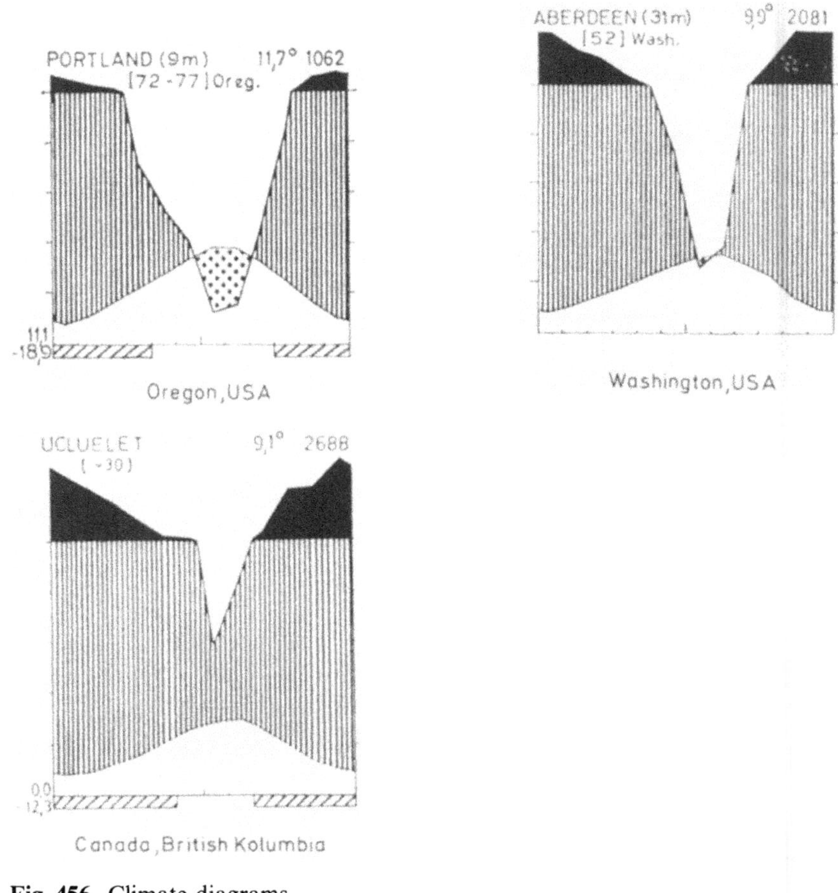

Fig. 456. Climate diagrams

Distribution

Horizontal: along the pacific coast from 40° to 57 °N and in Idaho and Montana from 46° to 54 °N.

Vertical: the northern area of distribution generally occurs at sea level. In Washington the species reaches 1200 m and occasionally as high as 2000 m.

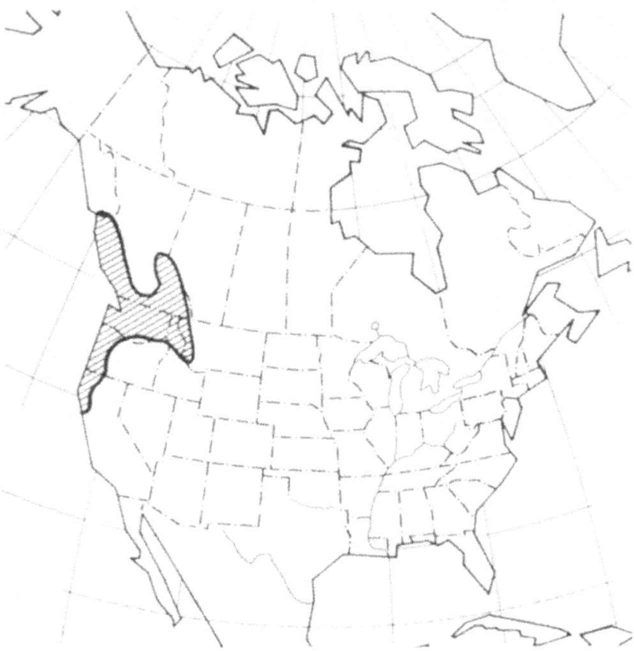

Fig. 457. Distribution of *Thuja plicata*

Wood

The heartwood is red to red-brown. The sapwood is yellow. Typical is a sweet cedar odor. The wood has a density of 350–400 kg/m^3. Stems are frequently lobed.

Cross-section: there are no resin canals. The parenchyma cells contain dark-colored substances. There is only a small amount of latewood. The transition from earlywood to latewood is continuous (Fig. 458a, b)

Tangential section: the average ray height is 6–12 cells. The cross-walls of the axial parenchyma cells have nodules.

Radial section: the rays are without ray tracheids. The pits in the cross-fields are cupressoid and have moderately slit-formed openings. The tangential cell walls of the rays are smooth.

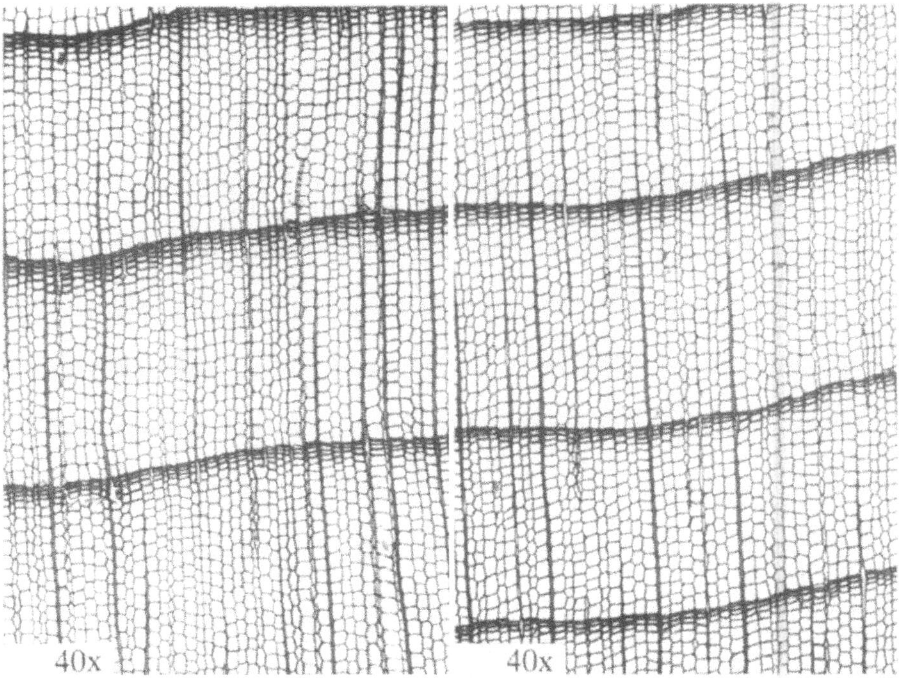

Fig. 458a,b. Cross-sections of *Thuja plicata*

It is possible to distinguish between *Thuja plicata* and *T. occidentalis* only on the basis of the color of the heartwood.

Thuja plicata	Heartwood red to red-brown
T. occidentalis	Heartwood pink to purple

It is impossible to differentiate microscopically between the species.

Use in Dendrochronology

Suitability of the Species

Redcedar is mainly found on sites located in super-humid Pacific climates with optimal water supply. The tree-ring sequences are, therefore, of minimal variability. The stems are lobed and the tree rings are irregular. Synchronization is practically impossible. Substantial climatological data are not to be expected using current methods. Forest dynamic processes can be reconstructed with the aid of growth changes and individual indicator years.

Previous Uses

Parker (pers. comm.) attempted to formulate chronologies by radiodensitometric means. Synchronization is problematic because the ring widths and latewood densities show only minimal variation while the stems are irregular in size.

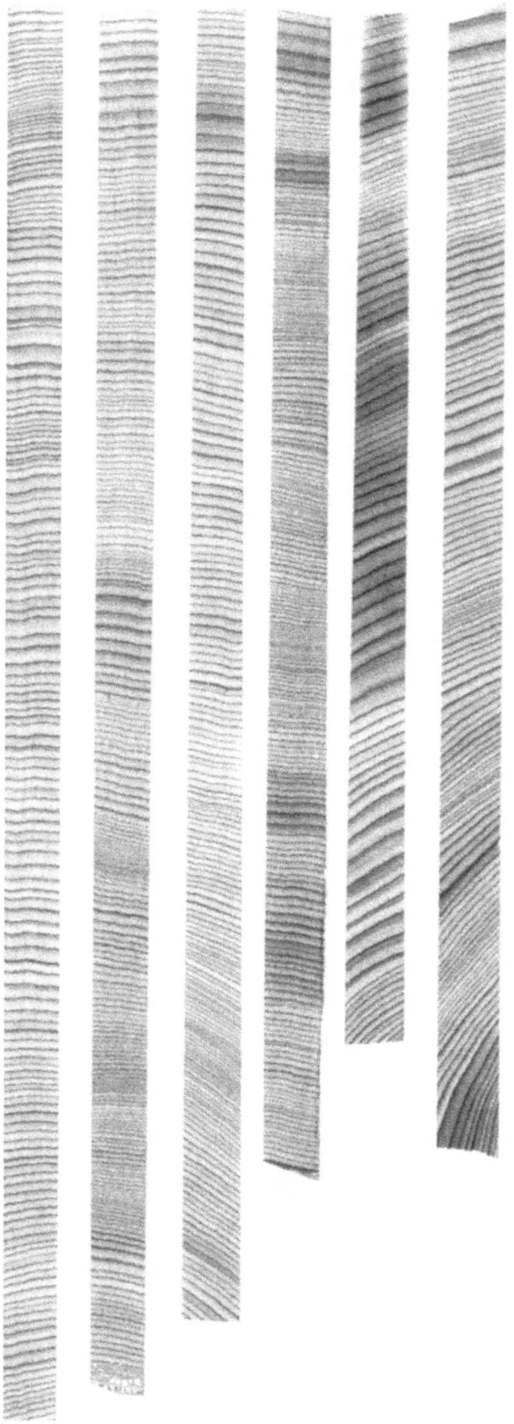

Fig. 459. Radiographs of *Thuja plicata* from Vancouver Island, Canada

Tsuga canadensis (L.) Carr.

Eastern hemlock

Morphology

This tolerant conifer has widespreading branches, an open crown, and a shallow root system. The tops of older trees are often tilted. The trees grow to an average height of 20–30 m, and in rare cases over 40 m. Stem diameters in managed forests measure 40–60 cm and in natural stands occasionally 1.5 m. Maximum age is close to 1000 years.

Fig. 460. *Tsuga canadensis* (Drawing by Hirzel)

Site, Site Conditions, Characteristics

Eastern hemlock grows mainly in mild, wet climates. Temperatures vary in January from −10 to 5 °C, and in July from 15 to 25 °C. The frost-free period lasts for 3–6 months. Annual precipitation is 800–1200 mm. The species flourishes in podzol. *Tsuga canadensis* either forms pure stands or, more commonly, is associated with Eastern white pine, birch, or poplar. Trees over 300 years old are common. The growth rate of shallow-rooted trees declines during periods of dryness.

Fig. 461. A young, cultivated *Tsuga canadensis*

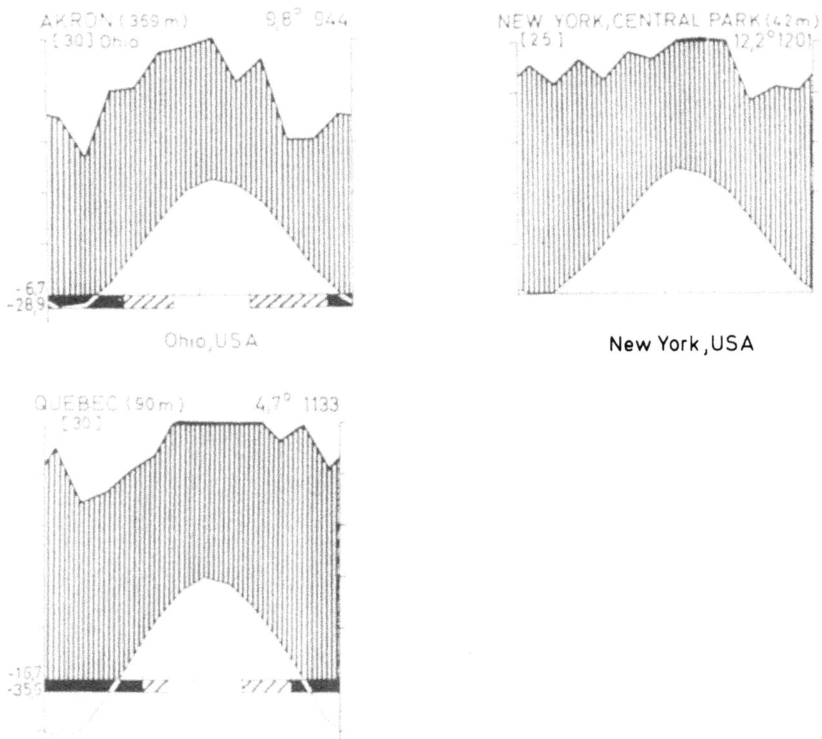

Fig. 462. Climate diagrams

Fig. 463a–c. Cross-sections of *Tsuga canadensis*

Distribution

Horizontal: in eastern and southeastern North America.

Vertical: in the northeast from sea level up to 650 m, in the southern Appalachians at an elevation of 600–1500 m. See *Tsuga mertensiana* for distribution map (Fig. 475).

Wood

See *Tsuga mertensiana* for description.

Use in Dendrochronology

Suitability of the Species

In humid, relatively mild climates cell wall growth in the latewood is limited. On shallow sites, the tree rings may reflect precipitation levels. The species is suitable for radiodensitometrical research if the powerful absorbing material is first removed by washing. The wood is extremely tough, making boring difficult, and the heartwood breaks easily.

Previous Uses

Only a few dendrochronological studies involving *Tsuga candensis* exist. Huges et al. (1982) formulated a few chronologies for dendroclimatological purposes. According to Hornbeck et al. (1987), *Tsuga canadensis*, contrary to *Abies balsamea* and *Picea rubens*, is quite useful for dendrochronology in northeastern USA. Lyon (1936, 1943) studied the pointer years of *Tsuga canadensis* in seven stands in Massachusetts and found a relation to summer precipitation. He identified the regional importance of the pointer years.

Tsuga heterophylla (Raf.) Sarg.

Western hemlock

Morphology

This tall slender conifer has slightly drooping branches and a shallow root system. It normally reaches a height of 40–45 m, sometimes as high as 70 m. The trunk has a diameter of 80–100 cm, in some cases 150 cm.

Fig. 464. *Tsuga heterophylla.* (Drawing by Hirzel)

Site, Site Conditions, Characteristics

Western hemlock flourishes in very wet, foggy marine climates. Temperatures average around 0 °C in January and around 10 °C in July. Annual precipitation is 1000–3000 mm. The frost-free period lasts for 3–7 months. The species grows the best in acidic soils with a thick cover of raw humus. Pure stands develop along the coast. The trees grow to be 500 years old. *Tsuga heterophylla* frequently associates with *Picea sitchensis*, *Thuja plicata*, and *Pseudotsuga*

Fig. 465. A conifer rain forest in the Olympic National Park, Washington

Fig. 466. A *Tsuga heterophylla-Picea sitchensis* forest near a fjord at Valdez in southern Alaska

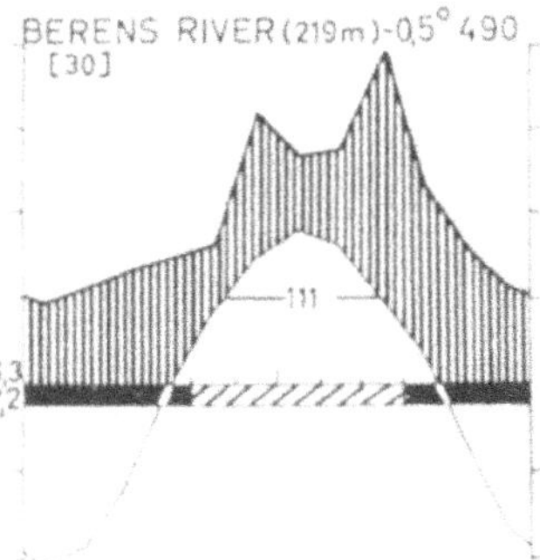

Canada,
British Kolumbia.

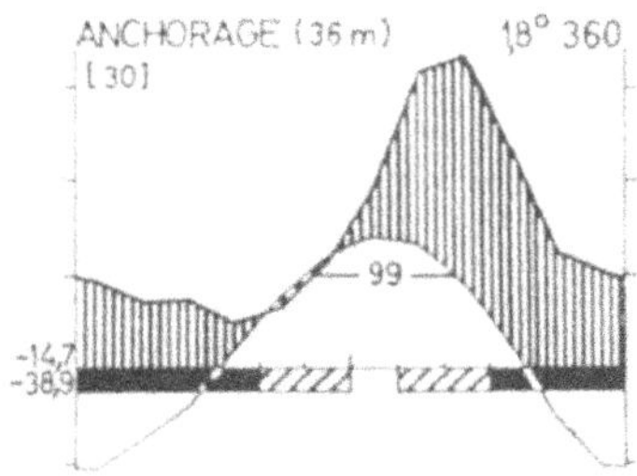

Alaska, USA

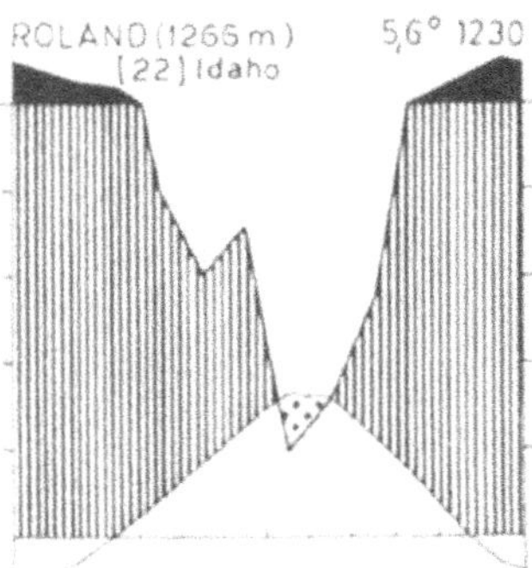

Idaho, USA

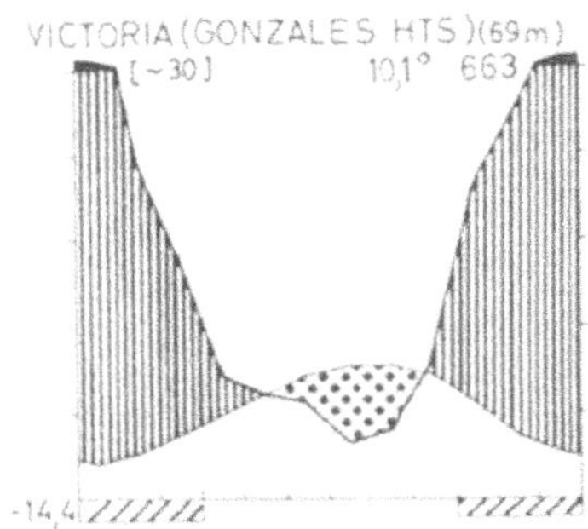

Canada,
British Kolumbia.

Fig. 467. Climate diagrams from regions where *Tsuga heterophylla* grows

menziesii. Young stands are often damaged by snow. The stands are prone to attack by the black-headed budworm.

Distribution

See *Tsuga mertensiana* for distribution map (Fig. 475).
Horizontal: practically only along the northwest coast of North America.

Vertical: along the coast of Alaska from sea level to 600 m, in Montana up to an elevation of 1600 m. The distribution is very similar to that of *Tsuga mertensiana* and *Thuja plicata*.

Wood

It is not possible to distinguish the wood from that of *Tsuga heterophylla* on the basis of its anatomic characteristics. The transition from earlywood to latewood is often more abrupt than in *Tsuga heterophylla*.

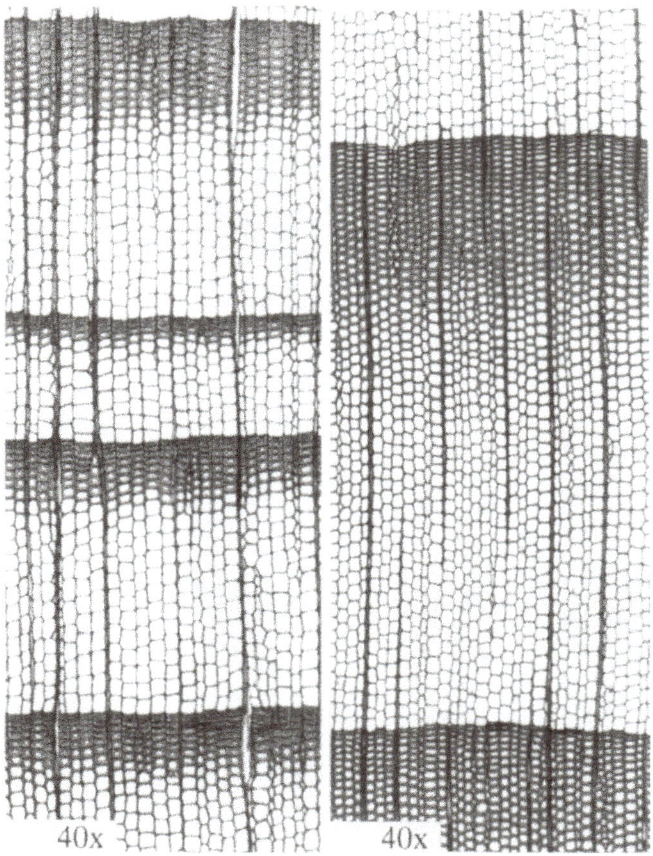

Fig. 468a,b. Cross-sections of *Tsuga heterophylla*. (Courtesy of P. Baas, Leiden)

Tsuga species cannot be microscopically differentiated from one another. See *Tsuga mertensiana* for description.

Use in Dendrochronology

Suitability of the Species

In temperate, humid climates the width and density fluctuations are minimal. Despite its old age, the wood is not very well suited for dendrochronological research. It remains unclear if the species can be used for densitometric purposes. Currently, only a few dendrochronological studies exist. Graumlich and Brubaker (1987) has demonstrated a relationship between summer temperatures and ring widths.

Tsuga mertensiana (Bong.) Carr.

Mountain hemlock

Morphology

The trees of this species generally have irregular, sparsely set crowns. On optimal sites they grow to be 35 m in height and have stem diameters of 1 m. At the timberline, the trees are prostrate shrubs. The root system is deep.

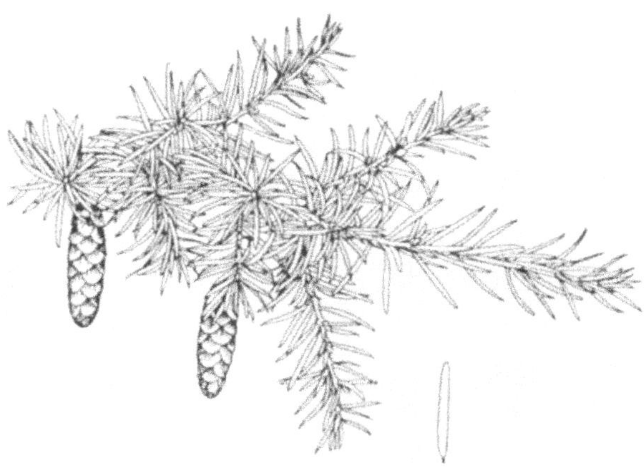

Fig. 469. *Tsuga mertensiana.* (Drawing by Hirzel)

Site, Site Conditions, Characteristics

Tsuga mertensiana is found in Pacific mountain climates. It can tolerate cold, snowy winters and short vegetation periods. Precipitation on nearly all sites is high, averaging between 1000–3000 mm annually. The species grows on practically all foundations except limestone and flourishes in deep soil with normal drainage. Mountain hemlock associates with many other conifers. In the north it forms stands together with *Picea sitchensis* and sometimes *Picea glauca*. In Idaho, near the forest border, it is accompanied by *Picea engelmannii* and *Pinus monticola*. On the central Pacific coast, *Tsuga mertensiana* is mainly associated with *Tsuga heterophylla*. In California, especially on volcanic rock, the species forms the forest border and mixes with *Pinus albicaulis* and sometimes with *Abies amabilis* and *A. lasiocarpa*. Virgin stands are often 300–500 years old.

Fig. 470. *Tsuga mertensiana* at the timberline in Lassen National Park

Fig. 471. 200-year-old trees at the timberline at an elevation of 800 m near Sitka in southern Alaska (Photo Noble)

Fig. 472. A wind-blown *Tsuga mertensiana* on Mt. Lassen, California

Fig. 473. *Tsuga mertensiana* on an optimal site in Oregon. The presence of lichens indicates a high humidity

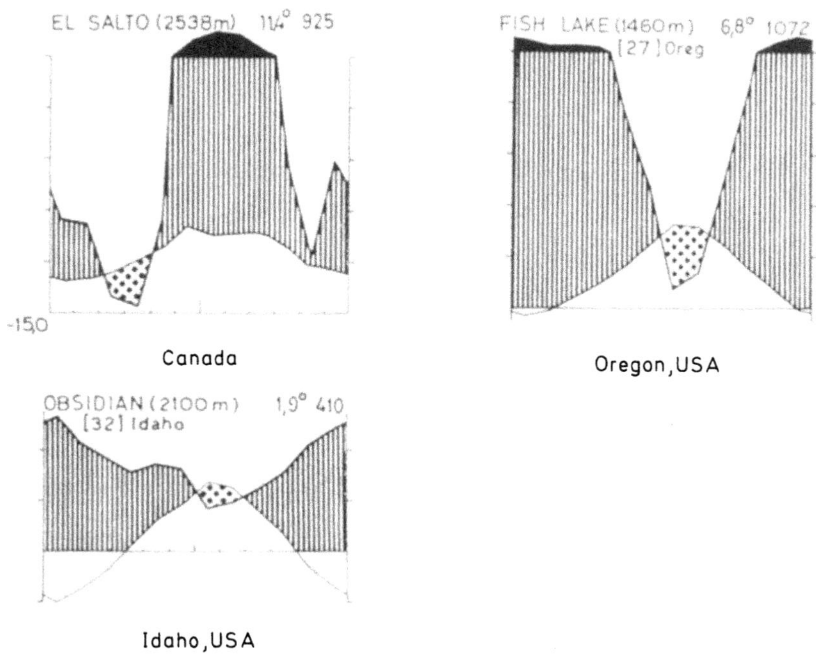

Fig. 474. Climate diagrams

Distribution

Horizontal: along the Pacific coast, from California to Alaska, as well as in areas of Montana and Idaho.

Vertical: at the sea level in Alaska. In Oregon usually at an elevation around 1500–1800 m. In California the trees are found at elevations as high as 3300 m.

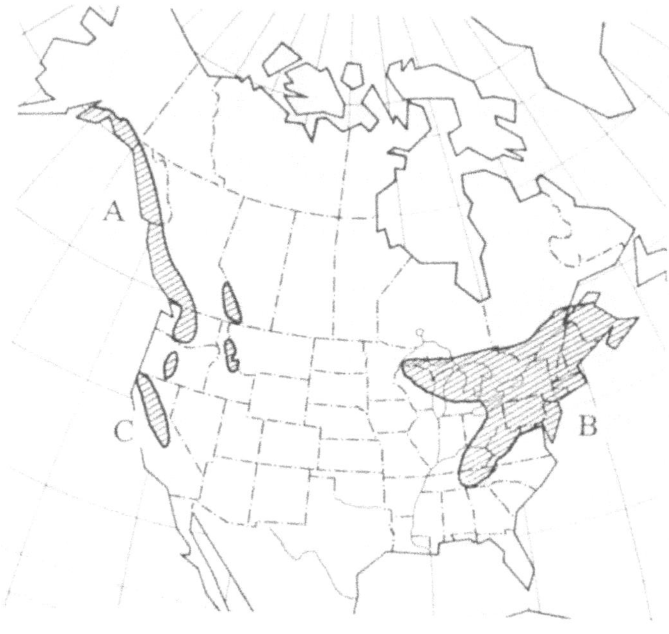

Fig. 475. Distribution of *A Tsuga heterophylla*; *B Tsuga canadensis*; *C Tsuga mertensiana*

Wood

The wood is yellow-brown with a white shimmer. The latewood zone is often slightly red. The wood has no actual heartwood, but the outside tree rings are usually lighter in color. Insects cause dark spots. The wood has a fresh acid odor and a density of 450–500 kg/m^3.

Cross-section: normally without resin canals, but sometimes the tangential rows have traumatic resin canals. The transition from earlywood to latewood is continuous (Fig. 476) to almost abrupt. The extractives in the parenchyma are colorless.

Tangential section: the average ray height is 8–16 cells.

Radial section: rays with ray tracheids. Pits in the earlywood are piceoid to cupressoid (narrow to wide slit-formed openings).

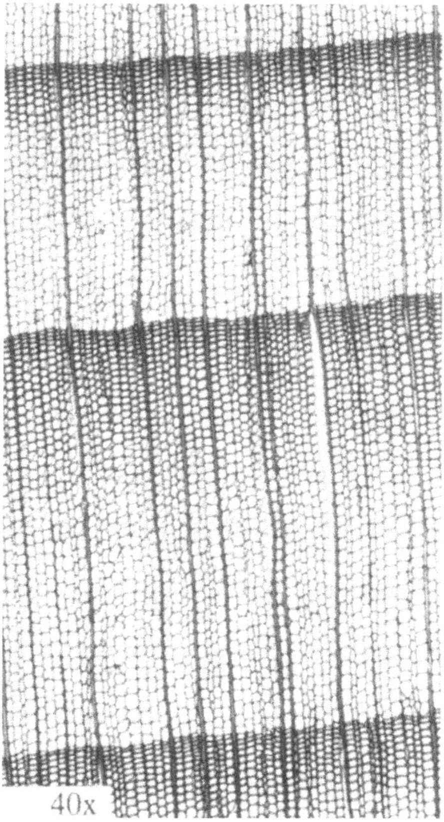

Fig. 476. Cross-section of *Tsuga heterophylla*

It is impossible to differentiate clearly between the various *Tsuga* species on the basis of their anatomic characteristics.

Use in Dendrochronology

Suitability of the Species

This mountain tree lends itself well to dendroecological research as well as to studies on wind patterns, volcanic explosions, and pathological events. It is also suitable for dendroclimatological purposes.

Previous Uses

The species has, until recently, rarely been used. It is possible to reconstruct past summer temperatures from the maximum densities of the tree rings. Graumlich and Brubaker (1987) have demonstrated the growth-stimulating effects of high summer temperatures.

Fig. 477. Radiographs of *Tsuga mertensiana*

Acer saccharum Marsh.

Sugar Maple

Morphology

Reaching heights of 30 to 40 m, this deciduous tree has deep roots and a broad, open crown. The largest trees can have maximum stem diameter of 1 m.

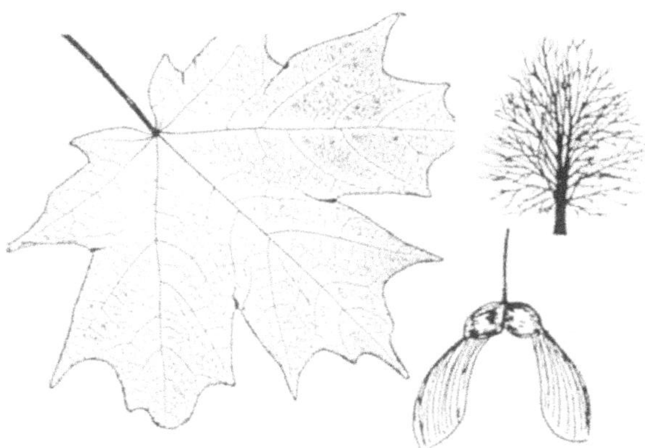

Fig. 478. *Acer saccharum.* (Krüssmann 1972)

Fig. 479. A sugar maple stand subjected to sap (maple syrup) collection. The younger, more vigorous trees from planted, pure stands are the most productive. Notice the plastic collection lines running between the trees

Site, Site Conditions, Characteristics

The demanding *Acer saccharum* is the main species of the northern hardwood zone. It flourishes in regions with annual precipitation of 700–1200 mm. January temperatures within the distribution area average between −10 to −20 °C. Sugar maple is threatened especially by early and late frosts. The species grows predominantly in deep, nutrient-rich soil. It avoids bogs and dry sites. Within its main area, it most commonly associates with *Betula alleghanensis, Fagus grandifolia, Pinus strobus, Tsuga canadensis, Picea rubens,* and *Abies balsamea.* Normally, sugar maple has a life expectancy of 120–150 years, and in rare cases lives for 250 years.

Fig. 480. Bored sugar maple with a tapping devise. Each year the trees are freshly bored

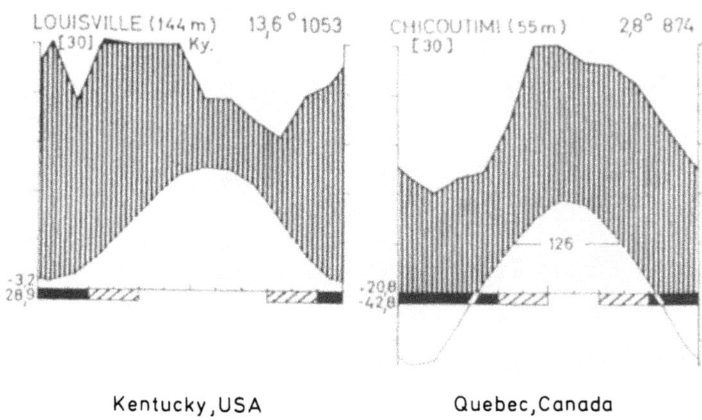

Fig. 481. Climate diagrams

Distribution

Horizontal: sugar maple grows only in northeastern North America. The northern border lies at approximately 47 °N, which corresponds to the 0 °C isotherm. In the west, the area seldom extends beyond the Mississippi. The southern border lies at 35 °N.

Vertical: in the north, distribution extends from sea level to 750 m. In the southern Appalachians, the area reaches elevations of 1600 m.

Fig. 482. Distribution of *Acer saccharum*

Wood

With heartwood. The sapwood is yellow-white, the heartwood is red-brown to brown. The longitudinal section is smooth. Rays are difficult to detect with the naked eye. The density is 600–750 kg/m^3.

Cross-section: diffuse-porous. The vessels are either single or arranged in short, radial rows. The tree-ring boundary is distinguished by a row of elements with a small lumen.

Tangential section: the ray width averages 3 to 4 rows.

Radial section: vessels with distinct helical thickenings. The rays are homogeneous.

Use in Dendrochronology

Suitability of the Species

As the main species of the hardwood zone in North America, sugar maple is particularly well suited for forest dynamic studies. Equally suited are *Betula*

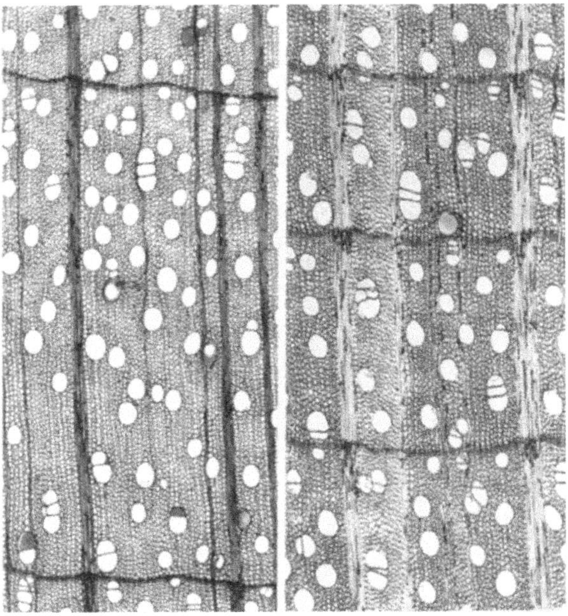

Fig. 483. Cross-sections of *Acer saccharum*

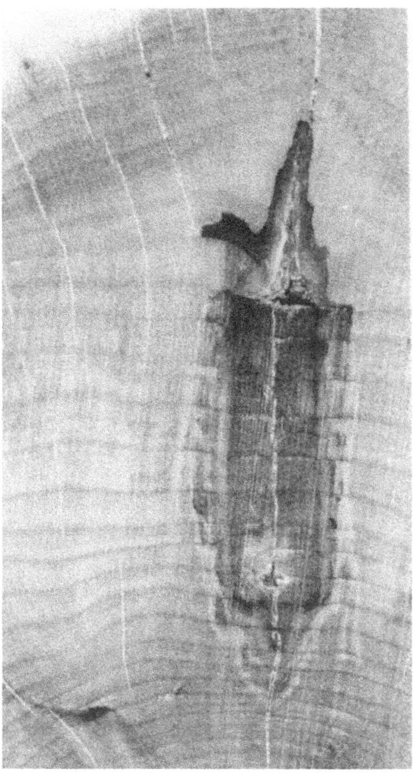

Fig. 484. Compartmentalized bore area. The bore hole is located under the dark spot. After 5 years the wound was healed over

alleghanensis and *Betula papyrifera.* In areas with manufacture of maple syrup, sap is repeatedly collected from the trees in the spring. The vitality of the tree is maintained through its exceptional compartimentalization capacity.

Previous Uses

Payette et al. (1990) studied a maple-beech old growth forest. They noted abrupt growth changes in trees within a 0.25 ha area and were able to determine 30 phases of altered canopy conditions since 1860. According to Payette (pers. comm.), local growth reductions caused by maple decline are discernable in the tree-ring sequences. Houle (1990) determined that since the beginning of the maple decline other species have become more common.

Populus spp.

Poplar

***Populus balsamifera* L. Balsam poplar**
***Populus tremuloides* Michx. Quaking aspen**
***Populus fremontii* Wats. Fremont cottonwood**
***Populus trichocarpa* Torr et Gray Black cottonwood**
***Populus deltoides* Bartr. ex Marsh Eastern cottonwood**

Morphology

Poplar from moist, southern sites generally grow to heights of 30–40 m with trunk diameters between 1 and 2 m. Poplars in northern latitudes or on dry sites reach heights of around 20 m and have trunk diameters of 40–70 cm. All species are deciduous.

Fig. 485. *Populus trichocarpa.* (Viereck and Little 1972)

Site, Site Conditions, Characteristics

With the exception of *Populus tremuloides*, most *Populus* species are found growing along river banks. They require moist, deep, nutrient-rich, clay soils. Poplar species grow in subartic as well as semiarid climates and are generally the pioneering species on recently flooded or burnt sites and on abandoned farmland. In wet soils along streams, poplar is often accompanied by willow, alder, maple, and ash species. In mountain sites *Populus tremuloides* generally forms pure stands or mixes with, for example, *Pinus contorta*. Poplars rarely grow to be very old. In most cases they barely live more than 100 years, although Dunnwiddie and Edwards (1984) report finding 230-year-old poplars in Alaska. The species are prone to attack by many leaf- and wood-damaging pests, which can strip large stands of its foliage. Such events are reflected in the tree rings.

Fig. 486. *Populus fremontii* in an Arizona flood plain

Fig. 487. *Populus* sp. infested with dwarf mistletoe, Poland

Fig. 488. Radiographs of *Populus tremula* (left two cores) and *P. trichocarpa*

Distribution

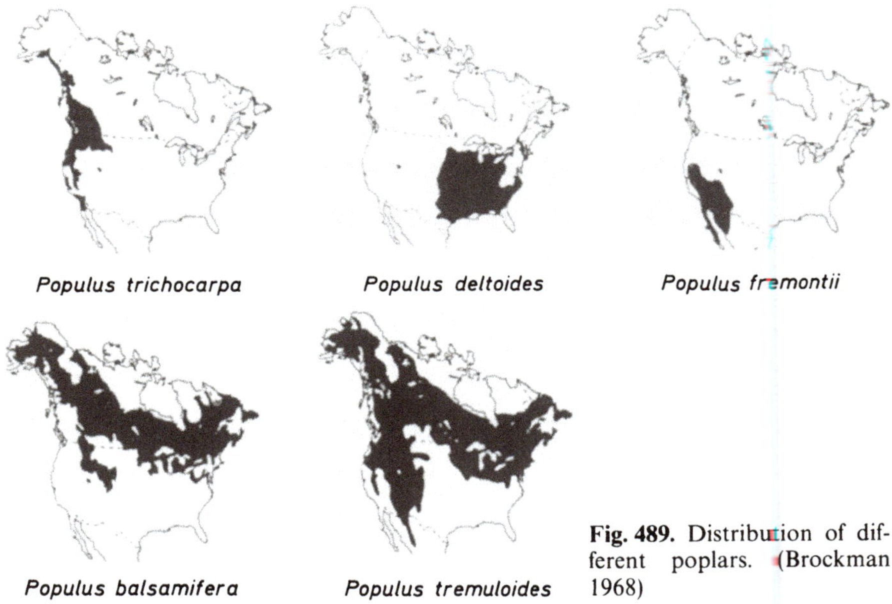

Fig. 489. Distribution of different poplars. (Brockman 1968)

Use in Dendrochronology

Suitability of the Species

Poplars are of minimal use in chronological and climatological studies. The wood is rarely used in construction, it does not preserve well in dry or wet environments, and never grows to be very old. On the other hand, poplars are eminently suited for dendroecological research. They allow, above all, the reconstruction of riparian forest dynamics, be it the determining of the age of the trees in relationship to geomorphological structures in riverbeds, the dating of wounds caused by ice-flows, fire, or by damage inflicted by beaver, deer, bears, or porcupines. Poplars also allow the dating of abrupt growth fluctuations as a result of sediment build-up, exposure of the root system or displacement of the tree by flooding. Using poplars located in delta regions, it is possible to reconstruct fluctuations in the level of sea water following ground swelling, ground sinkage, or changes in the tide.

Previous Uses

Dunwiddie and Edwards (1984) documented temperature reconstructions using tree-ring sequences from *Populus balsamifera* in northern Alaska. By determining the ages of *Populus deltoides* trees in the bed of the Missouri River, Everitt (1968) was able to date the dynamics of the river in relationship to river flow and sediment build up. Clark (1987) made available a flood statistic from Twentyfive Wash in northern Arizona by using live *Populus fremontii* trees that had, at one time, been covered with sediment.

Poplar has yet to be used in Europe for dendrochronological purposes.

Quercus alba L.

White oak

Morphology

On optimal sites, this large, slow-growing, deciduous tree can reach a mature height of 45 m. In open stands it has a broad, rounded crown and a straight trunk. The thickest stems have diameters of up to 2 m.

Fig. 490. *Quercus alba.* (Elias 1980)

Site, Site Conditions, Characteristics

Quercus alba has a diverse habitat. While on southern sites precipitation measures 2000 mm annually, in the north it is 750 mm. In the northeast, average yearly temperatures reach only 7 °C, while in the south they are 21 °C. *Quercus alba* can survive in the snowless region of northern Florida just as well as in the extremely snowy region of Maine. It grows optimally in nutrient-rich soils with normal drainage but can also tolerate dry soils made of sand and gravel. *Quercus alba* never forms pure stands and is in optimal regions associated with many

Fig. 491. A free-standing white oak on an optimal site

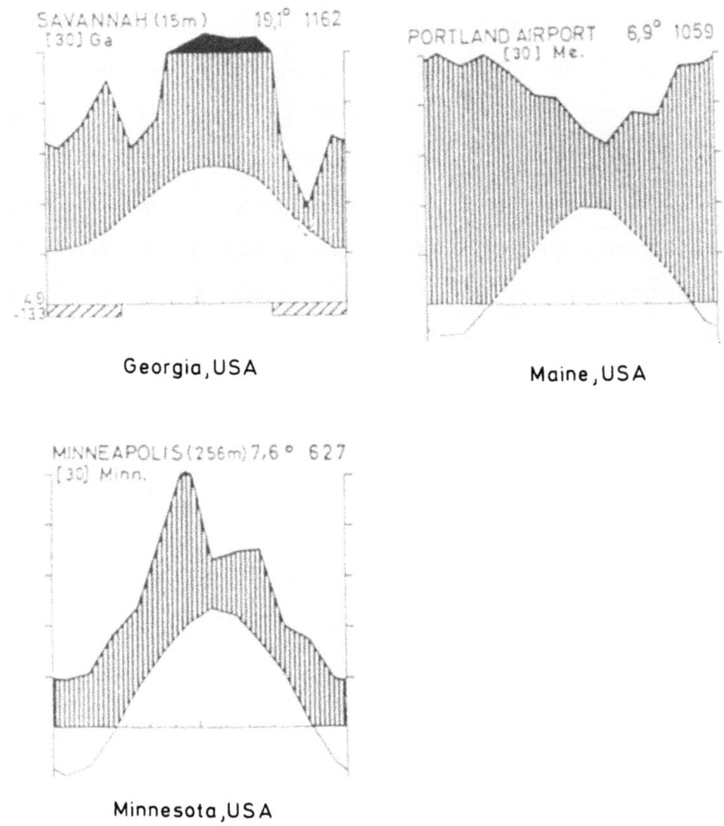

Fig. 492. Climate diagrams

different deciduous species, such as *Carya*, or *Fagus grandifolia* to name only a few. The oldest living *Quercus alba* trees are believed to be around 600–800 years old.

Distribution

Horizontal: *Quercus alba* is the most widely distributed oak species in America. It occurs in practically the entire eastern part of the USA. In the west, the species borders the great plains and in the east the Atlantic ocean. *Quercus alba* does, however, avoid the subtropical regions of the Atlantic Southeast.

Vertical: in the north, the area of distribution is mainly at elevations under 200 m, but in the southern Appalachians it can be found growing at elevations of up to 1500 m.

Fig. 493. Distribution of *Quercus alba*

Wood

The wood cannot be distinguished from that of *Quercus robur*, *Q. petraea*, or *Q. pubescens*. The ring-porous cross-sections are characterized by very wide rays, an abrupt transition from early- to latewood and flame-like groupings of vessels in the latewood of larger growth rings.

Use in Dendrochronology

Suitability of the Species

Corresponding to its wide ecological spectrum, *Quercus alba* is suitable for all forest dynamic studies.

Previous Uses

According to Cook (1982), there exist several recent 300-year chronologies from sites in southern USA. The oldest chronology reaches back to 1588. Fritts (1959) used dendrometers to study the relationship between growth, phanology, photosynthesis, and environmental factors. He also used the technique of response function for the first time in dendrochronology on this material. Stahle et al. (1985) integrated several chronologies in the network of south-central USA. Phipps and Whiton (1987) studied growth of white oaks from 89 localities to reveal the causes of the oak decline.

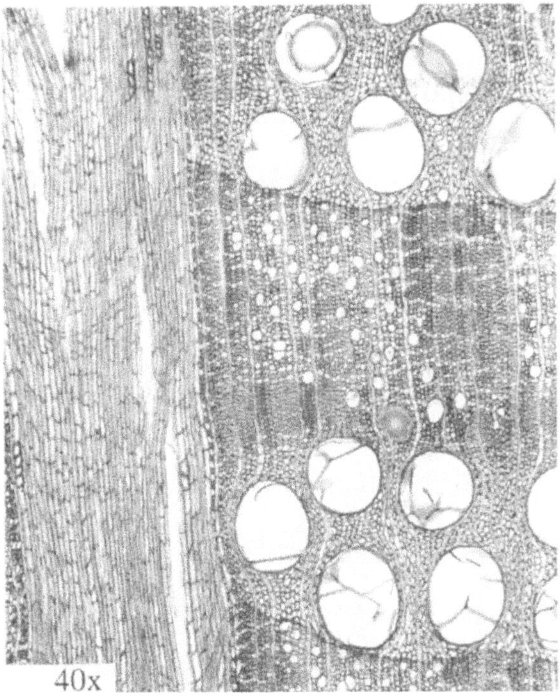

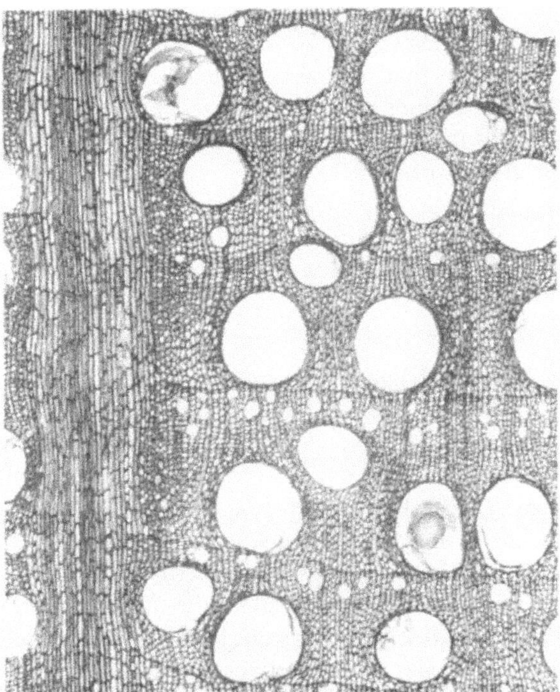

Fig. 494a,b. Cross-sections of *Quercus alba*

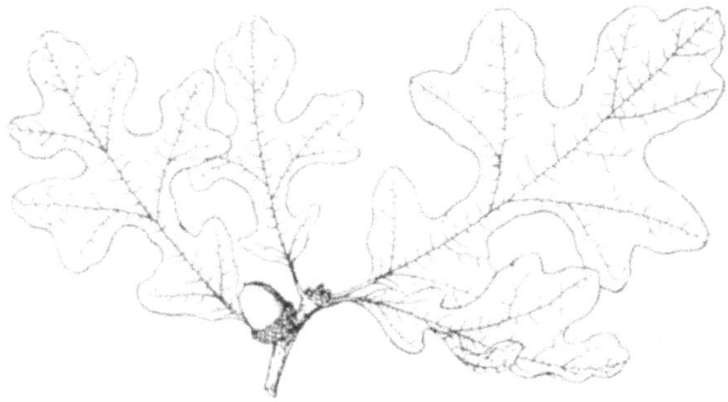

Fig. 495. *Quercus stellata.* (Elias 1980)

Quercus stellata Wangenh.

Post oak

Morphology

This deciduous species grows to a height of 8–20 m, has a broad, open crown and a usually crooked stem. On extremely dry sites, the species is bushy with several boles.

Fig. 496. Old *Quercus stellata* trees in Arkansas. (Photo Stahle)

Site, Site Conditions, Characteristics

Post oak grows slowly and requires abundant light. Its area of distribution covers a wide ecological spectrum. The species is found in the warm, humid climate of Florida with an annual precipitation level of 1500 mm as well as in the semiarid climate of Texas with a precipitation level of 500 mm. In Iowa, post oak withstands heavy snowfall and average January temperatures of −5 °C, while in Texas, it tolerates average July temperatures of 31 °C. It grows along river banks in moist clay soils as well as in dry, sandy soils. Along river banks, post oak is accompanied by a variety of deciduous species, such as *Quercus alba*, *Q. prinus*, *Q. marilandica*, and *Carya* spp. On the arid western border, it forms sparse stands together with *Quercus marilandica*. In dry regions, the trees grow to be 200 to 300 years old.

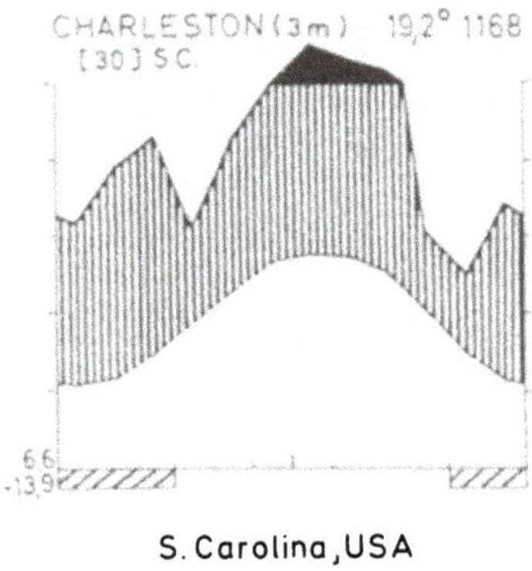

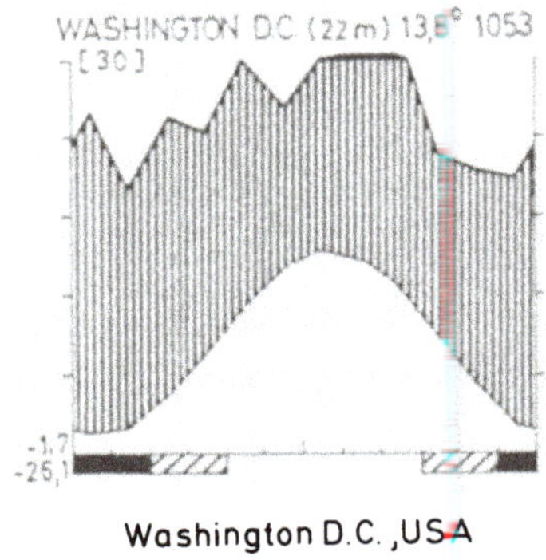

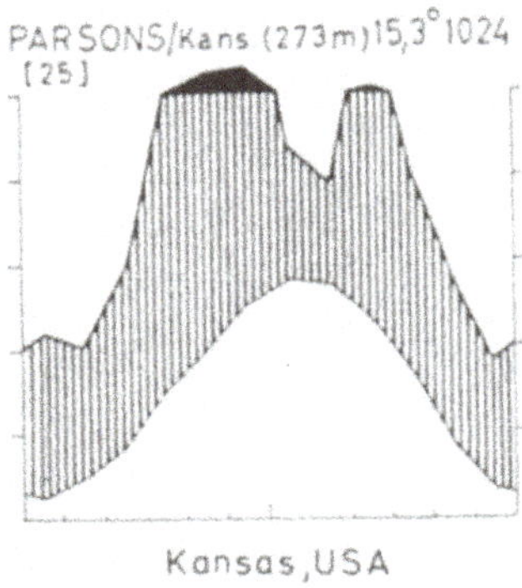

Fig. 497. Climate diagrams

Distribution

Horizontal: mostly found in southeastern and south-central USA, excluding the subtropical regions of Florida.

Vertical: on low sites near sea level up to a maximum elevation of 900 m in the southern Appalachians.

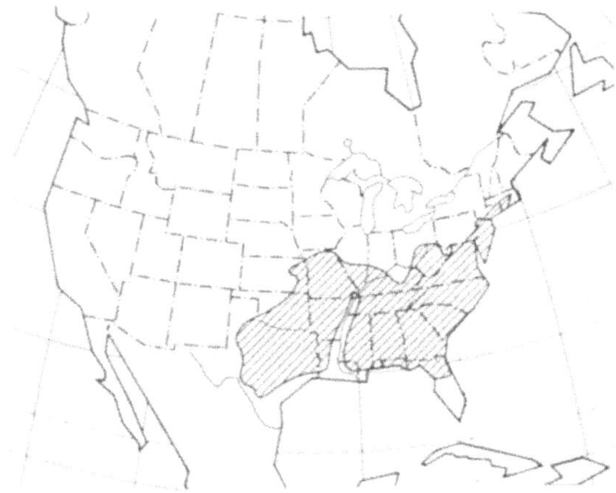

Fig. 498. Distribution of *Quercus stellata*

Wood

It is impossible to distinguish *Quercus stellata* from *Q. alba* on the basis of its anatomic characteristics. See *Quercus alba* for details.

Use in Dendrochronology

Suitability of the Species

Quercus stellata is very well suited for dendrochronological purposes. It grows relatively old and in a wide geographical and ecological spectrum. *Quercus stellata* is also quite common and perhaps the dendrochronologically most important species in southeastern USA.

Previous Uses

Recently, Stahle and co-workers have formulated a network using almost 50 different chronologies from southeastern USA. Previous analysis has yielded the following results.

Precipitation periods from August of the preceding year to July of the growing year can be reconstructed for the last 200–300 years. (Stahle and Hehr 1984; Stahle et al. 1985).

Using frost rings it is possible to determine the late-frost history for the past 300 years.

Blasing et al. (1988) established a 15–25-year periodic of drought spells.

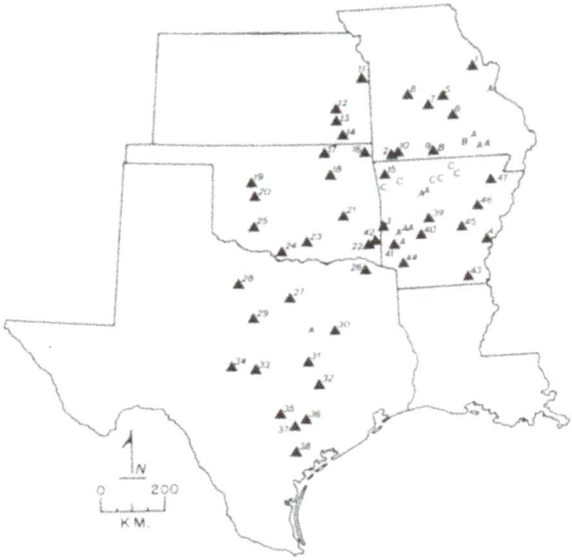

Fig. 499. The dendroclimatological network (Stahle et al. 1985) in southeastern USA. *1–3 Quercus alba*; *5–37 Quercus stellata*; *38–43 Taxodium distichum*

With increasing aridity from east to west (1200 to 600 mm annual precipitation), the ring width sensitivity declines, as well as the correlation between individual curves within a site and precipitation.

In certain phases distinct positive and negative pointer years are present which make dating easier and provide information about extreme events.

The network combines with the already established network in western USA.

Tree Species in the Southern Hemisphere

Agathis australis (D. Don) Salisb.

Kauri Pine

Morphology

This evergreen conifer has 2–3-cm-long, lanceolate leaves. The crowns of older trees are dense, and their stems are straight and branchless. The largest trees grow to be 50 m in height and have stem diameters of over 3 m. The thick bark exudes a rubbery substance.

Fig. 500. *Agathis australis.* (Drawing by Sebek)

Site, Site Conditions, Characteristics

Agathis australis is a marine, subtropical climate species. It flourishes on sites with an annual precipitation level of 1000–2500 mm and average yearly temperatures of 13–16 °C. In rare cases, smaller trees can be found growing under

Fig. 501. An over 1000-year-old Kauri Pine. The bushy crown is typical. (Salomon 1980)

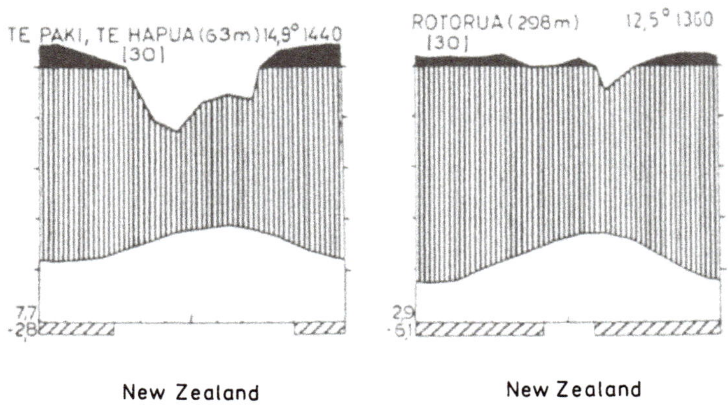

Fig. 502. Climate diagrams

extreme conditions. Kauri pine is a mixed forest species associating mainly with angiosperm species such as *Akamae rosaefolia*, *Beilschmiedia tarairi*, *B. tawa*, and conifers such as *Phyllocladus trichomanoides* and *Dacrydium cupressinum.*

Distribution

Horizontal: endemic to the New Zealand's North Island. Until the last century Kauri pine mixed forests covered areas of over 1.2 10^6 ha. The original area of distribution has been decimated by uncontrolled cutting.

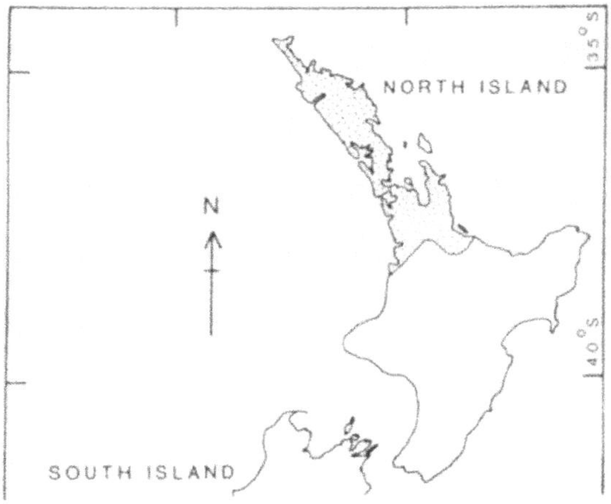

Fig. 503. Distribution of *Agathis australis* in New Zealand (North Island)

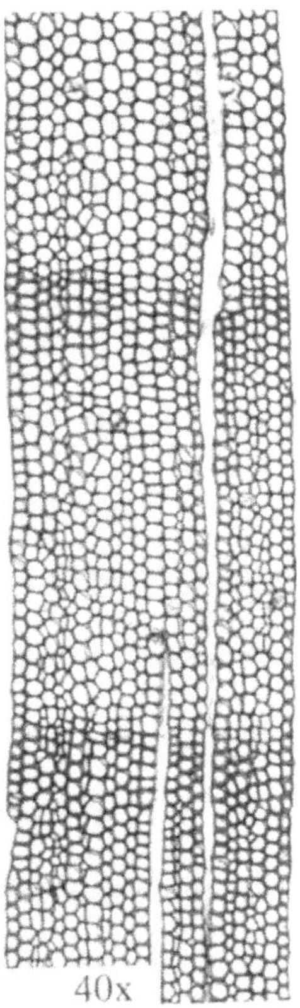

Fig. 504. Cross-section of *Agathis australis*

Fig. 505. Of interest here are the many wedging rings. (Dunwiddie 1979)

Vertical: predominantly on low sites up to an elevation of around 450 m. Smaller trees can be found at elevations up to 800 m.

Wood

There is only a minimal difference between the heartwood and sapwood. The wood is light brown to gray-brown and has a density of 600 kg/m³.

Cross-section: resin canals are absent. The tree ring boundaries are often indistinct (Fig. 504).

Tangential section: small rays, with a height of only two to about ten thin-walled cells.

Radial section: Multiple rows of alternate intervessel pits (araucarioid type). Pits in the cross-field are small and frequently slit-formed, cupressoid to taxodioid.

Use in Dendrochronology

Suitability of the Species

Agathis australis is basically well suited for dendrochronological research. Like other Araucariaceae in regions without distinct seasons, the tree-ring boundaries are often indistinct. Trees on dry northern slopes (S. Hemisphere) are those best suited for synchronization, although only a few event years usually exist to make synchronization possible. Event years occur in low temperature regions as well as in regions with minimal precipitation. It has yet to be determined whether the noticeable long-term trends are climatic in origin. Forest destruction by fire might also be involved.

Because the species grows to be very old, the first studies, along with isotope research, will most likely entail the collection of paleoclimatic data.

Previous Uses

Recent material has been examined by Dunwiddie (1979) and Ahmed and Ogden (1985). From the 11 synchronized chronologies, the oldest reaches back to 1580. Bridge and Ogden (1986) synchronized subfossile trunks from the high swamps on the North Island. The floating chronology, which was dated using radiocarbon methods, comes from the period of 3500 to 300 B.C.

Araucaria angustifolia (Bertoloni) O. Kuntze

Paraña pine

Morphology

This large conifer reaches heights of up to 35 m and has scaly, lanceolate leaves. The crown is broad and umbrella-shaped. The twigs of older trees are set on the

Fig. 506a,b. **a** Fruit and **b** twig from *Araucaria angustifolia*

branches as on a bush. The maximum trunk diameter is over 1 m. The thick, fissured bark protects the trunk from fire.

Site, Site Conditions, Characteristics

Paraña pine grows in tropical and subtropical mountain climates and requires deep soils on high plateaus and slopes (Hueck 1966). It frequently grows along rivers in alluvial soils or moderately swampy areas. The species flourishes in regions with an average annual temperature of around 13–17 °C. In Campos do Jordao (northern Sao Paolo), temperatures in January, the warmest month, average around 16 °C and those in July, the coldest month 8 °C (Lamprecht 1984). The species tolerates minimal temperatures of −10 °C (Lamprecht 1980). The annual precipitation level is 1500–2000 mm; there are no distinct periods of drought. Snowfall is common in the highest areas.

In contrast to most trees in the tropics, *Araucaria angustifolia* forms almost pure stands. *Cedrela* species and representatives of the Myrtaceae and Proteaceae families often accompany *Araucaria angustifolia*. The shrub layer is frequently dominated by tree ferns such as *Dicksonia* spp. and *Alsonia* spp. A smaller tree, *Podocarpus lambertii*, frequently occurs in the undergrowth but does not form tree rings. Grassy camps above the forest border created by *Araucaria* are common. The oldest trees live for around 300 years.

Distribution

Horizontal: the main area of distribution occurs between the 23rd and 30th southern latitudes on the Atlantic coast of Brazil. Isolated mountain areas reach the 18th latitude.

Vertical: in the south, the lowest area is at 500–600 m, and in the north at 1200 m. The species can be found at elevations up to 1800–2000 m, but its main area of distribution is between 1200 and 1600 m.

Fig. 507. Distribution of *Araucaria* forest (*hatched*). (Hueck 1966)

Fig. 508. Canyons in the Serra Geral, Rio Grande do Sul. (Hueck 1966)

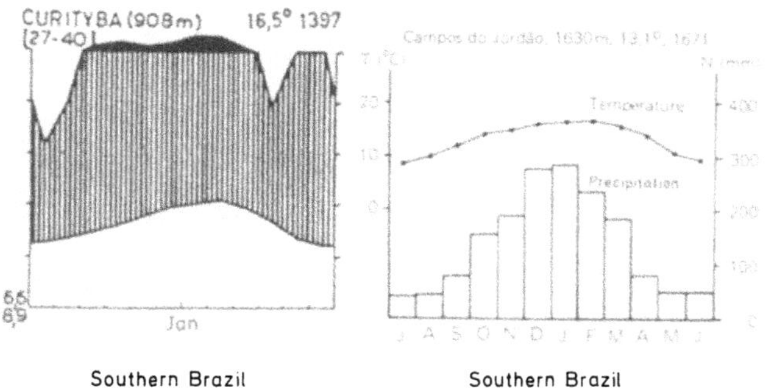

Fig. 509. Climate diagrams. (Lamprecht 1984)

Wood

The sapwood is yellow-white. The heartwood is gray-brown to light brown. The wood has a density of 480–640 kg/m^3.

The wood from *Araucaria angustifolia* is anatomically similar to that of *A. araucana*. Brazilian species do not have pronounced tree rings.

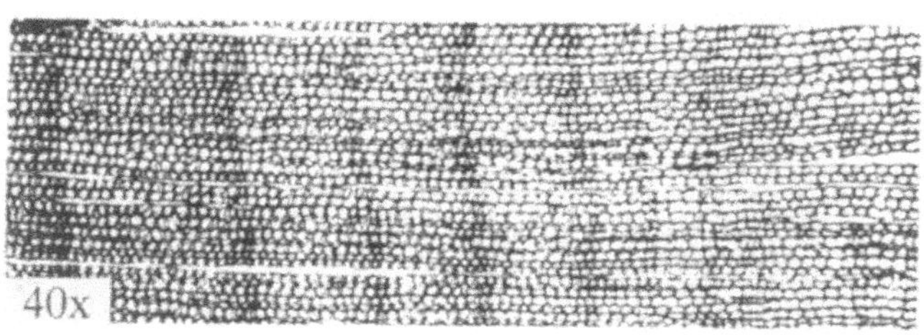

Fig. 510. Cross-section from *Araucaria araucana*. (Lamprecht 1984)

Use in Dendrochronology

Suitability of the Species

According to Lamprecht (1984), *Araucaria angustifolia* growing in the northern subtropical part of its distribution area is very limited in its suitability for dendrochronological purposes of any kind. Neither temperature nor precipitation appears to have any clear restricting effects on its growth. A recent synchronization of the trees was made possible by following the event years over a long period of time. In this dating process, however, mistakes in age determinations are frequent.

Previous Uses

Lamprecht (1984) completed a detailed study on *Araucaria angustifolia* using radiodensitometric procedures. This study helped to draw attention to the problems of dendrochronological research in tropical regions.

Fig. 511. Radiographs of *Araucaria angustifolia*, Brazil

Araucaria araucana (Molina) K. Koch

Chile pine, monkey puzzle

Morphology

Araucaria araucana grows to heights of 30 m and has scaly, thorny leaves and rod-shaped branches. The crowns of older trees are spread out like flat umbrellas. The thick, fissured bark serves as fire protection. At optimal sites the stems can have diameters as great as 1.5 m.

Fig. 512. *Araucaria araucana.* (Drawing by Hirzel)

Site, Site Conditions, Characteristics

Araucaria araucana covers a wide ecological spectrum. The species is found growing in swamps as well as in deep, dry basalt sites. In regions of Argentina, *Araucaria araucana* withstands annual precipitation levels of only 600 mm, while in the coastal regions of Chile it tolerates an annual precipitation of up to 2000 mm. In Argentina the lowest winter temperatures can drop to −20 °C and in Chile to −10 °C. The trees are mainly found in Chile along the coast but can also withstand the summer droughts of the savannah regions of Argentina. In South America, the species is frost-resistant, while in central European regions the trees are prone to freeze during an extremely cold winter, such as that of 1956 (Hueck 1966). *Araucaria araucana* forms pure stands at the forest border. On lower sites with moist soils, the stands are often mixed with *Pilgirodendron uviferum.* On central sites various *Nothofagus* species, such as *N. dombeyi* or *N. pumilio*, accompany, and on certain sites even suppress, *Araucaria araucana.*

Fig. 513. A forest border site on loose, very permeable volcanic rock

Fig. 514. *Araucaria* on a dry, deep soil site (Chile)

Fig. 515. A swamp site in Naheulbuta, Chile

Fig. 516. Thick, fissured, fire-resistant bark of an old *Araucaria* in Chile

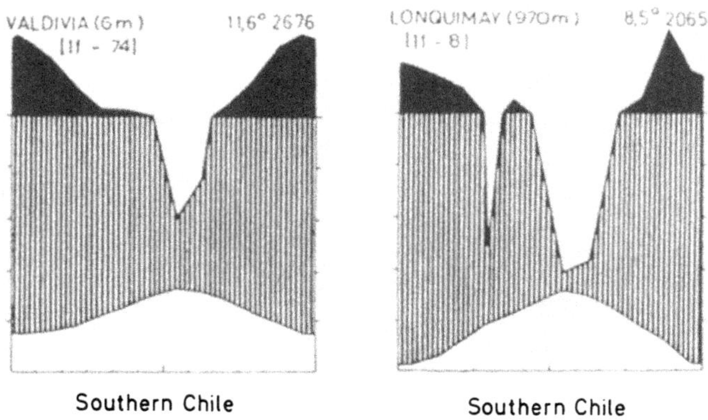

Fig. 517. Climate diagrams. A climate diagram for areas close but outside the natural distribution of *Araucaria araucana*. Winter temperatures in this region are probably 2–5 °C lower

Distribution

Horizontal: between the 37th and 40th southern latitudes in the Andes of Chile and Argentina.

Vertical: the lower border lies in the Argentine pampas region at an elevation of 600–800 m. The upper border is established at 1800–1900 m at the Andes volcanoes.

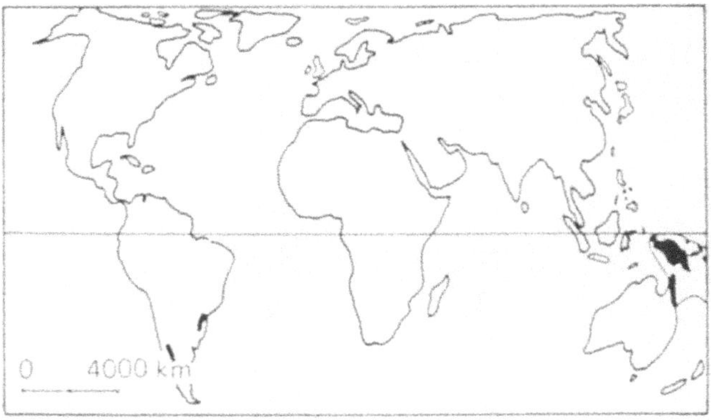

Fig. 518. The worldwide distribution of *Araucaria* (Hueck 1966). *Araucaria araucana* grows in Chile and *A. angustifolia* in Brazil

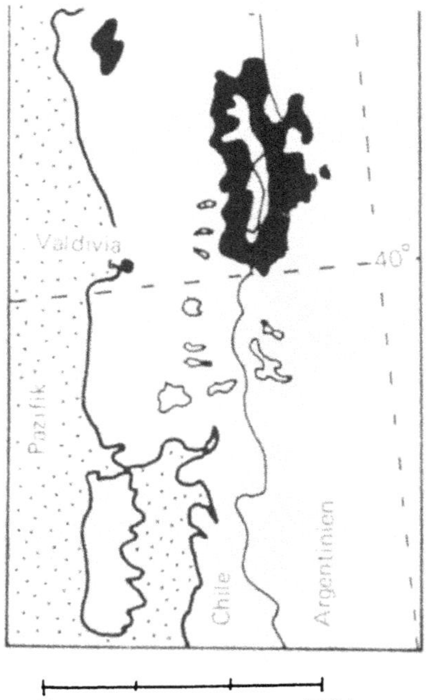

Fig. 519. Distribution of *Araucaria araucana* according to Lamprecht (1984)

Wood

The sapwood is yellow and the heartwood brown. The wood has a density of 550 kg/m^3.

Cross-section: there are no resin canals. The tree rings are indistinct (Fig. 520c) to distinct (Fig. 520a, b). A wave pattern is common in the tree-ring boundaries. The cell walls of the earlywood are thick. The difference between earlywood and latewood is small. The tracheids are round; there is no axial parenchyma.

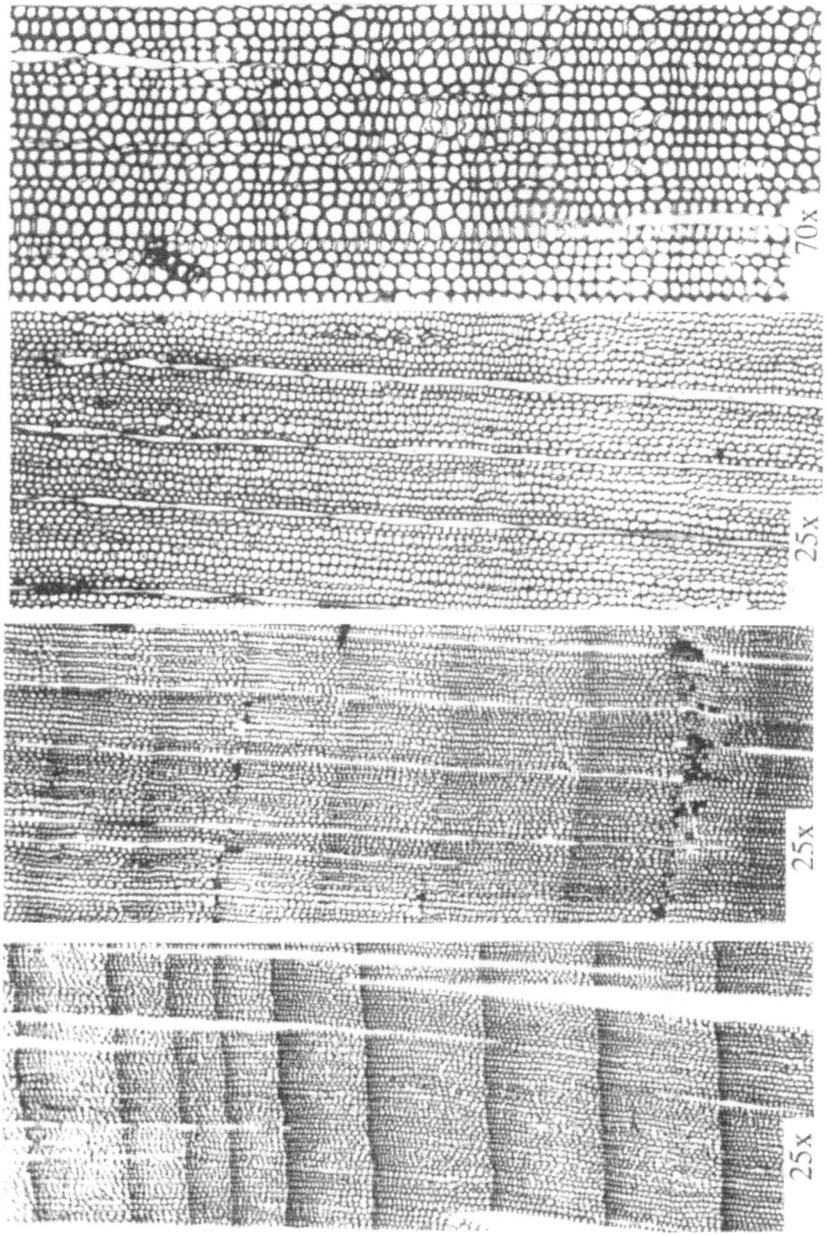

Fig. 520a–d. Cross-sections of *Araucaria araucana*

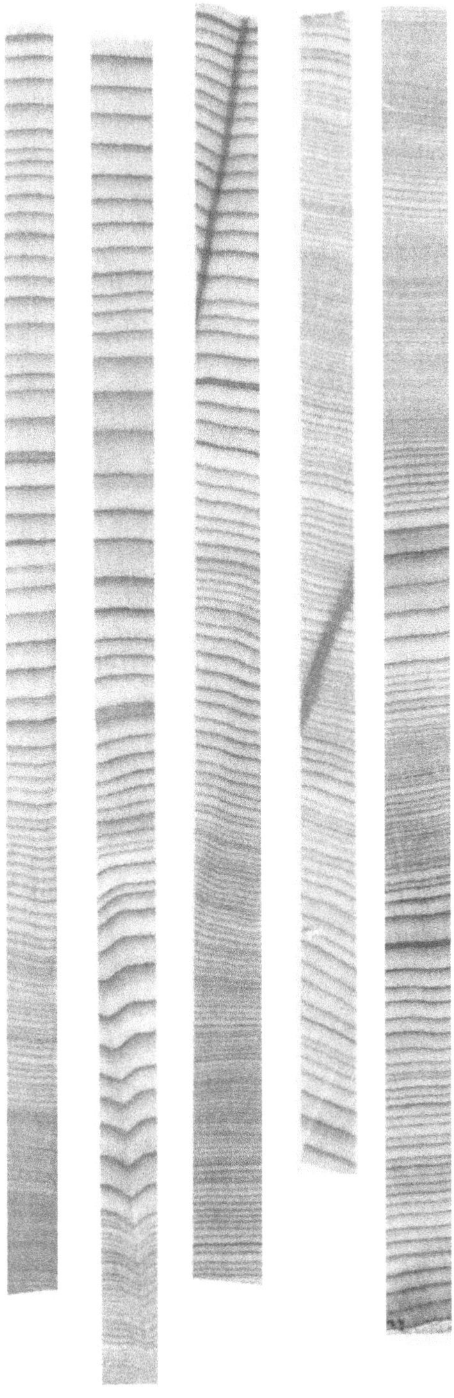

Fig. 521. Radiographs of *Araucaria araucana*, Chile

Tangential section: the rays are uniseriate, rarely biseriate. The average ray height is generally 8–15 cells. The ray cells are thin-walled and round.

Radial section: the rays are homocellular. Resin substances are frequent in the ray cells. In the cross-field of the earlywood there are generally four to eight abietoid to cupressoid pits. The ray cell walls are smooth. The tracheid pits are often in several alternating rows, and frequently of a honeycomb shape. The border pits are round, usually in pairs and touching one another axially.

Use in Dendrochronology

Suitability of the Species

Because *Araucaria araucana* covers a wide ecological spectrum and grows to be quite old, it is basically suited for all dendroecological and dendroclimatological studies. *Araucaria araucana* was seldom used throughout history for building material and is, therefore, only of minimal use for purely chronological problems. Dating is also problematic in that the indistinct tree-ring boundaries and many missing rings usually only have narrow sequences. Synchronization is, however, possible by using the event years, namely wide/narrow tree rings, distinct/indistinct latewood, and a continuous earlywood to latewood transition. Synchronization based solely on tree-ring curves is almost impossible. Using abrupt growth fluctuations and fire scars, the forest fire history of specific regions can be reconstructed. Volcanological studies are presumably also possible.

Previous Uses

LaMarche et al. (1979) have shown that synchronization of an *Araucaria araucana* tree-ring series is possible. The longest Argentine chronology reaches back to 1140 A.D. the longest Chilean chronology to 1242 A.D. Using trees from dry sites in low precipitation areas on the eastern slopes of the Andes, Holmes et al. (1979) reconstructed the past riverflow course of the Rio Negro in Argentina (Province Neuquen). Radiodensitometrical analysis on trees from various sites in Chile brought little climatological information. Even the cell growth of trees from forest borders showed no signs of any mid-summer temperature restriction (January to March), which would, for example, be typical in Pinaceae at the forest border in the northern hemisphere.

Austrocedrus chilensis (D. Don) Florin et Boutelje

Chilean cedar, Chilean incense cedar

Morphology

This species grows to a height of 25 m with a maximum stem diameter of 1 m. Its bark is striped and its narrow, cypress-like crown is frequently blown into a flag shape by the wind.

Fig. 522. *Austrocedrus chilensis.* (Drawing by Hirzel)

Site, Site Conditions, Characteristic

Austrocedrus is found in the southern Andes at low altitudes with minimal precipitation. It grows best in regions with an average annual precipitation level

Fig. 523. A dry site on a basalt foundation in Chile. The tree-ring sequence in trees from these sites is very sensitive. (Holmes et al. 1979)

Fig. 524. Hundreds of years old Cypress on a rocky cliff in Chile. (Holmes et al. 1979)

Fig. 525. Young *Austrocedrus* stand from a temperate site in Lanin National Park, Argentina. (Hueck 1966)

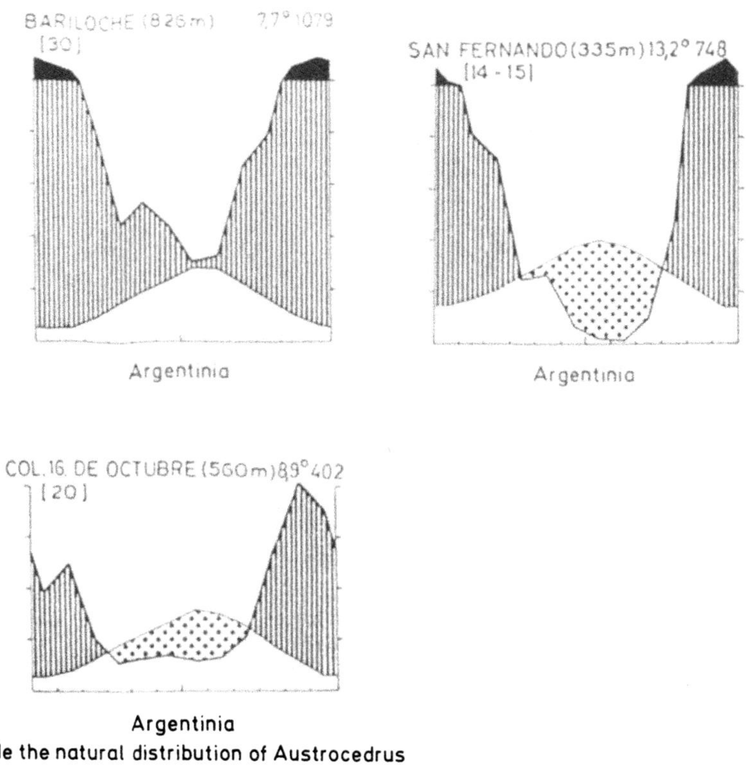

Fig. 526. Climate diagrams

of 1000 mm. Its distribution is restricted to sites with annual precipitation above 600 mm. On most sites temperatures in July average above 0 °C. *Austrocedrus chilensis* flourishes in loose, easily rooted soils. On extreme sites on basalt rock, the species is suppressed, especially the older trees. Most of the sites have been degraded by overuse and grazing. *Austrocedrus* forms pure stands in the pampas regions of Patagonia, Argentina or, in drier areas, is associated with *Schinus crenatus, Lomatia hirsuta, Diostea juncea,* and *Maytenus boaria.* Sometimes the stands are mixed with *Araucaria araucana* (Hueck 1966).

Distribution

Horizontal: in the southern Andes between the 32nd and 44th latitudes, mostly on the Argentine side.

Vertical: the upper border is located in the southern area at an elevation of around 1400 m and in the northern at 2000 m. The lower border is in the south at 500 m, and in the north at 900 m.

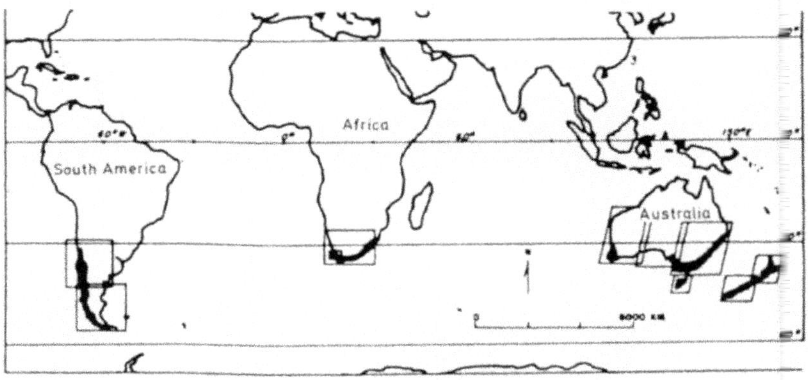

Fig. 527. Distribution of the species *Austrocedrus* in South America. *Widdringtonia* in South Africa. *Libocedrus* and *Arthrotaxis* in Australia and New Zealand. (Hueck 1966)

Wood

The sapwood is yellow-white and the heartwood chestnut brown. The wood has a density of 500 kg/m^3. The tree-ring boundaries are distinct and narrow latewood zones are typical. Variability in the widths of the tree rings and latewood is high. Density fluctuations are common in trees from relatively moist sites (Lamprecht 1984).

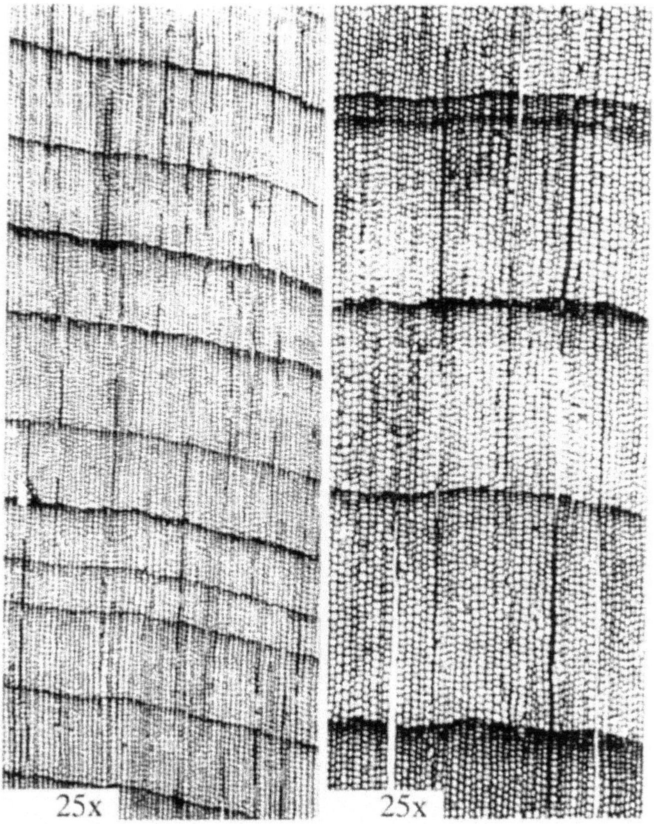

Fig. 528a,b. Cross-sections of *Austrocedrus chilensis*

Use in Dendrochronology

Suitability of the Species

Austrocedrus is suitable for dendroecological and dendroclimatological purposes because the tree grows to be quite old and has distinct tree-ring boundaries, as well as numerous event rings, thereby making it possible to date the trees. Trees from shallow soils allow for the reconstruction of annual precipitation.

Previous Uses

The chronology made available by Holmes et al. (1979) is the most comprehensive. The oldest Argentine chronology reaches back to 1418 A.D. and the oldest from Chile begins in 1011 A.D.

Fitzroya cupressoides (Mol.) Johnst.

Patagonian cypress, alerce

Morphology

Fitzroya cupressoides is the largest tree in the Andes. It grows to a height of 50 m and can have a trunk diameter of 4–5 m. The oldest trees are generally damaged and deformed.

Fig. 529. *Fitzroya cupressoides.* (Drawing by Hirzel)

Site, Site Conditions, Characteristics

Fitzroya is found in the moist Valdivian rain forest with a marine climate. The average temperatures in the coldest month (July) remain constantly above 0 °C and the warmest temperatures (December) vary around 18 °C. Precipitation is high with 3000–4500 mm falling annually. According to Veblen and Ashton (1982), *Fitzroya* can be found growing on very wet, marsh sites, where it is mostly accompanied by *Podocarpus nubigenus*, *Pilgirodendron uvifera*, *Saxegothea conspicua*, and *Drimys winteri*. *Fitzroya* can, however, also be found on relatively dry sites with podzol. On these sites the stands are mixed with *Nothofagus betuloides*. *Fitzroya* grows to be very old. The oldest trees are believed to be over 5000 years old. The largest trees have diameters of 4.5 m (see Fig. 531). In a radius of 35 cm were 1300 tree rings counted on the peripheral section of fallen neighboring trees. Allowing that the tree pictured in the photo grew faster when it was younger, even conservative estimates of its germination date go as far back as 3000 to 4000 B.C.

The existence of the stands is threatened by fire, a result of deforestation.

Fig. 530. *Fitzroya cupressoides* on a moist site in the Cordilliera Pelada, Chile

Fig. 531. *Fitzroya* with a diameter of 5 m in a Cordilliera Pelada gorge in Chile. This could be one of the world's oldest living trees

Distribution

Horizontal: mainly on the west slopes of the southern Andes in Chile between 40° and 43° latitude south.

Vertical: in the coastal ranges from 700 up to 1000 m above sea level.

Wood

The sapwood is light yellow; the heartwood dark brown-red. The density varies between 300 and 650 kg/m^3.

Cross-section: resin canals are absent. The tree-rings are distinct (Fig. 533). The latewood zones are generally narrow. There is an abrupt transition from earlywood to latewood. Parenchyma in the heartwood contain colored substances.

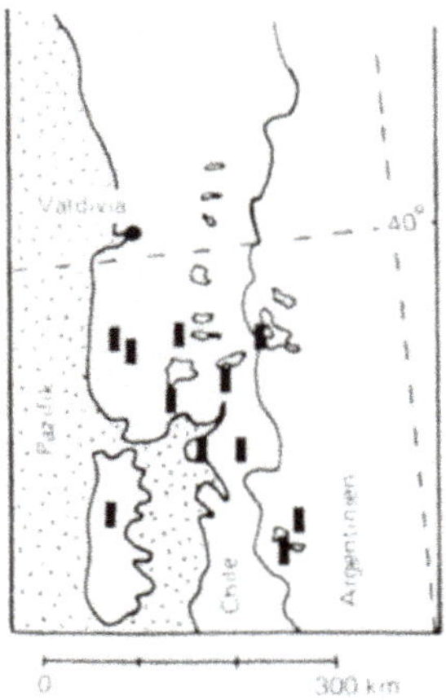

Fig. 532. Distribution of *Fitzroya cupressoides*. (Lamprecht 1984)

Tangential section: the ray height is 36 cells.

Radial section: the rays are without ray tracheids. As a rule, there are two to three cupressoid to piceoid pits in the cross-fields, and six in the border cells. The ray walls have nodules.

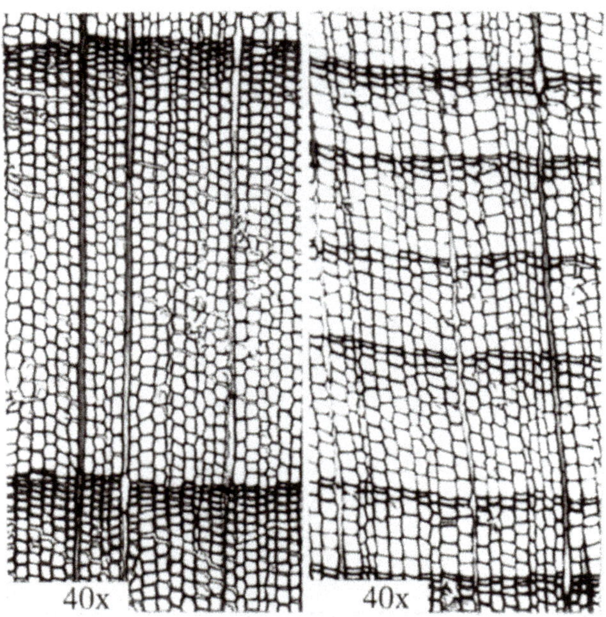

Fig. 533. Cross-sections of *Fitzroya cupressoides*. (Lamprecht 1984)

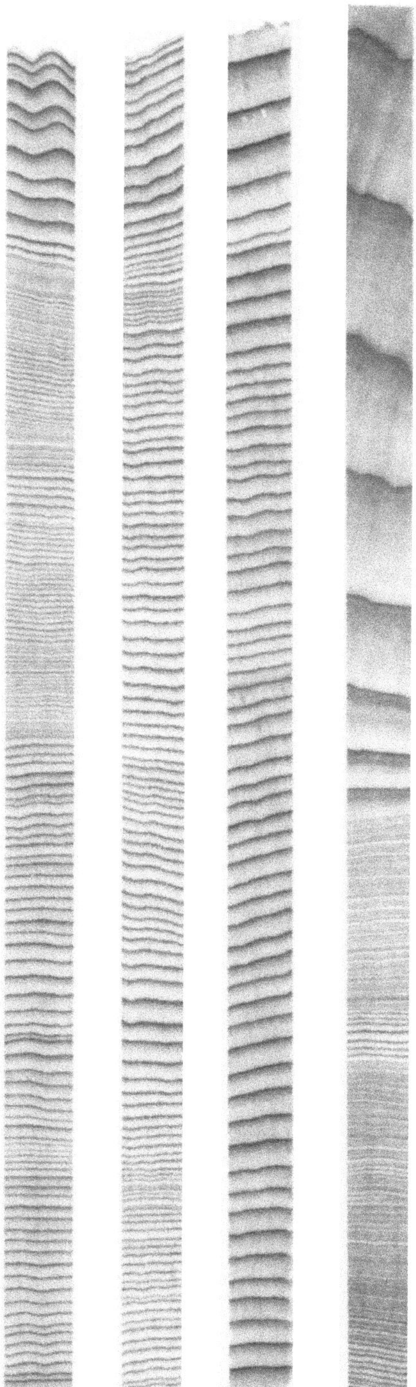

Fig. 534. Radiographs of *Fitzroya cupressoides*, Chile

Use in Dendrochronology

Suitability of the Species

Fitzroya is basically suited for dendrochronological research. Its tree rings are very distinct, it is distributed over a broad ecological spectrum, and the trees grow to be very old. Site differences are indicated in the tree rings. Forest dynamic processes are revealed in the frequently long-lasting variations in tree-ring width. Climatological information is for this reason difficult to ascertain. According to Boninsegna and Holmes (1985), the trees can be dated to a limited extent.

Previous Uses

This species has very seldom been used in dendrochronology. The longest chronology reaches back to 441 A.D. (Boninsegna and Holmes 1985).

Libocedrus bidwillii Hook

New Zealand cedar (cypress), Pahautea

Morphology

This large evergreen conifer can grow to a height of 20 m with a stem diameter of 1.2 m. Its crown is pyramid-shaped.

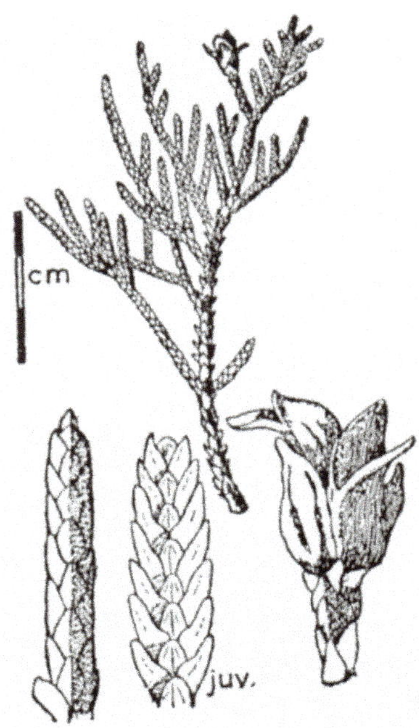

Fig. 535. *Libocedrus bidwillii.* (Poole and Adams 1980)

Site, Site Conditions, Characteristics

Libocedrus bidwillii grows in mountain regions with cool marine climates and high precipitation. It can withstand short-term periods of frost in the winter (July through August). Annual precipitation in its area of distribution is 2500–4500 mm.

Fig. 536. An old stand on the northeast coast of the South Island, New Zealand. (Salomon 1980)

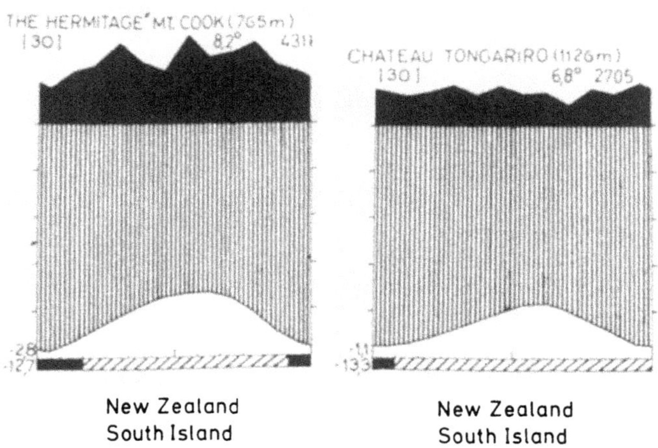

Fig. 537. Climate diagrams

Libocedrus generally builds stands mixed with *Nothofagus menziesii*, *Podocarpus* spp., *Dacrydium cupressinum*, *Beilschmiedia tawa*, etc. Regeneration takes place practically only after windfall and landslides (Norton 1983c).

Distribution

Horizontal: the species is endemic to New Zealand's South Island, where it is mostly found growing on the northeast side.

Vertical: at high altitudes at 250–1200 m.

Wood

The sapwood is yellow; the heartwood dark brown-red. The wood has a density of 400–500 kg/m^3.

Cross-section: resin canals are absent. The transition from earlywood to latewood is gradual (Fig. 538a,b). Density variations between earlywood and latewood are small. Intra-annual density fluctuations occur frequently. The parenchyma in the heartwood contain colored substances.

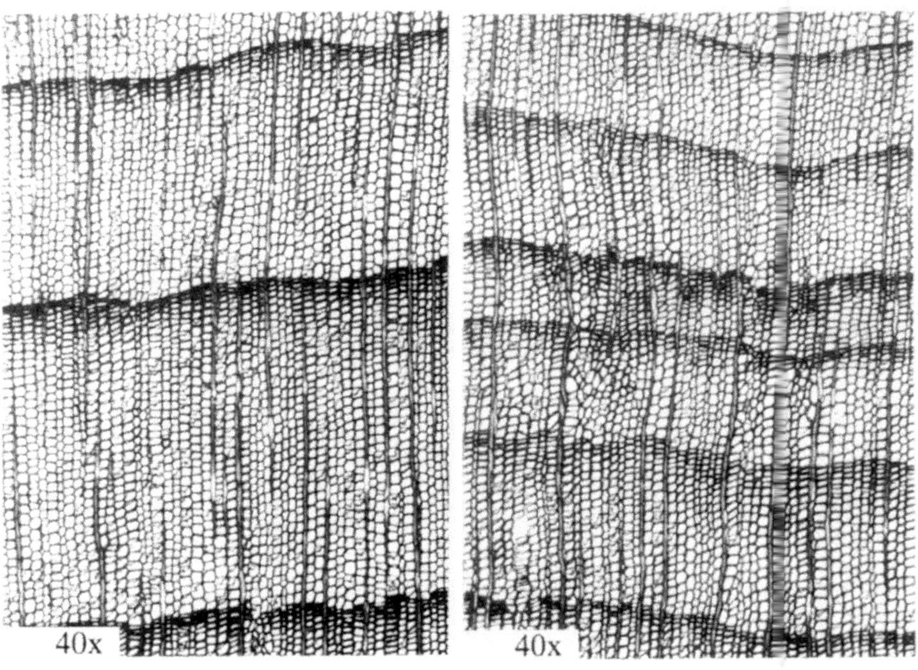

Fig. 538a,b. Cross-sections of *Libocedrus bidwillii*

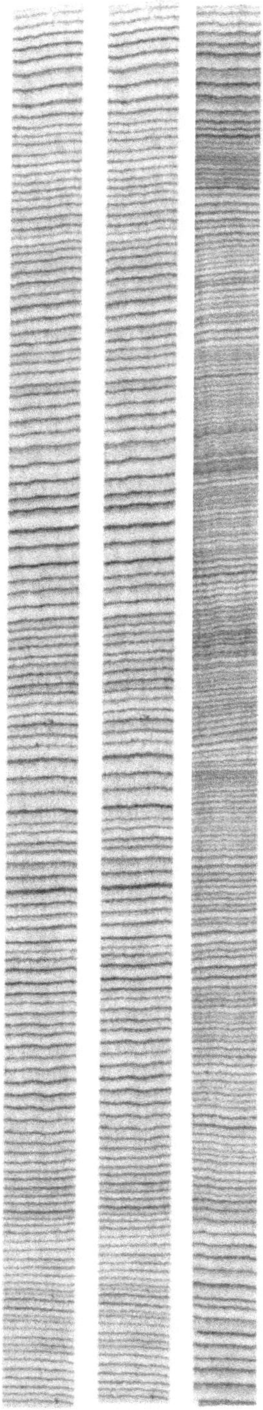

Fig. 539. Radiographs of *Libocedrus bidwillii*, New Zealand

Tangential section: the ray height is low, averaging three to five cells.

Radial section: the rays are without ray tracheids. There are many small cupressoid pits in the cross-fields. The walls of the parenchyma cells are smooth (ray and axial parenchyma).

Use in Dendrochronology

Suitability of the Species

The tree rings are distinct, but their widths vary only slightly. The tree-ring sequence can, however, be dated. Multiple, uneven rings can be found in the lobed stems of older trees (LaMarche et al. 1979). It remains to be determined whether density values can provide information adaptable to climatological studies. The species is especially suited for studies on forest dynamic processes.

Previous Use

Norton and Ogden (1987) recorded a chronology of recent trees, reaching back to 1346. According to Druce (1966), a volcanic eruption occurring in 1655 is indicated in the tree-ring sequences. A high autocorrelation appears to restrict the annual climatologic information content of the tree-ring widths. The few indicator years indicate cold summers (December).

Cedrela angustifolia Ses. et Moc.

Cedro salteño

Juglans australis Grieseb.

Argentine walnut

Morphology (*Cedrela angustifolia*)

This deciduous species grows to be extremely large with a height of over 20 m and a trunk diameter of up to 1.5 m. It has a widespreading crown.

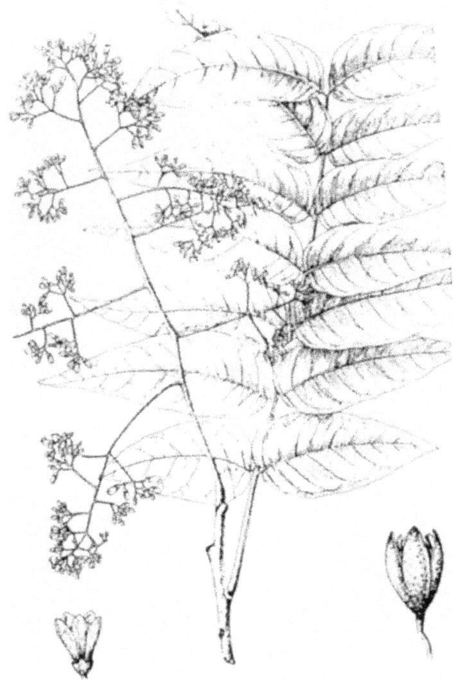

Fig. 540. *Cedrela angustifolia.* (Digilio 1966)

Fig. 541. *Juglans australis.* (Digilio 1966)

Morphology (*Juglans australis*)

A large deciduous species.

Site, Site Conditions, Characteristics

These two species are found in the subtropical climates of the so-called Nogal Pino forest of the Tucumanic-Bolivian forest region. *Cedrela* and *Juglans* generally form mixed stands together with *Podocarpus curvifolius*, *Blepharocalyx giganteus*, and *Phoebe australis*. In low, dry sites the stands border on the

Fig. 542. Old, free-standing *Cedrela angustifolia* in the Province of Salta, Argentina. (Courtesy of P. Boninsegna)

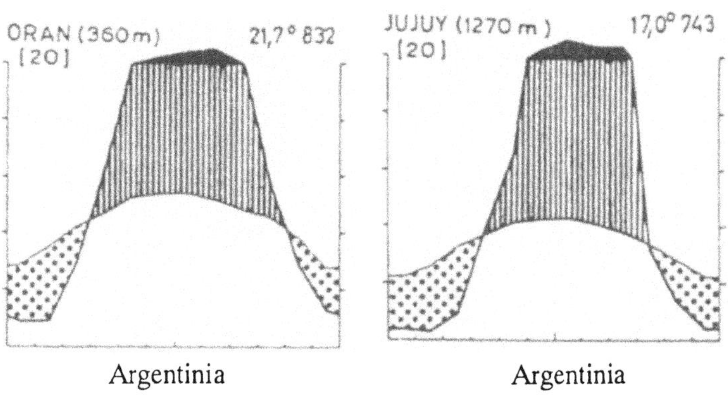

Fig. 543. Climate diagrams

Myrtacean hardwood forest and in higher, moister sites they border on the deciduous Aliso forest (*Alnus acuminata* and *Alnus jorullensis*). Annual precipitation is 600–1200 mm. The vegetation period is from September to May. *Cedrela angustifolia* can grow to be 400 years old; *Juglans australis* 330 years old (Legname 1982).

Distribution

Horizontal: on the eastern slopes of the Andes in Argentina (Tucuman and Salta provinces) and possibly also in Bolivia, from around 20 to 26 °S.

Vertical: from the base of the mountains at 600 m to an elevation of 1700 m. Only rarely do the two species reach the timberline.

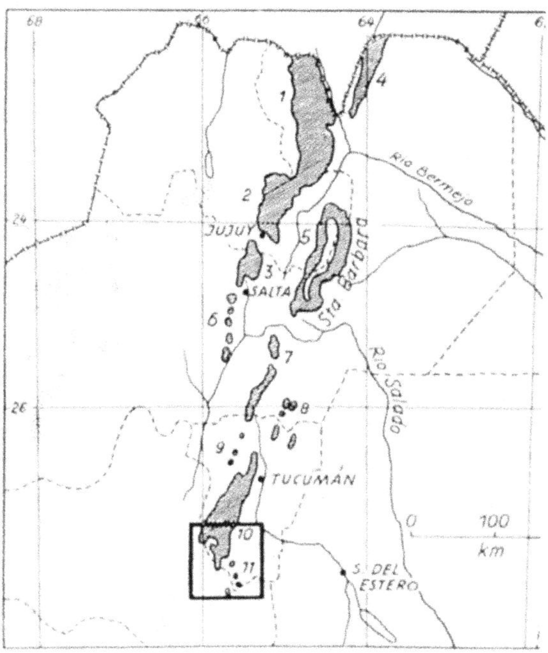

Fig. 544. Southern limit of the tucumanic–bolianic forests with *Cedrela* and *Juglans*

Wood (*Cedrela angustifolia*) (Fig. 545)

The heartwood is brown and the sapwood yellow. Diffuse-porous, large solitary vessels or short, scattered radial files. Terminal parenchyma in multi-rows in the earlywood, paratracheal. Tree-ring boundary is distinguished by the earlywood band, sometimes by the slightly flared vessels in the earlywood.

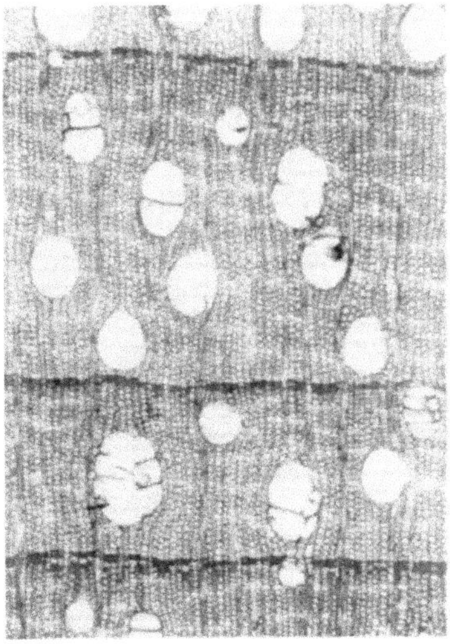

Fig. 545. Cross-section of *Cedrela angustifolia*

Wood (*Juglans australis*) (Fig. 546)

The heartwood is dark brown to violet. The sapwood is gray. Diffuse-porous. Solitary or short scattered radial files of vessels. Apotracheal parenchyma in single-rowed, tangential bands.

The tree-ring boundaries are distinguished by two to four tangential, thick-walled latewood cells. Density fluctuations and sometimes double rings are present. Crystals occur in the axial parenchyma cells.

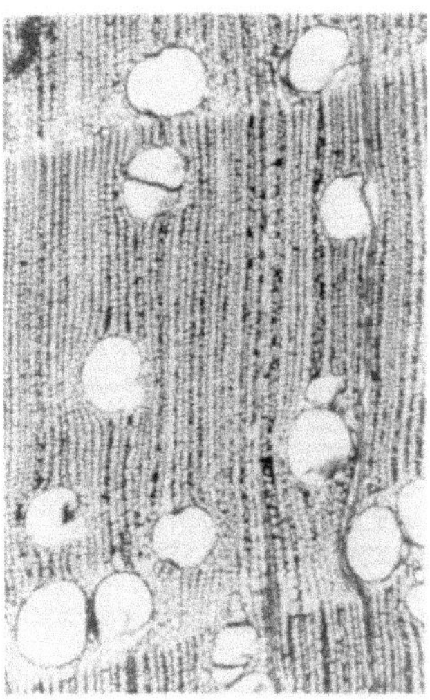

Fig. 546. Cross-section of *Juglans australis*

Use in Dendrochronology

Suitability of the Species

Due to the wide distribution, of the two species, their distinct tree-ring sequences and their long lifespans, both are quite suitable for dendrochronological purposes. In subtropical regions, the two species are likely to become important in ecological and climatological studies.

Previous Uses

Villalba and Boninsegna (1985) recorded chronologies of trees from three different sites in Argentina (23–24° southern latitude). *Cedrela* has been dated back 171 years and *Juglans* 294 years. Low temperatures and high precipitation during the vegetation period (September to November) appear to induce radial growth. Both species are quite sensitive.

Eucalyptus pauciflora

Snow gum, cabbage gum

Morphology

The size and shape of this evergreen species vary. It can grow to heights of 20 m; its stem is often twisted and bent. It has green-brown spotting on its bark and a broad, open crown.

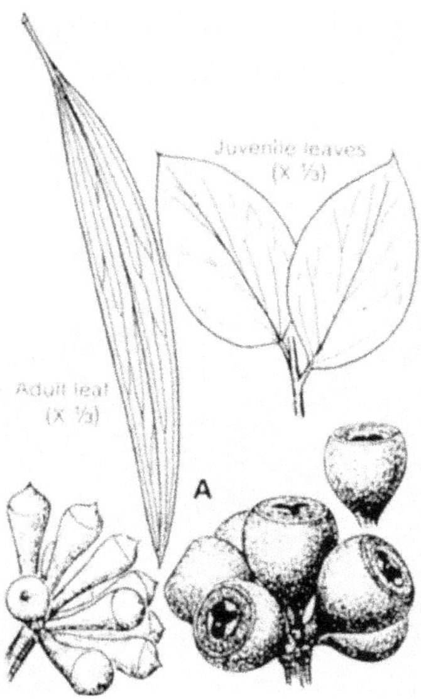

Fig. 547. *Eucalyptus pauciflora.* (Costermans 1983)

Site, Site Conditions, Characteristics

Snow gum is found in the cool, moist climate of mountain and subalpine zones where annual precipitation, distributed equally throughout the months, is 1500–1600 mm. Average temperatures in January (summer) are around 10 °C and in winter near the freezing point. The species can withstand minimum temperatures of 5 °C. At high elevations, *Eucalyptus pauciflora* forms pure stands, while at low elevations it mixes with other *Eucalyptus* species. For the most part the species grows on granite or basaltic foundations. In forest border regions, the stands are interspersed with open meadows, decreasing the risk of fire. Scattered old stands and bush forests are therefore common. As a rule, stems of trees which are twisted or bent over are covered with numerous resin pockets (Kino veins) making the wood suitable for local demands (fence posts, fire wood).

Fig. 548. *Eucalyptus pauciflora* at the timberline in Victoria, Australia. Short and bushy trees as well as tall and slim ones are subjected to various environmental factors

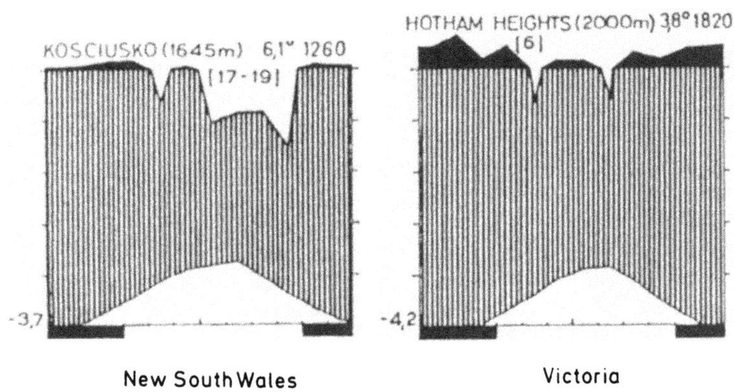

Fig. 549. Climate diagrams

Distribution

Horizontal: Australia. In the mountains of Queensland to the southern Australian border.

Vertical: mainly on high altitude mountain sites at 700 m in the north and 1000 m in the south to the timberline at 1600–2000 m.

Fig. 550. Distribution of *Eucalyptus pauciflora* (Costermans 1983)

Wood

Without heartwood, red-brown wood with a density around 800 kg/m^3.

Cross-section: diffuse-porous. Vessels are exclusively solitary and scattered with tyloses. The tree rings are characterized by distinct earlywood and latewood zoning. In the more or less vessel-free latewood bands, density fluctuations occur which, following damage, can lead to false rings.

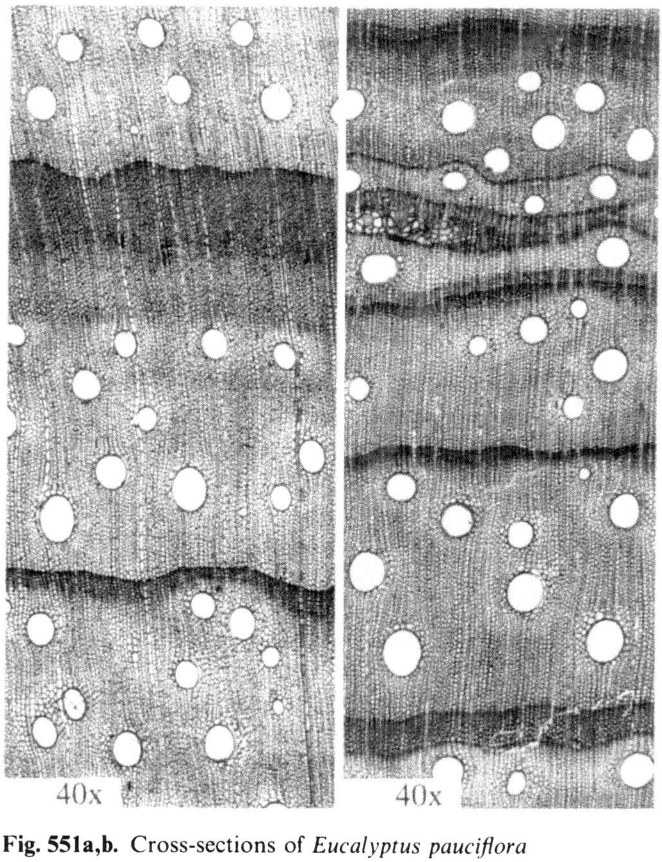

Fig. 551a,b. Cross-sections of *Eucalyptus pauciflora*

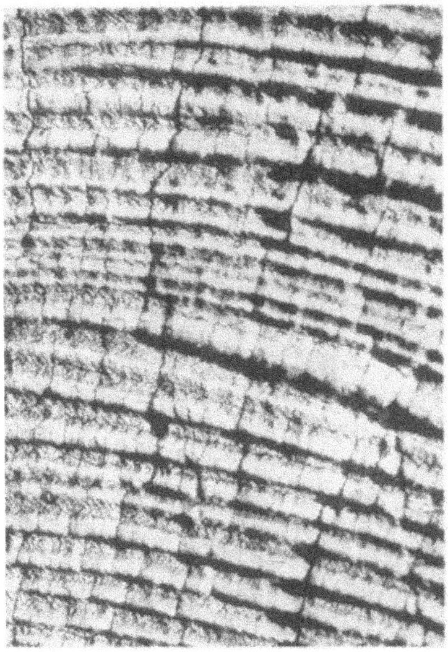

Fig. 552. Cross-section of a beam of *Eucalyptus pauciflora*. The weathering patterns bring out the nonuniform areas of different density within each annual ring

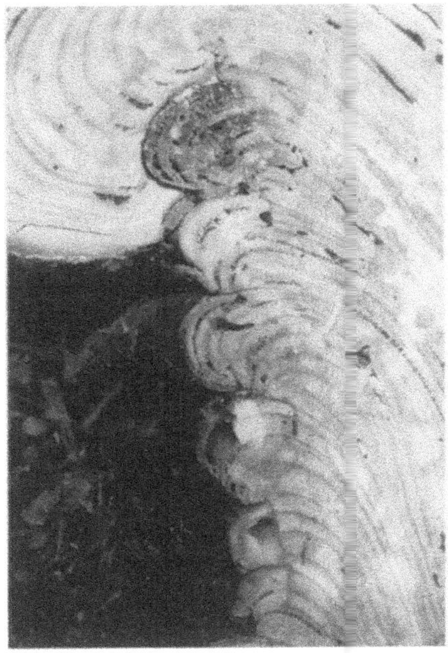

Fig. 553. Cross-section of a stem of *Eucalyptus pauciflora* with several healing layers that developed after fire injury. Material courtesy by J. Banks

Use in Dendrochronology

Suitability of the Species

Eucalyptus pauciflora forms very distinct tree rings. It has yet to be proven, but it is generally presumed, that this is the case in all *Eucalyptus* spp. from mountain sites. The ordered anatomical structure and the weathering pattern of beams suggest that the *Eucalyptus* is suitable for radiodensitometric research. It is uncertain what climatological information might be obtained. Tree-ring sequences from *Eucalyptus* trees could be especially suited as ecological indicators (stand dynamics, the effects of wind, fire, and water).

Previous Uses

In Australia, dendrochronological research has been minimal (Ogden 1978, 1982). Banks (1986) and Glasby et al. (1988) studied tree-ring-forming *Eucalyptus oreades* in subalpine and mountain zones of the Australian Alps. Using fire scars, they dated forest fires before and after European settlement.

Nothofagus menziesii Oerst.
Nothofagus solandri Oerst.

Silver beech, Black beech

Morphology

These evergreen deciduous trees usually are heavily branched and have narrow crowns. On optimal sites they both reach heights of 25–30 m while on timberline sites, with heavy snowfall, the species are bush-shaped.

Fig. 554. *Nothofagus menziesii.* (Poole and Adams 1980)

Fig. 555. *Nothofagus menziesii.* (Drawing by Sebeck)

Fig. 556. *Nothofagus solandri.* (Poole and Adams 1980)

Site, Site Conditions, Characteristics

Both species form forests in the temperate zone of super oceanic climate of the southern hemisphere. Average annual precipitation here is 1000–6000 mm. In

Fig. 557. Old *Nothofagus menziesii.* (Salomon 1980)

Fig. 558. Old *Nothofagus solandri.* (Salomon 1980)

Fig. 559. *Nothofagus solandri/N. menziesii* forest border on the South Island, New Zealand. (Salomon 1980)

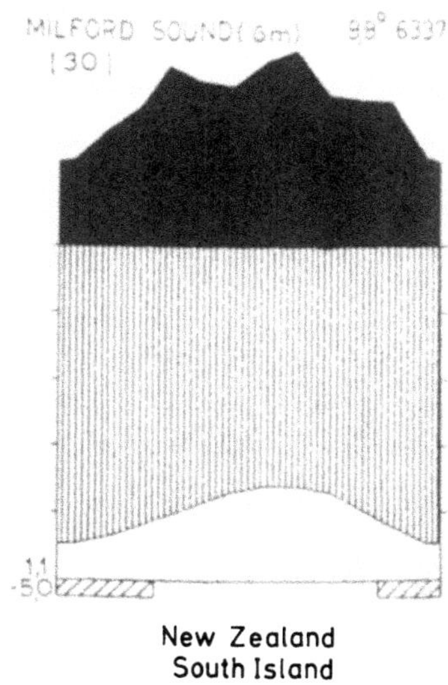

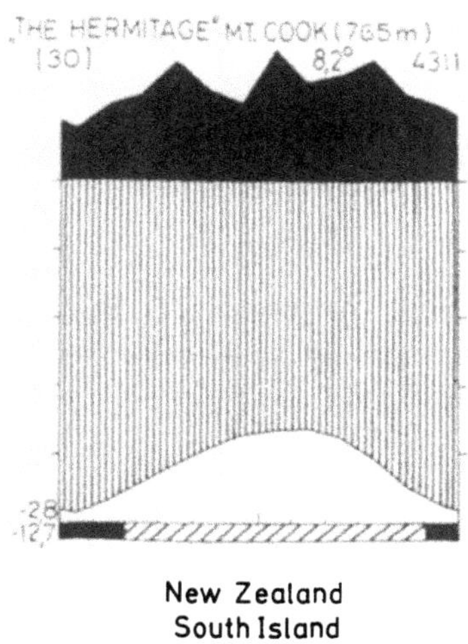

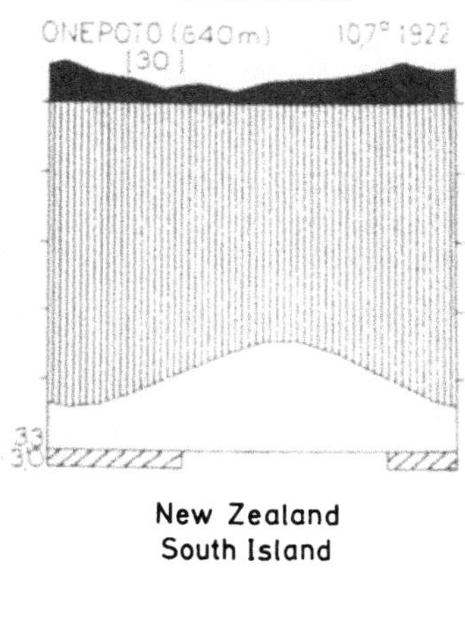

Fig. 560. Climate diagrams

the southwest of the South Island of New Zealand annual precipitation can be more than 4000 mm and mostly falls as rain. Winters are relatively mild. At sea level, temperatures average around 5 °C, and near the timberline at an elevation of 1200 m the average is 0 °C. The evergreen trees can tolerate snowy winter climates. They grow predominately on acid rock (granite, slate, sandstone) foundations in brown soil and podzol. In high altitudes they form pure stands. At the timberline, *Nothofagus solandri* stands are mixed with *Podocarpus nivalis*, and *P. alpinus*. On subalpine sites, *Pseudowintera colerata*, *Pseudopanax simplex*, and *Phyllocladus alpinus*, as well as evergreen species are found growing under the canopy of *Nothofagus*. In mountain zones, *Nothofagus fusca*, *Dacrydium cupressinum*, *Podocarpus ferrugineus*, *P. hallii*, *Weinmannia racemosa*, along with other species accompany the two *Nothofagus* species. Both have long life expectancies, *Nothofagus menziesii* to 600 years and *N. solandri* to around 400 years. The growing period falls in the warmer seasons between November and May and becomes increasingly short the higher the altitude (Norton 1984). Thus, at an elevation of 800 m, the radial growth of *Nothofagus* begins in November and ends in May, while with trees at the timberline at an elevation of 1290 m, it begins in December and ends in April. The growth rate is for a large part restricted by the temperature, yet brief dry spells can also stunt the growth.

Distribution

Nothofagus menziesii

Horizontal: the species is endemic to New Zealand and is widely distributed on both islands. On the South Island it is mainly found growing west of the main mountain chain.

Nothofagus solandri

Horizontal: sparsely distributed on the North Island with two large separate areas on the Southern Island.

Vertical: from sea level to the timberline. The species forms stands in the mountain zone and constitutes the forest border at the subalpine level. The forest border on the South Island is at around 1200–1300 m (Norton 1983b, 1984).

In the southern hemisphere, predominately in temperate marine climates, *Nothofagus* (with around 100 different species) substitutes for the *Fagus* genus of the northern hemisphere.

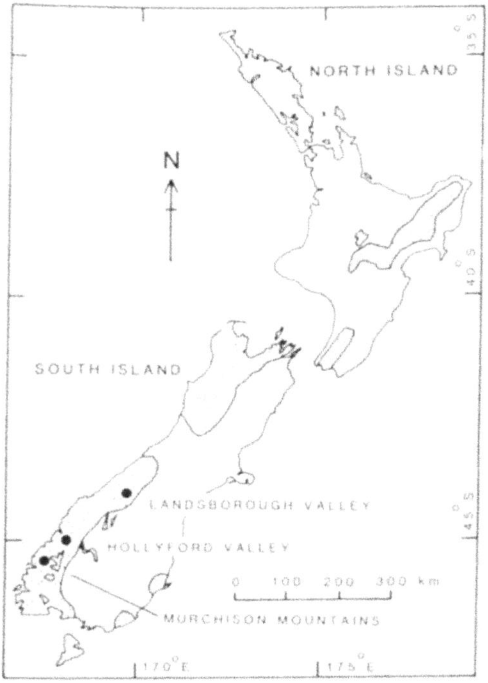

Fig. 561. Distribution of *Nothofagus menziesii*. (Norton 1983b)

Fig. 562. Distribution of *Nothofagus solandri*. (Norton 1983b)

Wood

The heartwood is colorless. The sapwood is similar to that of *Fagus sylvatica*: red-brown to pale yellow. The density is 500–600 kg/m^3.

Cross-section: diffuse- to semi-diffuse porous. The vessels are either single or in short radial files. Tyloses are rare. Parenchyma is sparse and apotracheal.

Tangential section: rays commonly uniseriate, sometimes biseriate.

Radial section: simple perforation plates, sometimes scalariform. Heterogeneous rays. Predominately libriform fibers, albeit sometimes occasionally septate.

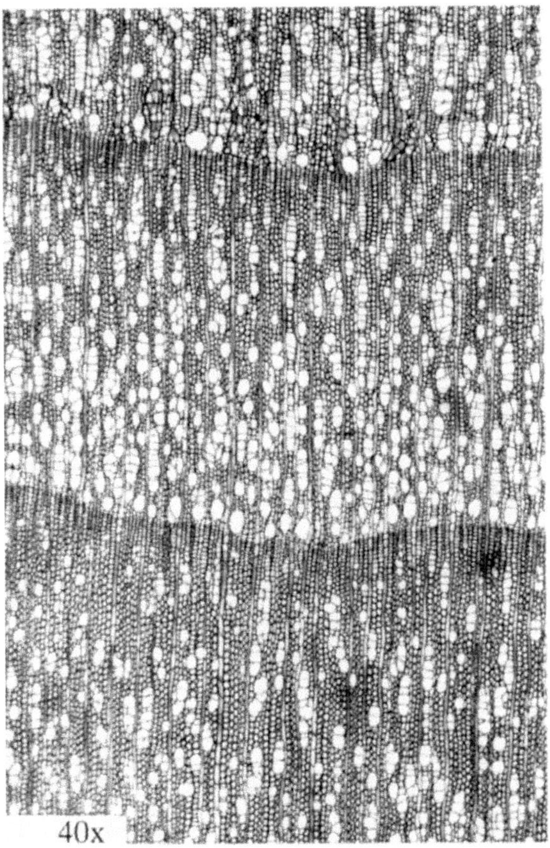

Fig. 563. Cross-section of *Nothofagus solandri*

Both species are indistinguishable on the basis of their anatomical characteristics. For the most part, the genus *Nothofagus* differs from *Fagus* only by the absence of large rays (Meylan and Butterfield 1978).

Use in Dendrochronology

Suitability of the Species

Both species are suitable for ecological and climatological analysis. Their tree-ring boundaries are distinct, they grow to be quite old, and their distribution area covers a broad ecological spectrum. The trees are also suitable for historical-chronological purposes, because European settlers often used the wood for lumber. There are no preserved wood constructions built by the original inhabitants.

Previous Uses

Norton and Ogden (1987) made chronologies available by the use of recent materials.
Nothofagus menziesii: 5 chronologies, first tree ring 1580.
Nothofagus solandri: 25 chronologies, first tree ring 1630.

According to Norton (1983a, 1984), the chronologies of trees from subalpine zones contain, above all, information on summer temperatures. Chronologies of trees on low-altitude sites, on the other hand, provide information on precipitation during the growing seasons. The possibilities for forest dynamic research have by no means been fully exhausted.

References

Aas B (1969) Climatically raised birch lines in southeastern Norway 1818–1968. Norsk Geogr Tidskr 23:119–130

Ahmed M, Ogden J (1985) Modern New Zealand Tree-ring chronologies III. Agathis australis Salis. Tree-Ring Bull 45:11–24

Alexander V, McCord S (1987) Late holocene sediment yield and transport in a northern Arizona drainage basin reconstructed by tree-ring analysis. In: Proc Int Symp Ecol Asp Tree Ring Anal, Tarrytown, NY, pp 213–223

Anon (1981) Forestry in Japan. For Ag Agric For Fish Tokyo, 95 pp

Artman A (1949) Jahrringchronologische und -klimatologische Untersuchungen an der Zirbe und anderen Bäumen des Hochgebirges. Diss Univ München

Attolini MR, Galli M, Nanni T, Ruggiero L, Znanni F (1988) Preliminary observations of the fossil forest of Dunarobby (Italy) as a potential archive of paleoclimatic information. Dendrochronologia 6:141–149

Badré L (1983) Histoire de la forêt française. Arthand, Paris, 309 pp

Baillie MGL (1982) Tree-ring dating and archaeology. Croom, Helm, 274 pp

Baillie MGL, Pilcher JR, Pearson GW (1983) Dendrochronology at Belfast as background to high-precision calibration. Radiocarbon 25:171–178

Baillie MGL, Bartholin T, Eckstein D, Jones PD, Karlen W, Schweingruber FH, Zetterberg P (1990) A 1400-year tree-ring record of summer temperatures in Fennoscandia. Nature 346:434–439

Balchunas VP, Brukshtus VE (1983) Dendrochronological scale of eastern spruce (Picea orientalis) of the eastern Caucasus. In: LaMarche VC Jr (ed) Soviet publications in dendrochronology. Translations of recent articles and books. Tree-Ring Soc, Tucson, Ar, pp 142–143

Banks JCG (1986) Fire and stand histories in subalpine forests on the Thredbo ski slopes, Koscuisko National Park, N.S.W. Australia. In: Proc Int Symp Ecol Asp Tree Ring Anal, Tarrytown, NY, pp 163–174

Bartholin T (1983) Dendrokronologi i Lapland. Dendrokronologiska Sällskapet, Medellanden 5:3–16

Becker B (1983) Dendrochronologie und Paläoökologie subfossiler Baumstämme aus Flussablagerungen. Ein Beitrag zur nacheiszeitlichen Auenentwicklung im südlichen Mitteleuropa. Mitt Komm Quartärforsch Österr Akad Wiss 5:1–12

Becker B, Giertz-Siebenlist V (1970) Eine über 1100jährige mitteleuropäische Tannenchronologie. Flora 159:310–346

Becker B, Billamboz A, Egger H, Gassmann P, Orcel A, Orcel Chr, Ruoff U (1985) Dendrochronologie in der Ur- und Frühgeschichte. Die absolute Datierung von Pfahlbausiedlungen nördlich der Alpen im Jahrringkalender Mitteleuropas. Antiqua 11:68 pp

Becker M, Bräker OU, Kenk G, Schneider O, Schweingruber FH (1990) Kronenzustand und Wachstum von Waldbäumen im Dreiländereck Deutschland-Frankreich-Schweiz in den letzten Jahrzehnten. Allg Forstz 11:263–274

Bednarz Z (1984) The comparison of dendroclimatological reconstructions of summer

temperatures from the Alps and Tatra mountains from 1741–1965. Dendrochronologia 2:63–72
Bednarz Z (1986) An example of the application of tree-ring chronology of dwarf mountain pine (Pinus mugo var. Mughus Zenari) for the dating of geomorphological processes in the Tatra mountains. Dendrochronologia 4:75–78
Begemann HF (1981–1987) Das große Lexikon der Nutzhölzer, 7 vols. Deutscher Betriebswirte Verlag, Gernsbach
Bégin C, Filion L (1985) Analyse dendrochronologique d'un glissement de terrain de la région du Lac à l'Eau Claire (Québec nordique). Can J Earth Sci 22:175–182
Bégin C, Filion L (1988) Age of landslides along the Grande Rivière de la Balaine estuary, eastern coast of Hudson Bay, Québec (Canada). Boreas 17:290–299
Bégin Y, Arsenault S, Lavoie J (1989) Dynamique d'une bordure forestière par suite de la hausse récente du niveau marin, rive sud-ouest du Golfe du Saint Laurent, Nouveau Brunswick. Geogr Phas Quat 43:355–366
Bhattacharyya A (1987) Tree-ring chronologies from Nepal. Res Rep Lab Tree-Ring Res Univ Ar, Tucson, 26 pp
Bircher W (1982) Zur Gletscher- und Klimageschichte des Saastales. Glazialmorphologische Untersuchung. Phys Geogr Univ Zürich 9:1–233
Bitterli P (1987) Jahrringe in Legföhren (Pinus mugo Turra, ssp. mugo) als Ausdruck lokaler mechanischer Faktoren. Dendrochronologia 5:77–92
Bitvinskas TT (1974) Dendroclimatic research. Gydrometeoizdas, Leningrad, 172 pp
Bitvinskas TT (1983) Dendroclimatic research. In: LAMARCHE VC Jr (ed) 1983: Soviet publications in dendrochronology, I: Translations of recent articles and books, Tree-Ring Soc, Tucson, Ar, pp 197–336
Blasing TJ, Stahle DW, Duvick DN (1988) Tree ring-based reconstruction of annual precipitation in the South-Central United States from 1750–1980. Water Resourc Res 24:163–171
Boninsegna JA, Holmes RL (1985) Fitzroya cupressoides yields 1534-year long. South American chronology. Tree-Ring Bull 45:37–42
Borscheva NM (1983) Dendrochronological scales of Picea schrenkiana in the mountains of Zailivysky (north Tien Shan). In: LAMARCHE VC Jr (ed) 1983: Soviet publications in dendrochronology. In: Translations of recent articles and books. Tree-Ring Soc, Tucson, Ar, pp 157–159
Borscheva NM (1987) Ecological and biological pecularities of Shrenka fir tree rings in the high zones of the northern part of Tien Shan. In: Proc Symp Dendrochronol Meth For Sci Ecol Forecasting, Irkutsk 1987, pp 103–104
Bosshard HH (1982) Holzkunde. Bd 1, 2. Aufl. Birkhäuser, Basel 224 pp
Bradley RS, Jones DP (eds) (1992) Climate since 1500 A.D. Routledge, London, 679 pp
Bräker OU, Schweingruber FH (1984) Standorts-Chronologien, Teil 1: Iberische Halbinsel. Eidgenöss Anst Forstl Versuchswes Birmensdorf, 73 pp
Brenninkmeijer CA (1983) Deuterium, Oxygen-18 and Carbon-13 in tree rings and peat deposits in relation to climate. Dissertation, Rijksuniv Groningen, NL 146 pp
Brett DW (1978) Elm tree rings as a rainfall record. Tree-Ring Bull 38:87–94
Bridge MC, Ogden J (1986) A sub-fossil Kauri (Agathis australis) tree-ring chronology. JR Soc NZ 16:17–23
Briffa KR, Jones PD, Schweingruber FH (1988) Summer temperature patterns over Europe: A reconstruction from 1750 A.D. based on maximum latewood density indices of conifers. Quart Res 30:36–52
Brockman CF (1968) Trees of North America. Golden, New York, 280 pp
Browicze K (1982) Chronology of trees and shrubs in southwest Asia and adjacent regions, vol 1. Pol Acad Sci Inst Dendrol, Warzawa, Poznan, 172 pp
Brubaker LB (1987) Forest disturbance and tree-ring analysis. In: Proc Int Symp Ecol Asp Tree Ring Anal, Tarrytown, NY, pp 101–118
Brukstus VJ, Balciunas VO (1983) Dendrochronological scales of Picea orientalis and

Abies nordmanniana in the West Caucasus. In: LaMarche VC Jr (ed) 983: Soviet publications in dendrochronology. Translations of recent articles and books. Arizona, Tree-Ring Soc, Tucson, AR, pp 28–30

Chebakova NM (1987) The relation between tree growth of Abies sibirica and climate in high mountains of Southern Siberia (Zapadny Sayan). In: Symp Dendrochronol Meth For Sci Ecol Forecasting, Irkutsk, pp 113–114

Christensen K (1987) Tree-rings and insects: The influence of cockchafers on the development of growth rings in oak trees. In: Proc Int Symp Ecol Asp Tree Ring Anal, Tarrytown, NY, pp 142–154

Clark C (1982) Überschwemmungen. Der Planet Erde. Time-Life Bücher, Amsterdam, 176 pp

Clark S (1987) Potential for use of cottonwoods as indicators of past floods. In: Proc Int Symp Ecol Asp Tree Ring Anal, Tarrytown, NY, pp 243–248

Conkey L (1979) Response of tree-ring density to climate in Maine, USA. Tree-Ring Bull 39:29–38

Conkey L (1986) Red spruce tree-ring densities in eastern North America as indicators of past climate. Quart Res 26:232–243

Cook ER (1982) Eastern North America. In: Hughes MK, Kelly PM, Pilches JR, LaMarche VC (eds) Climate from tree rings. Univ Press, Cambridge, London, New York, pp 126–133

Cook E, Kairiukstis L (eds) (1990) Methods of dendrochronology in the environmental sciences. Int Inst Appl Syst Anal. Kluwer, Dordrecht 394 pp

Costermans L (1983) Native trees and shrubs of south-eastern Australia, 4th edn. Weldon, Sydney, 422 pp

Dallimore W, Jackson AB (1974) A handbook of Coniferae and Ginkgoidae. Arnold, London. 729 pp

Dean JS (1969) Chronological analysis of Tsegi phase sites in southeastern Arizona. Papers of Lab. of Tree-Ring Res Univ Press, Tucson 3:1–207

Dechamps R (1983) Clé d'identification à la loupe des bois des regions temprées et tropicales disponibles en Belgique Musée royal de l'Afrique centrale. Tervuren Doc Écon 5:23 pp

De La Hera C, Figueroa ME, Garcia Novo F (1981) Control climatico del crecimiento del pino pinonero en el bajo Guadalquivir. Avances sobre la investigacion en bioclimatologia. Sevillia VII. In: Simp Bioclimatologia, pp 625–634

Delwaide A, Filion L (1987) Tree-ring reconstruction of Indian-Cree tree harvesting in the area, Quebec, Poste-De-La-Baleine. In: Proc Int Symp Ecol Asp Tree Ring Anal, Tarrytown, NY, 343 pp

Digilio APL, 1966: Los arboles indigenas de la provincia de Tucuman. Univ Nacl Tucuman

Dobremez J.-F (1976) Le Nepal. Ecologie et biogéographie. Cahiers népalais Cent Nat Rech Sci Paris, 355 pp

Donati P, Orcel AeC. (1988) Dendrocronologia e monumenti nell'area ticinese. Z Schweiz Archäol Kunstgesch 45:277–294

Druce AP (1966) Tree-ring dating of recent volcanic ash and lapilli, Mt. Egmont NZ J Bot 4:3–41

Duever MJ, McCollon JM (1987) Cypress tree-ring analysis in relation to wetlands and hydrology. In: Proc Int Symp Ecol Asp Tree Ring Anal, Tarrytown, NY, pp 249–260

Dunwiddie PW (1979) Dendrochronological studies of indigenous New Zealand trees. NZ J Bot 17:251–266

Eckstein D (1985) On the application of dendrochronology for the evaluation of forest damage. In: Proc inventorying and monitoring endangered forests. IUFRO Conf, Zürich, 1985. Eidgenöss Anst forstl Versuchswes Birmensdorf, pp 287–290

Eckstein D, Frisse E, Quiehl F (1977) Holzanatomische Untersuchungen zum Nachweis anthropogener Einflüsse auf die Umweltbedingungen einer Rotbuche. Angew Bot 51:47–56

Eckstein D, Wrobel S, Aniol RW (eds) (1983) Dendrochronology and archeology in Europe. Mitt Bundesforschungsanst Forst- und Holzwirtsch, Hamburg 141:1–249

Elias TS (1980) The complete trees of North America. Van Nostrand Reinhold, New York, 948 pp

Elling W (1966) Untersuchungen über das Jahrringverhalten der Schwarzerle. Flora 156:155–201

Everitt BL (1968) Use of cottonwood in an investigation of the recent history of a flood plain. Am J Sci 266:417–439

Federer CA, Hornbeck JW (1987) Red spruce diameter growth and Weibull functions for even-aged stand development. In: Proc Int Symp Ecol Asp Tree Ring Anal, Tarrytown, NY, pp 18–25

Fenaroli L, Bambi G, Alberi G (1976) Dendroflora italica. Museo Tridentino, Trento, 717 pp

Ferguson WC, Graybill DA (1983) Dendrochronology of bristlecone pine. A progress report. Radiocarbon 25:287–288

Filion L Morin P (1988) Modifications morphologiques d'epinette blanche soumise à la sedimentation eolienne en milieu dunarie, Québec subarctique. Can J Bot 66:1862–1869

Filion L, Payette S, Gauthier L, Boutin Y (1986) Light rings in subarctic conifers as a dendrochronological tool. Quart Res 26:272–279

Fletscher J (ed) (1978) Dendrochronology in Europe: principles, interpretations and applications to archaeology and history. BAR (Br Archaeol Rep) Int Ser 51:356 pp

Fowells HA (1965) Silvics of forest trees of the United States. US Dep Agr For Ser Agr Handb 271:762 pp

Frisse E (1977) Xylometrische und dendroklimatologische Untersuchungen über den Einfluss von Temperatur- und Niederschlag auf Eichen und Buchen. Diss Univ Hamburg, 213 pp

Fritts HC (1959) The relation of radial growth to maximum and minimum temperature in three tree species. Ecol 40:261–265

Fritts HC (1965) Tree-ring evidence for climatic changes in western North America. M Weather Rev 93:421–443

Fritts HC (1976) Tree rings and climate. Academic Press, New York, 567 pp

Genova R (1986) Dendroclimatology of mountain pine (Pinus uncinata Ram.) in the Central Plain of Spain. Tree-Ring Bull 46:3–12

Genova R, Garcia C (1984) Analisis dendrocronologica des castano (Castanea sativa) en el macizo des Montseny. Mediterr Ser Biol 7:67–82

Giddings JL Jr (1938) Recent tree-ring work in Alaska. Tree-Ring Bull 5:2–16

Glasby P, Selkirk PM, Adamson D, Downing AJ, Selkirk DR (1988) Blue Mountains ash (Eucalyptus oreades RT Baker) in the western Blue mountains. Proc Linn Soc N S W 110:141–158

Goers H (1981) Unsere Baum-Veteranen. Landbuch, Hannover, 151 pp

Graumlich LJ, Brubaker LB (1987) Increasing net primary productivity in Washington (USA) forests during the last 1000 years. In: Proc Int Symp Ecol Asp Tree Ring Anal, Tarrytown, NY, pp 59–69

Graybill DA (1987) A network of high-elevation conifers in western United States. In: Proc Int Symp Ecol Asp Tree-Ring Anal, Tarrytown, NY pp 463–474

Grosser D (1977) Die Hölzer Mitteleuropas. Ein mikrophotographischer Lehratlas. Springer, Berlin, Heidelberg, New York, 208 pp

Harrington TC (1986) Growth decline of wind-exposed red spruce and balsam fir in the White Mountains. Can J For Res 16:232–238

Hess E, Landolt E, Hirzel R (1967) Flora der Schweiz 2nd edn. 3 vols. Birkhäuser, Basel, Vol1:858 pp, Vol2:956 pp, Vol3:876 pp

Hollstein E (1978) Wood technology and the dating of oak. West German chronologies for oak and beech. In: Fletcher J (ed) Dendrochronology in Europe. BAR Int Ser 51:33–44
Hollstein E (1980) Mitteleuropäische Eichenchronologie. Zabern, Mainz, 273 pp
Holmes RL (1982) Argentina and Chile. In: Hughes MK, Kelly PM, Pilcher JR, LaMarche VC Jr., (eds) Climate from tree rings. Univ Press, Cambridge, pp 84–90
Holmes RL, Stockton CW, LaMarche VC (1979) Extension of river flow records in Argentina from living tree-ring chronologies. Water Resour Bull 15:1081–1085
Holzhauser HP (1984) Zur Geschichte der Aletschgletscher und des Findelengletschers. Phys Geogr 13:1–488
Hornbeck JW, Smith RB, Federer CA (1987) Extended growth decreases in New England are limited to red spruce and balsam fir. In: Proc Int Symp on Ecol Asp of Tree Ring Anal Tarrytown, NY, pp 38–44
Hosie RC (1973) Native trees of Canada. Can For Serv, Ottawa, 380 pp
Houle G (1990) Growth patterns of sugar maple seedlings and mature trees in healthy and in declining hardwood stands. Can. J For Res 20:894–901
Hueck K (1966) Die Wälder Südamerikas. Fischer, Stuttgart, 422 pp
Hughes M (1989) The tree-ring record. In: Bradley RS (ed) Global changes in the past. UCAR/off Interdisc Earth Stud, Bouldes, Co, pp 117–137
Hughes MK, Kelly PM, Pilcher JR, Lamarche VC Jr (eds) (1982) Climate from tree rings. Cambridge Univ Press, Cambridge, 233 pp
Informationsdienst Holz, (ed) incomplete (1980) Monographische Baumdarstellungen. Arbeitsgem Holz eV Düsseldorf
Iseli M, Schweingruber FH (1989) Sichtbarmachen von Jahrringen für dendrochronologische Untersuchungen. Dendrochronologia 7:145–157
Jacoby GC, Cook ER (1981) Past temperature variations inferred from a 400-year tree-ring chronology from Yukon territory, Can. Arct Alp Res 13:409–418
Jacoby GC, Hornbeck JW (eds) (1987) International symposium on ecological aspects of tree-ring analysis. In: Proc Int Symp on Ecol Asp of Tree Ring Anal Tarrytown, NY, pp 17–21
Jazewitsch W von (1953) Jahrringchronologie der Spessart-Buchen. Forstwiss Cbl 72:234–247
Johnson EA (1987) The relative importance of snow avalanche disturbance and thinning of tree populations. In: Proc Int Symp on Ecol Asp of Tree Ring Anal, Tarrytown, NY
Kairiukstis L (1983) Dendrochronology for the indication of background climate variation. In: LaMarche VC Jr (ed) (1983) Soviet publications in dendrochronology, I: Translations of recent articles and books. Tucson, Arizona, Tree-Ring Soc. Originaltitel 1974: Dendroklimaticeskie issledovanija, pp 75–97
Kairiukstis L, Bednarz Z, Feliksik E (eds) (1987) Methods of dendrochronology, In: Proc of the Task Force Meeting on Methodology of Dendrochronology, East/West Approaches (Krakow 1986). Warsaw, Polish Acad Sci Syst Res Inst
Kaiser KF, Eicher U (1987) Analysis of fossil pollen, molluscs and stable isotopes in the Dättnau Valley. Boreas 16:293–303
Kaiser KF, Kaiser-Bernhard Ch (1987) Der Ausbruch des Katmai 1912 und das Erdbeben von Alaska 1964 in den Jahrringen der Sitkafichten auf der Insel Kodiak. Mitt Naturwiss Ges Winterthur 38:69–88
Katayama M, Kurata S et al. (eds) (1968) Illustrated important forest trees of Japan. Tokyo, 3 vols
Kienast F (1985) Dendroökologische Untersuchungen an Höhenprofilen aus verschiedenen Klimabereichen. Diss Univ Zürich
Kienast F, Schweingruber FH, Bräker OU, Schär E (1987) Tree-ring studies on conifers along ecological gradients and the potential of single-year analysis. Can J For Res 17:683–696

Kirchner O, Loew E, Schröter C (1906) Die Coniferen und Gnetaceen Mitteleuropas. Ulmer, Stuttgart, 343 pp

Klein P (1983) Dating of art historical objects. In: Eckstein D, Wrobel S, Aniol RW (eds) Dendrochronology and archaeology in Europe. Mitt Bundesforschungs anst Forst-Holzwirtsch (Hamburg)14:209–222

Klein P, Eckstein D (1988) Die Dendrochronologie und ihre Anwendung. Spektrum der Wissenschaft, D 6179, pp 56–68

Kojo Y (1987) A dendrochronological study of Cryptomeria japonica in Japan. Tree-Ring Bull 47:1–21

Kolishchuk VG (1983) Dendroscales of Swiss stone pine in the Ukrainian Carpathians. In: LaMarche VC Jr (ed) Soviet publications in dendrochronology. Translations of recent articles and books. Tree-Ring Soc, Tucson, Arizona, pp 171–173

Kolishchuk VG (1990) Dendroclimatological study of prostrate woody plants, 51–55. In: E.R. Cook L.A. Kairiukstis (eds) Methods of dendrochronology. 394 pp Kluwer, Dordrecht

Kontic R, Niederer M, Nippel CA, Winkler-Seifert A (1986) Jahrringanalysen an Nadelbäumen zur Darstellung und Interpretation von Waldschäden (Wallis, Schweiz) Eidgenöss Anst forstl Versuchswes Ber 283, 46 pp

Köstler JN, Brückner E, Bibelriether H (1968) Die Wurzeln der Waldbäume. Parey Hamburg, 284 pp

Krüssmann G (1972) Handbuch der Nadelgehölze. Parey Hamburg, 366 pp 2nd edn 1983, 396 pp

Kuhlmann L (1979) Change and stability of the altitude of birch-tree-limit in the northern swedish scandes 1915–1975. Acta Phytogeogr Suec 65:1–121

Kuivinen KC, Lawson MP (1962) Dendroclimatic analysis of birch in South Greenland. Arct Alp Res 14:243–250

Kuniholm PI, Striker CL (1983) Dendrochronological investigations in the Aegaen and neighboring regions. J Archeol 10:411–420

Kuniholm PI, Striker CL (1987) Dendrochronological investigations in the Aegaen and neighboring regions, 1983–1986. J Field Arch 14:385–398

Kunkel G (1987) Die Kanarischen Inseln und ihre Pflanzenwelt. Fischer, Stuttgart 202 pp

LaMarche VC, Fritts HC (1971) Tree rings, glacial advance, and climate in the Alps. Z Gletscher kd 7:125–131

LaMarche VC, Holmes RL, Dunnwiddie PW, Drew LG (1979) Tree-ring chronologies of the southern hemisphere. 3. New Zealand. Chronology Series V, Lab of Tree-Ring Res. Univ of Arizona, Tucson, AZ 85721 USA, 77 pp

LaMarche VC, Stockton CW (1974) Chronologies from the temperature sensitive bristlecone pines at upper timberline in western United States. Tree-Ring Bull 34:21–45

LaMarche VC Jr, Wallace RE (1972) Evaluation of effects of past movements on the San Andreas Fault, Northern California. Geol Soc Am Bull 83:2665–2676

Lamprecht A (1984) Dendroklimatologische Untersuchungen in Südamerika. Eidgenöss Anst forstl. Versuchswes Ber 263, 79 pp

Lamprecht H (1980a) Baumarten der Tropen. Verbreitung – Oekologie – Waldbau – Verwendung. Skriptum zur Vorlesung, Inst Waldbau, Univ Göttingen, 118 pp

Lamprecht H (1980b) Die einheimischen und die wichtigsten fremdländischen Baumarten. Skriptum zur Vorlesung, Inst Waldbau, Univ Göttingen, 273 pp

Laprise D, Payette S (1988) Evolution récente d'une tourbière à palses (Québec subarctique): une analyse cartographique et dendrochronologique. Can J Bot 66:2117–2227

Leavitt SW, Long A (1987) Influence of site disturbance on ∂^{13} C isotopic time series from tree rings. In: Proc Int Symp on Ecol Asp of Tree Ring Anal Tarrytown, NY, pp 119–129

LeBlanc DC, Raynal DJ, White EH, Ketchledge EH (1987) Characterization of historical growth patterns in declining red spruce trees. In: Proc Int Symp on Ecol Asp of Tree Ring Anal Tarrytown, NY, pp 360–371

Legname PR (1982) Arboles indigenas del noroeste Argentino. San Miguel de Tucuman.

Leuschner HH, Delorme A, Höfle HC (1987) Dendrochronological study of oak trunks found in bogs of northwest Germany. In: Proc Int Symp on Ecol Asp of Tree Ring Anal Tarrytown, NY, pp 298–318

Lingg W (1986) Dendroökologische Studie an Fichte (Picea abies) und Weisstanne (Abies alba) im subkontinentalen Klimagebiet Wallis, Schweiz. Eidgenöss Anst forstl. Versuchswes Ber 287, 81 pp

Liphschitz N (1986) Overview of the dendrochronological and dendroecological research in Israel. Dendrochronologia 4:37–58

Liphschitz N, Mendel Z (1987) Damage to Pinus halepensis xylem by the Israeli pine bast scale, Matsucoccus josephi. IAWA Bull 8:369–376

Little EL Jr (1971) Atlas of United States trees, vol 1. Conifers and important hardwoods. US Government Printing Office, Washington, 200 maps

Lyon CJ (1943) Supply and the growth rates of conifers around Boston. Ecology 24:329–344

Lyon CJ (1943)Supply and the growth rates of conifers around Boston. Ecology 24:329–344

Malokvasov DS (1987) Long-term radial growth fluctuations in Korean pine (Pinus koreansis) and kuril larch (Larix kurilensis) in the Soviet Far East: biological and geographical aspects. Proc Meet: Dendrochronol. methods in for. sci. ecol. forecasting, Irkutsk, pp 48–50

Marchand PJ, Goulet FL, Harrington TC (1986) Death by attrition: a hypothesis for wave mortality of subalpine Abies balsamea. Can J For Res 16:591–59[illegible]

Mayer H (1962) Abies pinsapo Boiss. in der Betischen Kordilliere Südspaniens. Forstwiss Cbl 81:48–62

Mayer H (1965) Marokkanische und numidische Tanne im nordafrikanischen Atlas. Forstwiss Cbl 84:265–328

Mayer H (1976) Gebirgswaldbau – Schutzwaldpflege. Fischer, Stuttgart, 435 pp

Mayer H (1980) Waldbauliche Probleme in Gebirgswäldern des Maghreb (Nordafrika). Symposiumsbericht «Waldbau unter ökologisch und wirtschaftlich extremen Bedingungen». IUFRO-Tagung 25.9.–3.10.1980 in Thessaloniki, pp 259–278

Mayer H (1984) Wälder Europas. Fischer, Stuttgart, 691 pp

McLean DA (1981) Impact of defoliation by spruce budworm populations on radial and volume growth of Balsam fir. Mitt forstl Bundesversuchsanst Wien 142: 293–306

Meusel H, Jäger E, Weinert E (1978) Vergleichende Chorologie der zentraleuropäischen Flora. Fischer, Jena 583 pp

Meylan BA, Butterfield BG (1978) The structure of New Zealand Woods. DSIR Bull 222, Wellington, 250 pp

Miehe G (1982) Vegetationsgeographische Untersuchungen im Dhaulagiri und Anapurna-Himalaya. Cramer, Vaduz. Diss Bot 66:1

Miehe G (1987) An annotated list of vascular plants collected in the valleys south of Mt. Everest. Bull Br Mus Nat Hist (Bot) 16:225–268

Mitsutani T (1990) Dendrochronology in Japan. Res Rep of the Nasa national cultural properties research 48 Institute, Nasa 166 pp

Morin A, Payette S (1984) Expension récente à la limite des forets (Québec nordique. Can J For Res 62:1404–1408

Morin H, Payette S (1986) La dynamique récente des combes à neige du golfe de Richmond (Québec nordique): une analyse dendrochronologique. Can J Bot 64:2113–2119

Mukamedsin KD (1977) Dendrochronological scale of a dendritic form of Turkestan Juniper in Bitvinskas TT, Dendroscales of Soviet Union I. In: LaMarche VC Jr (ed)

(1983) Soviet publications in dendrochronology. Translations of recent articles and books. Lab of Tree-Ring Res, Tucson, Arizona, 85721, USA, pp 172–173.
Müller E, Stierlin HR (1990) Sanasilva, Kronenbilder. Eidgenöss Anst forstl Versuchswes, Birmensdorf, 98 pp
Müller-Stoll H (1951) Vergleichende Untersuchungen über die Abhängigkeit der Jahrringabfolge von Holzart, Standort und Klima. Bibl Bot 122:1–93
Munaut AV, Serre-Bachet F (1982) The mediterranean area. In: Hughes et al. (eds) Climate from tree rings. Cambridge Univ Press, London, pp 151–155
Nobori Y, Nagata Y, Hojima R (1990) wood density analysis of picea glehnii plantation by X-ray densitometry. Kurri-TR-352, pp 3–11
Norton DA (1983a) Modern New Zealand tree-ring chronologies I. Nothofagus solandri. Tree-Ring Bull 43:1–17
Norton DA (1983b) Modern New Zealand tree-ring chronologies II. Nothofagus menziesii. Tree-Ring Bull 43:39–49
Norton DA (1983c) Population dynamics of subalpine Libocedrus bidwillii forests in the Crop River Valley, Westland, New Zealand. NZ J Bot 21:127–134
Norton DA (1984) Phenological growth characteristics of Nothofagus solandri trees at three altitudes in the Craigieburn Range, New Zealand. NZ J Bot 22:413–424
Norton DA, Ogden J (1987) Dendrochronology: a review with emphasis on New Zealand applications. NZ J Ecol 10:77–95
Noda M, Higashimura T (1990) An approach for determining non-climatic variation in width of annual tree rings in western Japan. Kurri-TR-352, pp 24–27
Ogden J (1978) On the dendrochronological potential of Australian trees. Aust J Ecol 3:339–356
Ogden J (1982) Australasia. In: Hughes MK et al. (eds) Climate from tree rings. Cambridge Univ Press, Cambridge, pp 90–104
Orcel A, Orcel Ch (1985) Etat de recherches en mars 1985. Dendrochronologia 3:153–176
Otha S (1987) The observation of tree-ring structure by soft X-ray densitometry. Mokuzai Gakkaishi 24:429–434
Panshin AJ, Zeeuw de C (1980) Textbook of wood technology. Fourth Ed. McGraw-Hill, New York, 722 pp
Parker ML, Henoch WES (1971) The use of Engelman spruce latewood density for dendrochronological purposes. Can J For Res 1:90–98
Parker ML, Jozca LA (1971) Dendrochronological investigation along the Mackenzie, Liard and south Nahanni Rivers, NWT Tech Rep 10 to Glaciol Div Water Resour Branch Dept of the Environ, 187 pp
Parsapajouh D, Bräker OU, Habibi H, Schär E (1986) Etude dendroclimatologique du bois de Taxus baccata du nord de l'Iran. Schweiz Z Forstwes 137:853–868
Payette S (1980) Les grandes crues glacielles de la rivière aux feuilles (Nouveau-Quebec): une analyse dendrochronologique. Nat Can 107:215–225
Payette S, Delwaide A (1991) Variations séculaires du niveau d'eau dans le bassin de la rivière Boniface (Québec nordique): une analyse dendrécologique. Geogr Phys Quart 45:59–67
Payette S, Filion L (1985) White spruce expansion at tree line and recent climatic change. Can J For Res 15:241–251
Payette S, Gagnon R (1990) Tree-limit dynamics in Ungava peninsula, northern Quebec. Holarct Ecol 2:239–248
Payette S, Morneau C, Sirois L, Desponts M (1989) Ecology 70:656–673
Payette S, Filion L, Delwaide A (1990) Disturbance regime of a cold temperate forest as deduced tree-ring patterns; the Tantaré ecological reserve, Quebec. Can J For Res 20:1228–1241
Peterson DL, Wakefield VA, Arbaugh MJ (1987) Detecting the effects of ozone pollution on growth of Jeffrey pine in the Sierra Nevada, California. In: Proc Int Symp on Ecol Asp of Tree Ring Anal, Tarrytown, NY, pp 402–409

Phillips R (1979) Trees in Britain, Europe and North America. Pan Books, London, 224 pp
Phipps RL, Whiton (1987) Decline in long-term growth trends of white oak. Can J For Res 18:24–32
Pignatti S (1982) Flora d'Italia. Bologna, Edagricole, 3 vols
Pividori M, Cherubini P (1990) Analisi degli anelli diaccrescimento per valutare d'eventuale influenza della vibratura meccanica sull'accrescimento diametrico in popolazioni di Pinus pinea L. Quaderni dell'istituto di assestamento e tecnologia forestale III, Univ studi di Firenze, pp 55–59
Poole AL, Adams NM (1980) Trees and shrubs of New Zealand. Wellington, 257 pp
Pruden MA, Fryer GI, Johnson EA (1987) Fire frequency and old trees in the southern Canadian Rockies. Proc Int Symp on Ecol Asp of Tree Ring Anal, Tarrytown, NY, pp 175–179
Richter K (1987) Dendrochronologische und dendroklimatologische Untersuchungen an Kiefern (Pinus sp.) in Spanien. Diss Univ Hamburg, 296 pp
Richter K, Eckstein D (1986) Estudio dendrochronologico en Espana. Dendrochronologia 4:59–74
Rol R (1962) Flore des arbres arbustes et abrisseaux vol I, plaines et colines La Maison rustique, Paris 95 pp
Ruiz-Flaño P (1988) Dendroclimatic series of Pinus uncinata R. in the central Pyrenees and the Iberian system, Spain. A comparative study. Pirineos 132:49–6[illegible]
Salomon JT (1980) The native trees of New Zealand Reed, Wellington Sydney 384 pp
Schär E, Schweingruber FH (1987) Nacheiszeitliche Stammfunde aus Grächen im Wallis. Schweiz Z Forstwes 138:497–515
Schenk CA (1939) Fremdländische Wald- und Parkbäume Parey, Berlin, 3 vols
Schmidt-Vogt H (1977) Die Fichte vol 1 Parey, Hamburg, 647 pp
Schöpfer W, Hradetzky J (1986) Zuwachsrückgänge in erkrankten Fichten– und Tannenbeständen – Auswertungsmethoden, Ergebnisse. Forstwiss Cbl 105:446–470
Schotterer U (1987) Klima – unsere Zukunft? Kümmerly und Frey, Bern, 156 pp
Schulman E (1956) Dendroclimatic changes in semiarid America. Univ of Arizona Press, Tucson, Arizona, 142 pp
Schweingruber FH (1978) Mikroskopische Holzanatomie. Eidgenöss Anst forstl Versuchswes, Birmensdorf, 226 pp
Schweingruber FH (1983) Der Jahrring. Paul Haupt, Bern, 234 pp
Schweingruber FH (1987) Flächenhafte dendroklimatische Temperaturrekonstruktionen für Europa. Naturwissenschaften 74:205–212
Schweingruber FH (1988) Tree-rings. Basics and applications in dendrochronology. Reidel, Dortrecht 276 pp
Schweingruber FH (1990) Anatomie europäischer Hölzer/Anatomy of European woods. Paul Haupt, Bern, 800 pp
Schweingruber FH, Kontic R, Winkler-Seifert A (1983) Eine jahrringanalytische Studie zum Nadelbaumsterben in der Schweiz. Eidgenöss Anst forstl. Versuchswes Ber 253, 29 pp
Scuderi LA (1984) A dendroclimatic and geomorphic investigation of late-Holocene glaciation, Southern Sierra Nevada, California. Diss Univ of Los Angeles, Dept of Geogr Microfilm, 225 pp
Sell J (1989) Eigenschaften und Kenngrössen von Holzarten, 2nd edn. Lignum, Zürich, 80 pp
Serre F (1969) Variation de l'épaisseur des annaux ligneux chez le thuya de Barbarie (Tetraclinis articulata) et le climat en Tunisie. Ann Fac Sc Marseille 42:193–204
Serre F (1973) Contribution a l'étude dendroclimatologique du pin d'halep (Pinus halepensis Mill.). PhD thèse Univ d'Aix, Marseille III, 244 pp
Serre-Bachet F (1982) The Mediterranean region. In: Hughes MK et al. (eds) Climate from tree rings. Cambridge Univ Press, Cambridge, pp 153–155

Serre-Bachet F (1985) Une chronologie pluriseculaire du sud de l'Italie. Dendrochronologia 3:45–66
Sheppard PR, Jacoby GC (1987) Dating earthquakes along the San Andreas fault system in California. In: Proc Int Symp on Ecol Asp of Tree Ring Anal, Tarrytown, NY, pp 281–289
Sheppard PR, Lassoie JP (1987) Fire history of lodgepole pine of Mt. San Jacinto, California. In: Proc Int Symp on Ecol Asp of Tree Ring Anal, Tarrytown, NY, pp 155–162
Siebenlist-Kerner V (1984) Der Aufbau von Jahrringchronologien für Zirbelkiefer, Lärche und Fichte eines alpinen Hochgebirgsstandortes. Dendrochronologia 2:9–29
Siren G (1963) Tree rings and climate forecasts. New Sci 364:18–20
Sprugel DG, (1976) Dynamic structure of wave-regenerated Abies balsamea forests in northeastern United States. J Ecol 64:889–911
Stahle DW (1987) Environmental factors involved in the growth and distribution of post oak in the south central United States. In: Proc Int Symp on Ecol Asp of Tree Ring Anal, Tarrytown, NY, pp 54–58
Stahle DW, Hehr JG (1984) Dendroclimatic relationships of post oak across a precipitation gradient in the southcentral United States. Ann Assoc Am Geogr 74:561–573
Stahle DW, Cleavland MK, Hehr JG (1985a) A 450-year drought reconstruction for Arkansas, United States. Nature 316 (6028):530–532
Stahle DW, Cook ER, White WC, (1985b) Tree-ring dating of baldcypress and the potential for millenia-long chronologies in the southeast. Am Antiquity 50 (4):796–802
Stahle DW, Hehr JG, Hawks GG Jr. Cleavland MK, Baldwin JR (1985c) Tree-ring chronologies for the Southcentral United States. Tree-ring Lab, an Off of the State Climatol. Dept of Geogr, Univ of Arkansas, Fayetteville, Arkansas 72701, 135 pp
Staschel R (1987) Dendroklimatologische und -chronologische Untersuchungen an Betula pubescens (Ehrh.) im Gletschervorfeld des Bodalsbreens (Jostedalsbreen / SW-Norwegen). Diplomarb, Geogr Inst Univ Göttingen, 94 pp (unpublished).
Stöckli V (1990) Dendrochronologische und ökologische Untersuchungen der Fir Wave Mortality am Mt. Washington, New Hampshire (USA). Unveröffentlichte Diplomarbeit, Botanisches Institut der Universität Basel, 158 pp
Stockton C (1985) Current research progress toward understanding drought. Proc Symp drought, water manage and food prod Agadir:21–35
Stockton CW (1976) Long term streamflow reconstruction in the upper Colorado river basin, usin tree rings. In: Falkenburg CG, Riley DH (eds) Colorado river basin modelling studies. Utah State Univ, pp 410–441
Stokes MA, Dietrich JH (eds) (1980) Proc fire history workshop Tucson, Arizona. General Tech Rep RM-81. Rocky Mountain For and Range Exp Stn For Ser, US Dept Agric, 142 pp
Sudworth GB (1967) Forest trees of the Pacific slope. New York, 455 pp
Swetnam TW (1987) Western spruce budworm outbreaks in northern New Mexico: Tree-ring evidence of occurrence and radial growth impacts from 1700 to 1983. In: Proc Int Symp on Ecol Asp of Tree Ring Anal, Tarrytown, NY pp 130–141
Tessier L (1982) Analyse dendroclimatologique comparée de six populations de Pinus sylvestris (L.) dans le Drôme. Ecol Mediter VIII (Marseille):185–202
Tessier L (1986) Approche dendroclimatologique de l'écologie de Pinus silvestris L. et Quercus pubescens Willd. dans le Sud-Est de la France. Acta Oecol 7:339–355
Tessier L, Pons A, Serre-Bachet F (1982) Analyse dendroclimatologique comparée de quelques populations de chêne pubescent (Quercus pubescens Willd.) de la Drôme et du Var. Ecol Mediterr VIII (Marseille):118–130
Till C (1984) A synthesis of response functions from eight cedar forests located in northern Africa. Dendrochronologia 4:73–82

Trendelenburg R, Mayer-Wegelin H (1955) Das Holz als Rohstoff. Hauser München, 541 pp
Treter U (1984) Die Baumgrenzen Skandinaviens. Steiner, Wiesbaden, 111 pp
Vaucher H (1986) Elseviers dictionary of trees and shrubs. Elsevier, Amsterdam, 413 pp
Veblen TT, Ashton DH (1982) The regeneration status of Fitzroya cupressoides in the Cordilliera Pelada, Chile. Biol Conserv 23:141–161
Viereck LA, Little EL (1972) Alaska trees and shrubs. Agriculture Handb 4 0, For Ser, US Dept of Agric, 265 pp
Villalba R, Boninsegna JA (1985) Cedrela angustifolia and Juglans australis: two new tropical species useful in dendrochronology. Tree-Ring Bull 45:25–35
Vinci Da, Leonardo, in Di Luigi XVIII, (1817) Trattato della Pittura. Rom., 396 pp
Vins B (1962) Die Auswertung jahrringchronologischer Untersuchungen in rauchgeschädigten Fichtenwäldern des Erzgebirges. Wiss Z Techn Univ Dresden 11 579–580
Vorreiter L (1949) Holztechnologisches Handbuch Bd 1. Fromme, Wien 543 pp
Walter H (1968) Die Vegetation der Erde II. Fischer, Stuttgart, 1001 pp
Walter H, Lieth H (1960, 1964, 1967) Klimadiagramm-Weltatlas. Fischer, Jena
Wickman BE (1978) A case study of Douglas fir tussock moth outbreak and stand conditions 10 years later. USA For Ser Res Pap PNW 244 22 pp
Wu X, Lin Z, Sun L (1988a) A preliminary study of the climatic change of the Hengudan Mountains area since 1600 A.D. Adv Atmos Sci:437–443
Wu X, Sun L, Cheng Z (1988b) Establishment of some tree-ring chronologies. Kexue Tongbao 33:1284–1289
Yamaguchi DK (1987) The development of old-growth Douglas-fir forests northeast of Mount St. Helens, Washington, following an A.D. 1480 eruption. In: Proc Int Symp on Ecol Asp of Tree Ring Anal Tarrytown, NY, pp 181–185
Yasuda H (1987) Characteristics in ring structure of Cryptomeria japonica grown under the surrounding area of a big town. In: Noda M (ed) Tree rings. Kurri-TR-300, 86 pp
Yokobori M, Otha S (1983) Combined air pollution and pine-ring structure observed xylochronologically. Eur J For Path 13:30–55
Z'graggen S (1992). Densitometrisch-klimatologische Unteruchung an Buchen (Fagus silvatica L.) Diss, Univ Basel, 167 pp
Zheng Sizhon, Wu Xiangding, Lin Zhenyo (1982) Status of dendrochronology in Asia. In: Hughes MK et al. (eds) Climate from tree rings. Cambridge Univ Press, Cambridge, pp 155–157
Zoller H (1981) Gymnosphermae, Nacktsamige Pflanzen. In: Hegi G (ed) Illustrierte Flora von Mitteleuropa. I/2, 3rd edn. Paul Parey, Berlin, 148 pp

Species Index – Latin and English

Species Index – English-Latin

www.ingramcontent.com/pod-product-compliance
Ingram Content Group UK Ltd.
Pitfield, Milton Keynes, MK11 3LW, UK
UKHW020131290926
15471UKWH00003B/423

* 9 7 8 3 6 4 2 7 7 1 5 9 0 *